FINITE MATHEMATICS
WITH STATISTICS FOR BUSINESS

FINITE MATHEMATICS WITH STATISTICS FOR BUSINESS

Ralph Crouch
PROFESSOR OF MATHEMATICS
DREXEL INSTITUTE OF TECHNOLOGY, PHILADELPHIA

McGraw-Hill Book Company
NEW YORK ST. LOUIS SAN FRANCISCO TORONTO LONDON SYDNEY

Finite mathematics with statistics for business

Library of Congress Catalog Card Number 67-20969

14519

1234567890 MPMM 7432106987

This Book Is Dedicated to C–,
Who Is a Little Better Than a Dee.

PREFACE

This book was written to serve as a text for a single course in mathematics for students of business. It should serve equally well for students of the social sciences or behavioral sciences. It is written for students who have little or no mathematical training and, perhaps of more importance, who are not necessarily mathematically motivated or inclined. A year of high school algebra is helpful but not crucial. The book is totally self-contained. In particular, it is expected that a large percentage of those students in junior or community colleges who take one or more courses in mathematics will find the text both suitable in content and appropriate in level of presentation.

The manuscript out of which this book was constructed was written and rewritten several times between 1963 and 1967. The selection of a text for a single course in mathematics for business, social science, or behavioral science students is often difficult, even painful, and in some cases, borders on disaster. This book grew out of the need for a text for such students. It has been class-tested for several years, and many changes in the selection of topics, in the level of presentation, and in the difficulty and quantity of examples, diagrams, and exercises have been made as a result of the experimentation. These changes should be reflected in the suitability of the material for its intended audience.

Some comments about the selection of content of the text are appropriate. Clearly, the text does not include all the mathematical topics that the students for whom it is intended should know. Nor are the mathematical needs of today's students easily predicted. The coverage of all such topics in a short text such as this is not possible. So a selection of topics from the many possible topics was necessary. The selection reflects the prejudices and beliefs of the author, and they no doubt can (and will) be challenged. But such factors as the limited preparation of the students, the application of mathematical ideas to real life, the sophistication of ideas that can be realistically taught, the number of teacher-student contact hours, the students' interest, etc., place boundary conditions on the topic selection that exclude many worthwhile topics. It is hoped that the variety of ideas presented here will offer a sufficiently wide choice to meet the demands of most teachers.

Since this text was written with a well-defined audience in mind, it was possible to design the book so that some features not present in other texts of this type are included.

Each section begins with a problem or problems which require for their solution some formula, idea, method, or technique (or combination of these) which is explained in the section. This is often called the *case method* of teaching and is fairly common for texts in some disciplines. But we know of no college mathematics textbook that employs this feature. There is an unusually large number of examples, over three hundred in all. This is in keeping with the thought that many of the students for whom this book is intended have difficulty with "abstract" ideas but can understand "concrete" examples of these ideas. These examples are always set in physical situations familiar to the student's environment and are oriented toward the business world. Because the mathematical ideas are relatively simple, the examples are not "deep." But an honest, sincere attempt has been made to make the examples meaningful. Experience teaches that students are better motivated by practical examples.

Chapter 1 is truly basic and very elementary. It should not be dwelt upon, for the applications of the basic ideas lie ahead. Chapter 2 has ideas in it that students with two years of algebra will probably have already studied. For those with one year of algebra, it will require a slower treatment. For some students in better-prepared classes, Chap. 3 may present the first really new ideas. In any case, Chap. 3 will serve to unify the concepts of sets, logic, counting, and probability. Chapter 4 is the capstone of the text, and it shows how ideas presented earlier in the book can be used for predictive purposes. If a later course is to deal with statistics and inference, then Chap. 4 will serve only as an introductory treatment. Chapter 5 is a self-contained introduction to the concepts of linear equations and linear programming. The treatment is elementary for both topics.

The list of credits for this book is long and hopefully all-inclusive. In the early experimental stages of the manuscript, Dr. Ralph Ball, who was teaching an early version of the manuscript, was particularly helpful. In its later stages, Professor Ray Graham and Professor Byron Newton assisted with a number of examples and exercises. Dr. Robert Wisner taught one version of the manuscript and made valuable suggestions. Mr. Robert Hausser contributed to the exercises. Many typists were involved in various stages of the manuscript, but the assistance of Dolores Gonzales and Peggy Gorski is hereby acknowledged. In the final draft, the assistance of Laura Graham was extremely valuable.

My wife typed at least four versions of this book, so she deserves *some* thanks.

Ralph Crouch

CONTENTS

1

STATEMENTS, SETS, AND DECISION MAKING

1.0 Introduction

Each of the chapters of this book has an introductory section, the primary purpose of which is to give to the reader some broad hints about the kinds of mathematical ideas that are to be presented in the chapter. Of course, the table of contents of a book is another device for presenting this same information in a more precise form. But in these sections an attempt will be made to be a bit more specific. In particular, special attention will be devoted to the sequence of ideas to be developed in the chapter and to the reasons behind the desirability—even the necessity—of assimilating these ideas.

The content of Chap. 1 largely comprises the development of some rather simple but basic mathematical ideas, with the necessary notation, which are to play key roles in all the rest of this book. But a motivation based on the threat "Learn this—you will need it later" leaves something to be desired. This is a mathematics book, and although a sincere effort is made to point out some applications of the mathematics as the ideas are developed, it is still true that one needs to know mathematics before it can be applied. So this is truly a foundations chapter, and in it, the nomenclature of the subject will be developed. Mathematics, by its very nature, is a precise subject which is conveyed by language. We shall begin by discussing precision in language.

Precision in language is essential to the conveyance of ideas, and *statements* are the means of conveyance. Hence a study of statements is the beginning point. But statements have to be interpreted with respect to logical possibilities, which leads to the study of sets of logical possibilities and, more generally, *sets*. This chapter is largely concerned with statements and their *truth sets* of logical possibilities, and the two concepts—statements and truth sets—are developed in parallel.

Certain kinds of statements are more involved than others and are called *compound statements* rather than *simple statements*. Compound statements arise through the use of connectives, and some connectives are *and, or, not, if . . . then,* and *if and only if*. There is a section devoted to each of these connectives. Variations and combinations of these connectives are also discussed. One of the most important uses of these mathematical ideas is in decision making, and to appreciate this one needs to understand the validity of an argument, *consistency* and *implication*. These subjects in their mathematical context are also discussed.

In the understanding of ideas, man is constantly seeking all the aids available. One source of help in understanding is geometry. This is exploited in this chapter by the use of Venn diagrams, which are sprinkled throughout where their use seems natural and desirable.

But by and large, the chapter is concerned with the development of mathematical tools which can be used in situations not yet easily described. Do not neglect the language or the notation developed, since much of what follows will be phrased in expressions first discussed here.

1.1 Statements

At the beginning of almost all the sections of this book, there is at least one special problem for your consideration. The question raised in each of these special problems has been selected with several ideas in mind. First and foremost, to determine the correct answer to the question requires some definition, formula, or fact (or a combination of these) that is to be developed in the section under discussion. Many of these problems are phrased in terms of a business situation and will quite often involve a hypothetical business firm which is created for purposes of illustration. Such special problems are by their nature a bit artificial. But they will have served their purpose if, as a result of these problems, the reader realizes that there are questions about the conduct of business that require mathematics for their successful resolution, and quite a bit of mathematics at that.

As the text is read, study these special problems and try to answer the questions asked. If you are unable to answer the questions posed, do not be disturbed. After all, that is one reason they are included. Just read the section and then try the problems again.

Problem

In a particular year, the annual earnings of four persons are, respectively, $10,000, $7,800, $4,300, and $15,400. Assume that each of the following sentences is expressed by one of these four persons. Determine which of the sentences is (1) a statement, (2) a tautology, (3) self-contradictory.

A. I made less than $16,000.
B. I made $8,750.
C. I paid $1,800 income tax.
D. I made less than $4,000.

One of the means of human communication is by the use of verbal expressions. These verbal expressions in turn lend themselves to written expression in a language. Some of these expressions are labeled *statements,* and the method used to choose from the set of all verbal expressions those which are statements is of primary concern in this section. To make this distinction, there will be a predetermined set of *logical possibilities* that will serve as a *universe* for the considerations, and each expression must be interpreted with respect to the universe of logical possibilities to determine if it "makes sense" and hence is a statement. Under these circumstances, the ones that can be labeled *true* or *false* are statements. The major concern in this section is with simple statements, although compound statements will be considered in much detail later. Examples will aid in understanding these remarks.

Listed below are 10 expressions that will be used for illustrative purposes for these ideas.

Example 1. There are fewer than three legs on my body.

Example 2. There are four heads on my shoulders.

Example 3. My income last year was $10,800.

Example 4. More than one-half the hair on my head is gray.

Example 5. All the clients have been contacted, and the reports have been mailed.

Example 6. The sales of automobiles are increasing, or production is decreasing.

Example 7. The shipment did not arrive on time.

Example 8. If I meet my sales quota, then the district sales manager will be pleased.

Example 9. Look, look.

Example 10. Good morning.

One of the characteristics that distinguishes this set of 10 expressions is that each might be heard in conversation. However, a little reflection shows that the first eight expressions differ in one rather important aspect from the last two. Each of the first eight sentences, if spoken by a given individual under a known set of circumstances, can be labeled as *true* or *false*. This is not so for the expressions "Look, look" and "Good morning." It is the property of possessing a connotation of "true" or "false" that is used as a selector for deciding which ones of a set of expressions are statements.

These considerations lead to the first definition of this book. Definitions in books that deal primarily with mathematics are used just as they are in books that deal with other subject-matter areas. Words (or phrases) will be given meaning in a definition by the use of words whose meanings are generally understood and also words which have previously been defined. Since there will be many definitions in this book that will often be referred to, they will be offset in the manuscript and also italicized so that they are available for quick reference.

Definition: (***Statement***) *An expression is a statement if its truth or falsity can be found through comparison with some predetermined set of logical possibilities.*

This definition will now be applied to each of the first 10 expressions in the examples.

Example 1 is a statement since, with any preconceived set of logical possibilities, it is possible to discover whether the expression is true or false by counting the limbs of the speaker. Since only humans will make such a remark, Example 1 is true regardless of who is the speaker.

A statement which is true for all logical possibilities is a tautology.

In a similar fashion, for any member of the set of persons who might make the statement in Example 2, it is possible to determine that the sentence is false. Because Example 2 is false regardless of who the speaker might be and because such statements are called *self-contradictory*, we arrive at another agreement.

A statement which is false for all members of the set of logical possibilities is self-contradictory.

If the sentence in Example 3 is uttered by any one of several members of a set of persons who pay income tax, then by the process of examining the income tax reports (the individuals or the government willing), the truth or falsity of the sentence can be established. The sentence in this example is another particular case of expressions that are called *simple statements*. It is to be expected that the statement is true for some of the persons in the set of taxpayers under consideration and false for other persons in a different subset.

If the expression in Example 4 is made by one of the persons in a set consisting of, say, Ralph, Jennie, and Jack, then its truth or falsity can be determined with respect to the preconceived set of logical possibilities. Given the set of logical possibilities, the expression becomes a statement, and its truth or falsity is determined by counting the hairs on the head of the speaker. It is to be expected that the sentence is true for certain of the logical possibilities and false for others.

Examples 5 to 8 are examples of *compound statements* (again with a prescribed set of logical possibilities), and they, along with many other such examples, will be discussed at some length in the sections which follow. The only important thing about these examples that need be noted now is that each, given a set of circumstances, is either true or false. Hence, each is an example of a statement. Methods and procedures for determining whether such statements are true will be discussed in Secs. 1.3 to 1.5 and 1.7.

It is reemphasized that Examples 9 and 10 are not statements, because the truth or falsity of the expressions is indeterminable with reference to any conceivable set of circumstances.

In what follows, statements will merit considerable attention, and it is inconvenient to rewrite a statement each time that it is necessary to refer to the statement. Therefore, a single letter of the alphabet will often be used to represent the entire statement. For no reason except that it seems to be customary, the letters p, q, r, etc., will most frequently be used to represent simple statements. When this notation is employed, each of the statements in Examples 1 to 4 can be represented by a single letter of the alphabet since they are examples of simple statements. Consider how this notation could be applied to Example 5. If the simple statement "All the clients have been contacted" is denoted by p and the simple statement "The reports have been mailed" is denoted by q, then the compound statement in Example 5 can be represented as "p and q." In a similar fashion, we could break down the statement in Example 6 and represent it, with the proper choice of statements, as "p or q."

Each of Examples 1 to 6 can be represented by either a single letter of the alphabet or two letters of the alphabet and a *connective*. In Example 5, the connective is *and;* in Example 6, the connective is *or*. Example 7 involves the connective *not*. If the letter t is used to represent the statement "The shipment arrived on time," then "not t" represents the statement "The shipment did not arrive on time." Use of a single letter to represent a statement, along with a little common sense applied to Example 8, and with the agreement that p represents the simple statement "I meet my sales quota" and q represents "The district sales manager will be pleased," becomes "if p, then q."

Since Examples 9 and 10 are not examples of statements and since considerations in the remainder of this book will be limited to statements, there will be no occasion to express them (or any other nonstatement sentences) in this notation.

Exercises

1. Label each of the following as a simple statement, a compound statement, or not a statement:
   a. The board of directors met today.
   b. The quality of the product has increased, and so have sales.
   c. A computer from IBM.
   d. Ford and Chrysler rank behind General Motors in numbers of automobiles sold.
   e. Allied Products and Studebaker decided against merger. (They were considering a merger at a joint meeting.)
   f. Wages must rise or the union will strike.

2. Let p represent the simple statement "Employees are happy," and let q represent the simple statement "Wages are high." Express the following in the notation of this section:
   a. Employees are happy.
   b. Wages are high.
   c. Employees are not happy.
   d. Wages are not high.
   e. Employees are happy, and wages are high.
   f. Employees are not happy, or wages are high.
   g. If wages are not high, then employees are not happy.
   h. Either wages are high, or employees are not happy and wages are not high.

3. For the following, label the simple statements which have been combined to form the compound statements. For example,

   If profits go up, then sales go up or costs go down.

   p: Profits go up.
   q: Sales go up.
   r: Costs go down.

   a. If profits are decreasing, then the market price of the stock goes down and the stockholders are unhappy.

b. If the gross national product increases, then production increases or prices increase.

c. A drop in revenues will cause a firm to cut advertising expenditures, and there will be a further drop in sales.

d. Poor working conditions or low wages will cause employees to be unhappy and strike.

4. Let p represent the simple statement "Sales have increased," and let q represent the simple statement "Working-capital requirements have increased." Express the following using the notation of this section:

   a. Sales have increased, and working-capital requirements have increased.

   b. Sales have not increased.

   c. If sales have increased, then working-capital requirements have increased.

   d. Working-capital requirements have not increased, or sales have increased.

   e. Sales have not increased, and working-capital requirements have not increased.

   f. If working-capital requirements have not increased, then sales have not increased.

   g. Sales have not increased, or working-capital requirements have not increased.

Answer to problem

(1) A, B, and D. C is not a statement because truth or falsity cannot be determined from the given information.

(2) A is true for all possibilities.

(3) D is false for all possibilities.

1.2 *Finite sets and subsets*

At the beginning of Sec. 1.1, we remarked that many of the sections of this book will begin with a problem or problems that will raise questions whose answers are obtainable with the aid of knowledge gained from a study of the section. This policy is executed for this section with the problem below.

Problem

The sales staff for the Ready Book Company in the Eastern region has seven members. The sales manager in the region must decide whether all, some, or none of the salesmen should attend the national sales meeting. How many choices does he have?

Three major concepts are presented in this section: set, subset, and the number of subsets in a set. There will not be any lengthy or deep discussion about theoretical set theory, but some mathematical symbols will be introduced to facilitate future communications. The concern voiced in Sec. 1.0 with regard to precision in the use of language should be kept in mind as these new methods of expression are adopted.

The mathematician (as well as anyone else) finds it necessary to use undefined words in his normal procedure of communication. One of the undefined words in this book is the word *set*. If you are observant, you have noticed that the word *set* has already been used in the previous section. But a definition of set was not given, nor will such a definition be given here. However, what is meant by the word will be made clear by the use of several examples.

Example 1. The Cozy Comfort Company makes sheets, blankets, and comforters for beds. The company likes to use its initials in its advertising, and so the business firm is designated by CCC. The chairman of the board of directors of CCC is Mr. Adams, and the person with the second greatest influence over the affairs of the company is the president, Mr. Beam. These two men, Mr. Adams and Mr. Beam, are the most influential men in the company, and they determine most of its policies. They are also an example of a set, which may be designated in one fashion (or notation) as

$$\{\text{Adams,Beam}\}$$

or simply $\{a,b\}$

Example 2. The board of directors of the company has four other members: Mr. Crow, Mr. Dann, Mr. Ealy, and Mr. Fain. These men also constitute an example of a set, which is designated by

$$\{\text{Crow,Dann,Ealy,Fain}\}$$

or simply as $\{c,d,e,f\}$

Example 3. Also employed by the company is a set of vice-presidents: Mr. Gumm, who is vice-president in charge of sales; Mr. Hand, who is the vice-president in charge of marketing; and Mr. Ivan, who is the vice-president in charge of advertising. In the notation adopted previously, the set of vice-presidents is expressed as

$$\{\text{Gumm,Hand,Ivan}\}$$

or $\{g,h,i\}$

Example 4. Further down in the management team (and a diagram is provided later to show the chain of command) are other members of the promotion department: Mr. Jett, who is division chief in charge of consumer response, and Mr. Klam, who is division chief in charge of production. These two men also constitute an example of a set, which is represented by

$$\{\text{Jett,Klam}\}$$

or

$$\{j,k\}$$

Since all the members of the executive team of CCC will frequently be used for illustrative purposes, it is convenient to give the set consisting of all the men mentioned a name, and we write

$$O = \{a,b,c,d,e,f,g,h,i,j,k\}$$

to represent the officials of the company. A diagram to show the chain of authority was promised, and it follows:

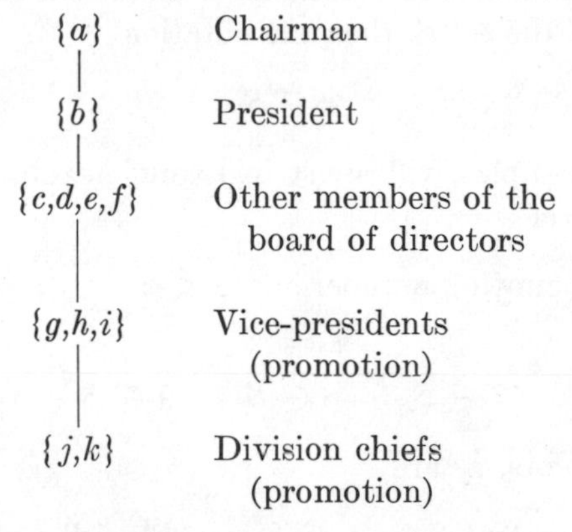

A word about notation: As can be seen from the examples, a set is a collection of objects. In these particular examples, the objects, or *elements*, are persons. The elements need not be persons, and you are familiar with such examples of sets as a set of dishes, a set of automobile tires, a set of chairs in a classroom. The symbols { } are called *braces* and are used to enclose the elements of a set. It is customary (and this custom will be observed here) to denote sets by uppercase letters and elements of the set by lowercase letters.

The method that has been used above to indicate the members of a set is referred to as the *tabulation method*. In this notation, the set is indicated by listing (tabulating) the members of the set. Another method used to describe sets is the *descriptive method*. The expression

$$\{x|x \text{ is a member of the board of directors of CCC}\}$$

is the descriptive way of expressing the set

$$\{\text{Crow,Dann,Ealy,Fain}\}$$

or

$$\{c,d,e,f\}$$

Perhaps you are more familiar with such sets as

$$\{x|x \text{ is a counting number and } 6 < x < 10\} = \{7,8,9\}$$

This is read "The set of elements x such that x is a counting number and x is greater than 6 and less than 10." The set of counting numbers whose elements are 7, 8, and 9 is first written in the descriptive notation and then tabulated.

The notation used to represent *membership* in a set is $\in$. To explain how this symbol is used, if x is a member of set S, then membership is expressed by the symbols

$$x \in S$$

which is read "The element x is a member of the set S." Similarly, if x is *not* a member of the set S, then the notation

$$x \notin S$$

is used. Some examples will assist in fixing meaning to the symbols. Refer back to the set

$$\{x|x \text{ is a counting number and } 6 < x < 10\} = \{7,8,9\}$$

and observe that

$$7 \in S \qquad 8 \in S \qquad 9 \in S$$

are all true statements, as are

$$1 \notin S \qquad 6 \notin S \qquad 143 \notin S$$

This concludes, for the present, the discussion of sets, their elements, and their notation, but the next idea is very closely related.

Perhaps you observed that in Sec. 1.1 the word *subset* was also used without a definition being provided. If it is now clear how the word *set* is to be used to refer to a collection of objects, then the word *subset* is also intuitively clear. To return to earlier examples, we speak of the subset of cups in a set of dishes, the subset of front tires in a set of tires, the subset of chairs in the first row in a set of classroom chairs. In most discussions, there will be some fixed set, which is called the *universal set* (and represented by U); and if it is desired to discuss parts (or portions) of U, then these parts of the universal set are subsets. To illustrate this, refer again to the set O of officials of CCC (page 9). As a particular case, it may be desirable to select from the set O the subset whose elements are the members of the board of directors or, for another example, the subset of the vice-presidents of CCC. Each of these is an example of a subset of the set of officials. Observe also that the set of vice-presidents is a subset of the set of members of the entire promotion department (which includes j and k). Written in tabular form, these three examples become

Example 5. $\{c,d,e,f\}$ is a subset of $\{a,b,c,d,e,f,g,h,i,j,k\}$.

Example 6. $\{g,h,i\}$ is a subset of $\{a,b,c,d,e,f,g,h,i,j,k\}$.

Example 7. $\{g,h,i\}$ is a subset of $\{g,h,i,j,k\}$.

A definition will assist in fixing these ideas and will lend precision to future discussions.

Definition: (**Subset**) *If S and T are sets, then S is a subset of T if and only if each element of S is also an element of T. The notation employed is $S \subseteq T$.*

The symbols $S \subseteq T$ are read "S is a subset of T, or equivalently, S is contained in T." Under many circumstances, T will contain an element or elements that are not members of S. (Examples 5 to 7 are all of this type.) Under these circumstances, S is called a *proper subset* of T, and we write

$$S \subset T$$

Given two sets S and T, there is no reason to expect that one is a subset of the other. To illustrate, $\{g,h,i\}$ is *not* a subset of $\{i,j\}$. The notation to represent this situation parallels the symbols $\in$ and $\notin$ chosen earlier to represent "member of" and "not a member of," respectively. If S is not a subset of T, then

$$S \nsubseteq T$$

represents this statement.

If the set of logical possibilities for a statement p is considered as a universal set U, then the set of logical possibilities for which the statement p is true is a subset of U—as is the set of logical possibilities for which the statement p is false.

Example 8. To illustrate, in Example 3 of Sec. 1.1, the statement

p: My income last year was $10,800.

is given. Suppose that a set of logical possibilities is chosen to be the set represented by $\{c,d,e,f\}$. Suppose further that information about the actual earnings of the persons involved is available and is

c, $19,600 *d*, $11,000 *e*, $10,800 *f*, $9,000

Then the subset of these logical possibilities for which statement p is true is $\{e\}$. The subset of logical possibilities for which the statement p is false is $\{c,d,f\}$.

Example 9. A similar analysis can be carried out for the statement in Example 4 of Sec. 1.1. Recall that the statement is

q: More than one-half the hair on my head is gray.

We chose as a set of logical possibilities the set

$$U = \{\text{Ralph,Jennie,Jack}\}$$

and we determine by inspection that the set for which statement q is true is

$$\{\text{Ralph}\}$$

The other elements of the set of logical possibilities are the subset of U for which q is false:

$$\{\text{Jennie,Jack}\}$$

In each of these examples, observe that the choice of the universal set is arbitrary and is a function of the person who makes the analysis. This observation will be of prime importance when this type of analysis is applied to decision making.

Example 10. Before leaving the subject of the set of logical possibilities and certain subsets thereof, there are two special cases that deserve careful attention. For the first of these two special cases, recall the statement in Example 1 of Sec. 1.1:

r: There are fewer than three legs on my body.

For any normal set of humans, which is the set U of logical possibilities for statement r, the subset R of U for which r is true is U; that is, $R = U$. For those special statements which are true for all members of the set of logical possibilities (*tautologies*), the subset of U selected as the truth subset is U.

One subset of U is U.

Example 11. In a similar fashion, we reason that statements such as the statement in Example 2 of Sec. 1.1,

t: There are four heads on my shoulders.

(which is *self-contradictory*) is true for none of the logical possibilities. Consequently, the subset T for which statement t is true has no members. From this example it can be concluded that another special subset of any set is the set with no members. The set which contains no elements is the *empty set*. The notation Φ is used to represent the empty set.

One subset of U is Φ.

The notation $\{\ \}$ is also used to denote the empty set. The braces with no elements enclosed is suggestive of emptiness and hence in some ways is "better." Both notations will be used in this book.

This section will be concluded by counting the number of subsets of a set. Counting is an important mathematical activity with a wide range of applications. Chapter 2 of this book is devoted solely to counting, and there are many other places in the book where counting is an essential part of the mathematical procedure. The first counting problem to be solved here is easily stated:

If the number of elements in a set U is known, say n, what is the number of subsets of U?

This question will be answered in the discussion which follows. Since this question involves counting and hence the use of counting numbers, recall that the *counting numbers* are

$$C = \{0,1,2,3, \ldots\}$$

Familiarity with this set, with such operations as addition and multiplication on the set, and with various properties of the operations is assumed of the reader.

A complete answer to the question posed will be given, but an example to point the way toward the answer will prove very helpful. (Indeed, this will be the procedure throughout the book.)

Example 12. Suppose that every morning some of the vice-presidents of CCC go to the company cafeteria for a cup of coffee. The coffee-drinking habits of Mr. Gumm, Mr. Hand, and Mr. Ivan vary, but we assume that each likes coffee and hence may or may not partake. Since the number of elements in the set is relatively small (three), a complete list of all possible subsets can be obtained easily. These subsets are tabulated in an orderly fashion as

$$\begin{array}{llll} \{g,h,i\} & \{g,h\} & \{g\} & \{\ \ \} \\ & \{g,i\} & \{h\} & \\ & \{h,i\} & \{i\} & \end{array}$$

The set consisting of all the vice-presidents is listed since all three may decide to get coffee. The next three sets represent those situations where two of the vice-presidents went to get coffee, and the next column of three sets represents the situations in which only one of the vice-presidents went to get coffee. Occasionally, of course, none of the three vice-presidents goes, and the empty set, denoted by $\{\ \ \}$, properly represents this occurrence. Notice that we have not listed both $\{g,h\}$ and $\{h,g\}$, since they represent the same set.

The order in which the elements of a set are listed does not change the set.

Notice also that we have not listed $\{g,h,g\}$, since we consider that $\{g,h,g\} = \{h,g\}$.

Elements of a set are listed but once in a tabulation.

With these agreements, the list is complete, and a count shows that the set of three vice-presidents has eight subsets.

Example 13. As another example, consider the set consisting of promotion heads, Mr. Jett and Mr. Klam. The subsets of this set are tabulated as

$$\{j,k\} \qquad \{j\} \qquad \{k\} \qquad \{\ \}$$

Why is the subset $\{k,j\}$ not in the list? The subset $\{j,k,j\}$? If a set S consists of two members, then the number of subsets of S is four.

Example 14. For another example, note that the subsets of the set consisting of a single member, President Beam, are

$$\{b\} \qquad \{\ \}$$

If a set consists of a single element, then the number of subsets is two.

These last three examples are sufficient to establish a pattern of the counting procedure. So that the pattern is more easily discernible, a summary is given:

Number of elements in the set	*Number of subsets*
1	$2 = 2$
2	$4 = 2 \cdot 2$
3	$8 = 2 \cdot 2 \cdot 2$

A simple observation is: The number of subsets is obtained as a product which uses 2 as a factor the same number of times as the number of elements in the set. We assert without any further justification (yet) that

The number of subsets contained in a set with n members is the product $2 \cdot 2 \cdot \cdots \cdot 2$ *(n factors).*

Example 15. To test this statement in one more special case, the subsets of the set $\{c,d,e,f\}$ are tabulated below.

$\{c,d,e,f\}$	$\{c,d,e\}$	$\{c,d\}$	$\{c\}$	$\{\ \}$
	$\{c,d,f\}$	$\{c,e\}$	$\{d\}$	
	$\{c,e,f\}$	$\{c,f\}$	$\{e\}$	
	$\{d,e,f\}$	$\{d,e\}$	$\{f\}$	
		$\{d,f\}$		
		$\{e,f\}$		

The number of subsets in this set of four elements is

$$1 + 4 + 6 + 4 + 1 = 16 = 2 \cdot 2 \cdot 2 \cdot 2$$

The assertion made above is verified for one more case (when the set contains four elements). (It is clear the listing would be essentially the same regardless of the nature of the elements in the set.)

These examples do not establish that the conjecture is true for *any* counting number n. Theorems about *all* counting numbers are not established by testing a few cases (although, had the investigation of the case for $n = 4$ led to a result contrary to the conclusion that there are $2 \cdot 2 \cdot 2 \cdot 2$ subsets, it could have been concluded that the conjecture is *not* true for all counting numbers). What is desired now is an argument to support (establish) the result for any set of n elements. Many arguments of this type will be given in this book, but such proofs (*proof* is what such arguments are called) will not be given in a formal, rigorous manner. The theorem (*theorem* is what such statements are called) is

If S is a set of n elements, then S has $2 \cdot 2 \cdot \cdots \cdot 2$ (n factors) subsets.

Suppose that a subset of set S (S has n elements) is to be formed. Then each individual element is considered, and a decision to select or reject the element as a member of the subset is made. For each element, there is a choice—it either belongs to the subset or does not belong to the subset. So, for the first element, there is a choice of two different things. The second element may or may not be in the subset that is being constructed, so again a choice of two things is made. This gives a total of $2 \cdot 2$ different choices for the first two elements. If this choice process is continued, then there are

$$2 \cdot 2 \cdot \cdots \cdot 2 \qquad (n \text{ factors})$$

choices to make. Stated differently, there are $2 \cdot 2 \cdot \cdots \cdot 2$ (n factors) *different* subsets of S. Another argument to establish this same result is given in Sec. 2.4.

We close with two observations. First, there is a shortcut method of indicating the product of n factors of 2. Such a product is written as

$$2^n = 2 \cdot 2 \cdot \cdots \cdot 2 \qquad (n \text{ factors})$$

The number 2 is called the *base,* and n is called the *exponent.*

Second, if the set S is empty, then the choice argument given above becomes nebulous. However, the empty set $\{\ \}$ has exactly one subset, $\{\ \}$. Recall that we observed earlier that every set contains itself as a (nonproper) subset. Since it is desirable that 2^0 be 1, such an agreement is made, and henceforth we write $2^0 = 1$.

Mathematical statements are often summarized in a special kind of symbolic expression called a *formula.* The main result of the remarks about counting subsets can be summarized in a formula. This is the first of several formulas to be presented in this book, and each of them will be set off so that they are readily available.

Formula: **(*Number of subsets in a set*)** *If S is a set such that the number of elements in S is n, then the number of subsets of S is 2^n.*

This formula can be used to compute the number of subsets of any set S with n elements. Particular instances are if S has 5 elements, then S has $2^5 = 2 \cdot 2 \cdot 2 \cdot 2 \cdot 2 = 32$ subsets; if S has 6 elements, then S has $2^6 = 2 \cdot 2 \cdot 2 \cdot 2 \cdot 2 \cdot 2 = 64$ subsets; and if S has 14 elements, then S has $2^{14} = 16{,}384$ subsets.

Exercises

1. Referring to the set O of the officials of CCC, let D represent the subset of board of directors and V the subset of vice-presidents. Determine which of the following are true:

   a. $a \in D$ *b.* $a \notin D$ *c.* $d \notin V$
   d. $j \notin V$ *e.* $k \in D$ *f.* $b \notin D$
   g. $e \in V$ *h.* $h \in V$ *i.* $h \notin D$
   j. $a \in V$ *k.* $D \subset V$ *l.* $V \not\subset D$
   m. $D \not\subset D$ *n.* $V \subset D$ *o.* $D \not\subset V$

2. Form five different subsets from each of the sets listed below. (x represents an element in the set of counting numbers, and $C = \{0,1,2,3,4, \ldots\}$.)

   a. $\{x|4 < x < 7\}$ *b.* $\{0,1,2\}$ *c.* $\{4,2\}$
   d. $\{x|7 > x\}$ *e.* $\{2,2^2,2^3\}$

3. Determine which of the following are true. For this exercise, let

   $$S = \{0,1,3,5,7\} \quad T = \{0,2,4,6,8\} \quad R = \{0,1,3\}$$

   a. $S \subset T$ *b.* $T \subset S$ *c.* $R \subset T$
   d. $S \neq T$ *e.* $T \subset R$ *f.* $0 \notin S$
   g. $4 \in T$ *h.* $R \notin S$

4. Refer to the result obtained in this section with regard to the number of subsets of a given set and determine the number of subsets of the set $\{1,2,3,4,5\}$.

   a. List three subsets such that each contains four elements.
   b. List 12 subsets such that each contains 10 elements.

5. The officials of CCC, $\{a,b,c,d,e,f,g,h,i,j,k\}$, are going to vote for one of their members to represent them at a district sales meeting.

   a. If only two members are under consideration, in how many ways could they vote? (Assume the candidates do not vote.)
   b. In how many ways could they vote if one member abstains?

6. The purchase of a large quantity of material may be made by a construction company. The purchasing agent may be allowed to decide from which company the purchase will be made. The companies bidding are denoted by A, B, and C. A table of logical possibilities is given below:

	Company buys	***Purchasing agent makes choice***	***Company A bids low***	***Company B bids low***	***Company C bids low***
P_1	No	Yes	Yes	No	No
P_2	No	Yes	No	Yes	No
P_3	No	Yes	No	No	Yes
P_4	No	No	Yes	No	No
P_5	No	No	No	Yes	No
P_6	No	No	No	No	Yes
P_7	Yes	Yes	Yes	No	No
P_8	Yes	Yes	No	Yes	No
P_9	Yes	Yes	No	No	Yes
P_{10}	Yes	No	Yes	No	No
P_{11}	Yes	No	No	Yes	No
P_{12}	Yes	No	No	No	Yes

The president of the company suggests that to analyze the company's position, all possible subsets of the set of possibilities should be studied. The purchasing agent says that there are over four thousand subsets of a set of 12 elements. Is he correct? Why?

7. The president of company B in Exercise 6 believes that the construction company will buy and that the purchasing agent will not be allowed to decide the seller.

   a. Write the subset of logical possibilities if he is correct.
   b. How many subsets has this set?

8. The president of company C in Exercise 6 believes that his company will be low bidder.

   a. Write the subset of logical possibilities if he is correct.
   b. How many subsets has this set?

9. Given $S = \{a,b,c\}$:

   a. List all subsets of S.
   b. List all proper subsets of S.

Answer to problem

$2^7 = 128$. The set has seven elements. The number of subsets possible is $2^n = 2^7 = 128$.

1.3 Conjunction of statements and intersection of sets

Problems

Given the following compound statement:

The membership of labor unions has increased, and wages have risen.

A. If the simple statement "Membership of labor unions has increased" is true and the simple statement "Wages have risen" is true, is the given statement true?

B. If the simple statement "Membership of labor unions has increased" is true and the simple statement "Wages have risen" is false, is the given statement true?

C. If the simple statement "Membership of labor unions has increased" is false and the simple statement "Wages have risen" is true, is the given statement true?

D. If the simple statement "Membership of labor unions has increased" is false and the simple statement "Wages have risen" is false, is the given statement true?

In the preceding sections, precise meaning has been given to such expressions as *simple statement, set, element, subset, contained in,* and *proper subset.* These concepts, with the accompanying notations, will serve as a foundation for the ideas to be discussed in the remainder of this chapter (and, for that matter, in the remainder of the book).

In this section, we shall consider one of the *connectives* used to form compound statements from simple statements. At least one example of a compound statement formed from two simple statements by the use of the connective *and* is already on exhibit.

Example 1. In Sec. 1.1, the following statement was considered:

All the clients have been contacted, and the reports have been mailed.

Recall also that in Sec. 1.1, a notation was developed for expressing such compound statements. If p represents the simple statement "All the clients have been contacted" and if q represents the simple statement "The reports have been mailed," then the notation that has been agreed upon can be used to express the statement in the example as

$$p \text{ and } q$$

It will be one of the primary concerns of this section to determine when a statement of the form "p and q" is to be labeled as true. Also recall that in order to decide whether statements are true, a set of logical possibilities is made and the statement is considered in terms of this set.

There are four logical possibilities for the statements p and q, which will now be discussed in an effort to arrive at some common agreements. Again, the goal is to decide when "p and q" is true.

For the first case, suppose that it is known that all the clients have been contacted and that all the reports have been mailed.

1. It may be that both p and q are true.

Under these circumstances, anyone who hears the statement "p and q" would undoubtedly agree that it is true. So it is agreed that if both statements are true, then "p and q" is true also.

Suppose, however, that it is known that all the clients have been contacted but that all the reports have not been mailed.

2. It may be that p is true and q is false.

Then, if told that "p and q," we would be inclined to disbelieve it and call the total statement false. So let us agree that if p is true and q is false, the compound statement "p and q" will be considered false.

As a third case, suppose it is known that all the clients have not been contacted but, in spite of this, all the reports have been mailed.

3. It may be that p is false and q is true.

Under these circumstances, if the compound statement "p and q" were made, we would again be inclined to disbelieve it and label the compound statement as false. So it is agreed that if statement p is false and statement q is true, the compound statement "p and q" is false.

The remaining possibility is that all the clients have not been contacted and the reports have not been mailed.

4. It may be that p is false and q is false.

Certainly in this case we would label the compound statement "p and q" false. So if p and q are false, the statement "p and q" is false.

If the word *and* is used to form a compound statement from other statements, then the new statement is called the *conjunction*. It is customary to denote the connective *and* by the symbol $\wedge$, a convention to be observed here. With this notation, the four cases are summarized in Table 1. Such tables are referred to as *truth tables;* T represents true, F represents false.

Some instructions about the way to read Table 1 may be helpful. The entries in each row refer to one of the four cases. Reading horizontally, if p is true (T) and q is true (T), then "p and q" ($p \wedge q$) is true (T). Similarly, the third row is read "p false, q true implies that (p and q) is false." To verify your understanding, read the second and fourth rows of Table 1.

Table 1

p	q	$p \wedge q$
T	T	T
T	F	F
F	T	F
F	F	F

The purpose of the first example was to lead to a *reasonable* agreement for the truth value to be assigned to compound statements of the form "p and q." This agreement forms a definition which is ordinarily taken for the truth value of such compound statements. Since this definition will be referred to in future considerations, it is restated below; again, it is set off for convenience in reference.

Definition: **(*Truth value of the conjunction of two statements*)** *If p and q are statements, the compound statement "p and q" is true if and only if both p and q are true; otherwise, it is false.*

The concepts of *statement* and *set* are very closely connected, and it will be the main purpose of the remainder of this section to show how set notation can be used to illustrate truth sets for conjunctive statements.

Perhaps it is in order to review the agreements of this section. For any statement which is compounded from two other statements by the use of the connective *and*, a set of logical possibilities is determined by considering all the possibilities for the statement "p and q." The only one of these four possibilities for which we label the compound statement true is when *both* p and q are true. In the remaining three cases, the compound statement is false.

Now a new notation will be used to illustrate the set of logical possibilities and some of its subsets. The following notation is adopted: The letter U represents the set of logical possibilities, the letter P represents the subset of logical possibilities for which statement p is true, and letter Q represents the subset of logical possibilities for which the statement q is true. Diagram 1 illustrates this notation. (Diagrams of this type are called *Venn diagrams.*)

The shaded area in Diagram 1 (the area which contains the number 1) represents the subset of logical possibilities for which the statement "p and q" is true. The areas labeled 2, 3, and 4 represent the remaining cases. The numbers have been chosen to correspond to the cases in the truth table: 2 corresponds to p true, q false; 3 corresponds to p false, q true; 4 corresponds to p false, q false.

Lest Diagram 1 mislead you into believing that each of the four

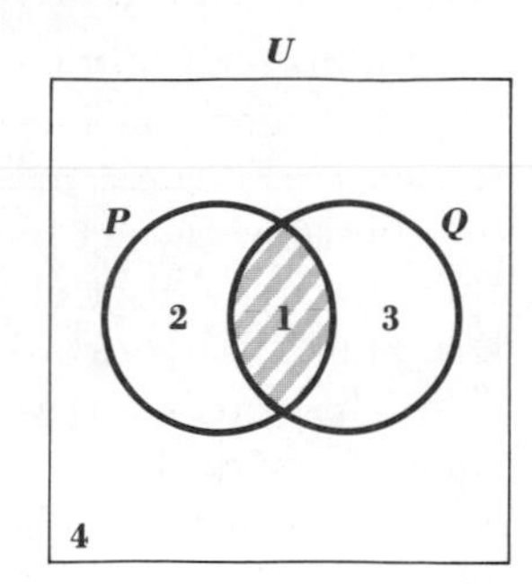

Diagram 1

subsets contains but one element of the set U, we hasten to give a second example of two statements, the conjunction of the statements, and the diagram which illustrates the statements.

Example 2. A salesman for a manufacturer of women's clothing travels in the Midwest. He finds it convenient to partition his set of customers into two categories: *occasional customers or regular customers.* Suppose that he also partitions the same set of customers into three other categories: (1) purchasers of large amounts, (2) purchasers of medium amounts, *or* (3) purchasers of small amounts. Now consider the compound statement

Burns Department Store is a regular customer and does not purchase small amounts.

This compound statement is divided into two parts: the statement

p: Burns Department Store is a regular customer.

and the statement

q: Burns Department Store does not buy small amounts.

The compound statement that we have under consideration can be written in terms of statements represented by p and q as "p and q." Following procedures that have been indicated before and using obvious shortcuts in notation, the set U of logical possibilities is given in Table 2.

Table 2

Logical possibilities	*Frequency*	*Quantity*
1	Regular	Large
2	Regular	Medium
3	Regular	Small
4	Occasional	Large
5	Occasional	Medium
6	Occasional	Small

A little reflection will show that these six possibilities are all that can arise as a result of the categories that have been used. As in Diagram 1, the set U of logical possibilities is divided into four subsets. With P as the subset of the set of logical possibilities for which statement p is true and Q as the subset of logical possibilities for which statement q is true, Diagram 2 illustrates the sets U, P, and Q.

The numbers that appear in Diagram 2 refer to the entries of logical possibilities in Table 2. Observe that logical possibilities 1 and 2 fall in the area for which both p and q are true. Logical possibility 3 is true for p but false for q. Logical possibilities 4 and 5 are true for q but false for p. Logical possibility 6 is false for both p and q. Because of the definition for the truth value of "p and q," the set consisting of logical possibilities 1 and 2 is the set for which "p and q" is true.

There is a new (but related) use to which the word *and* can now be put. Previously, *and* has been used as a connective for statements. But it can also apply to elements of sets. When *and* is used as in the statement "p and q," it means *both*. This leads to another definition. (You are reminded that definitions are especially set off for quick reference.)

Definition: **(*Intersection of sets*)** *If P and Q are subsets of the set U, then the intersection of the sets P and Q is the set of elements in P and in Q. Symbolically, we write*

$$P \cap Q = \{x | x \in P \textit{ and } x \in Q\}$$

Reread the definitions for "p and q" and "P and Q" to see the very close connection between the two usages of the word *and*. To make this even more apparent, the two notations that have been adopted, $\wedge$ and $\cap$, are chosen so that their similarity of appearance will serve as a reminder of the relationship.

Diagrams such as Diagrams 1 and 2 can be helpful in illustrating sets, but diagrams can also be misleading. Only one of three possible relations between P and Q has been illustrated, so two more examples are given to illustrate the other cases.

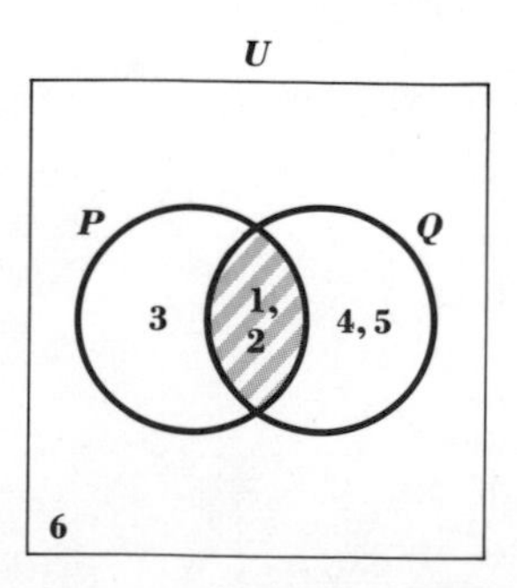

Diagram 2

Example 3. This example will use sets whose elements are counting numbers because (for one reason) it is desirable to become better acquainted with this set of numbers for use in later work. Let U be the set

$$U = \{0,1,2,3,4,5,6,7,8,9,10,11,12\}$$

let P be the set

$$P = \{0,1,2\} = \{x|x \in U \text{ and } x < 3\}$$

and let Q be the set

$$Q = \{0,1,2,3,4,5\} = \{x|x \in U \text{ and } 0 \leq x \leq 5\}$$

(The symbol $0 \leq x$ is read "0 is less than or equal to x." The symbol $x \leq 5$ is read "x is less than or equal to 5." More generally, the symbol $a \leq b$ is read "a is less than or equal to b.")

The definition for the intersection of sets can be used to compute $P \cap Q$. The definition states that the set $P \cap Q$ is the set of elements which are in *both* P and Q. By observation,

$$P \cap Q = \{0,1,2\} = P$$

A diagram like Diagrams 1 and 2 can be used to represent sets U, P, Q, and $P \cap Q$. Diagram 3 does just that.

Diagram 3 differs from Diagrams 1 and 2 in that set P is a subset of set Q.

The point of this example is that there are situations such that the set P is a subset (in this case, proper subset) of Q. Nevertheless, the intersection of the sets P and Q is still determined by identifying the elements in both sets. All such elements are in P, and this result is italicized below for easy reference.

If $P \subseteq Q$, then $P \cap Q = P$.

A little thought will show that this situation is *symmetric*, hence

If $Q \subseteq P$, then $Q \cap P = Q$.

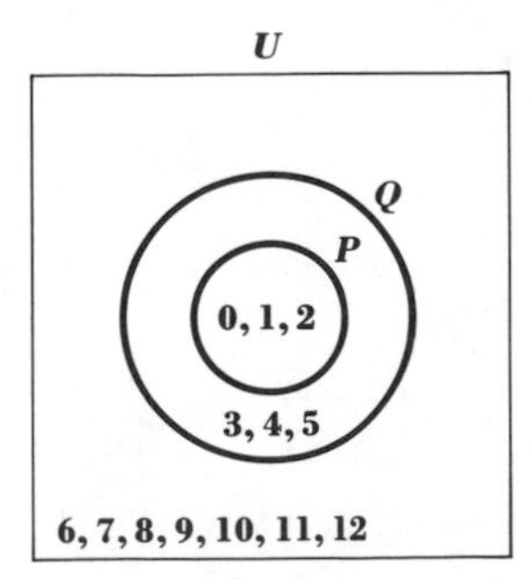

Diagram 3

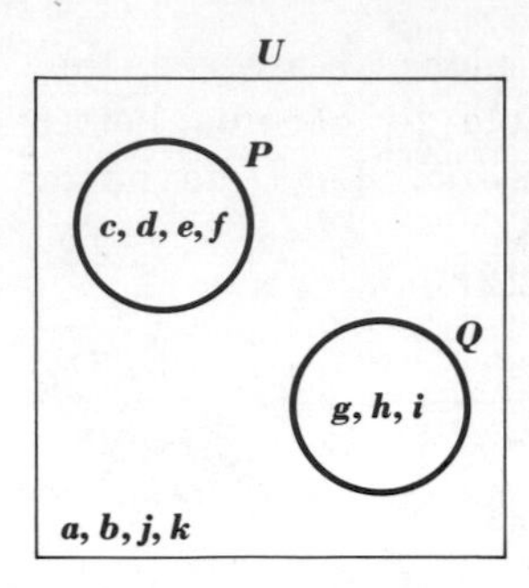

Diagram 4

There are no statements p and q given here to illustrate the case for $Q \subseteq P$, but such statements can be constructed, and Exercise 11 suggests that you do so.

Example 4. As another example, let $U = \{a,b,c,d,e,f,g,h,i,j,k\}$. Let $P = \{c,d,e,f\}$ and $Q = \{g,h,i\}$. Diagram 4 represents the sets.

In this example, the sets P and Q have no elements in common, and consequently their intersection is the empty set. Compare Diagrams 2, 3, and 4 to see the different formats that may occur.

In summary, (1) two sets may have some but not all elements in common; (2) one of the sets may be contained in the other; or (3) the two sets may have no elements in common.

We have shown how (given statements p and q) the connective *and* can be used to form the conjunction of the two statements "p and q." Obviously, the connective *and* can be used to form the conjunction of three (or more) statements. In what follows, the conjunction of three statements is defined. After the truth value for the conjunctions of three statements is determined, an example is given.

Suppose that p, q, and r are statements. The set of logical possibilities contains eight elements, as listed in Table 3.

Table 3

Logical possibilities	p	q	r
1	T	T	T
2	T	T	F
3	T	F	T
4	T	F	F
5	F	T	T
6	F	T	F
7	F	F	T
8	F	F	F

The definition for the truth value of $p \wedge q$ given by Table 1 can be used to compute Table 4.

Table 4

Logical possibilities	p	q	$p \wedge q$
1	T	T	T
2	T	T	T
3	T	F	F
4	T	F	F
5	F	T	F
6	F	T	F
7	F	F	F
8	F	F	F

The definition can be applied again for the results given in Table 5.

Table 5

Logical possibilities	$p \wedge q$	r	$(p \wedge q) \wedge r$
1	T	T	T
2	T	F	F
3	F	T	F
4	F	F	F
5	F	T	F
6	F	F	F
7	F	T	F
8	F	F	F

The entries in Table 5 are obtained in the order $p \wedge q$ and then $(p \wedge q) \wedge r$. Another way that the definition could be used would be to compute $q \wedge r$ and then $p \wedge (q \wedge r)$. The final column of truth values arrived at in the second fashion will agree with those in Table 5, but it is suggested that the reader perform this computation. The common result is taken as the definition.

Definition: **(*Truth value for the conjunction of three statements*)** *If p, q, and r are statements, then the compound statement "p and q and r" $[(p \wedge q) \wedge r = p \wedge (q \wedge r)]$ is true if and only if p, q, and r are all true; otherwise, it is false.*

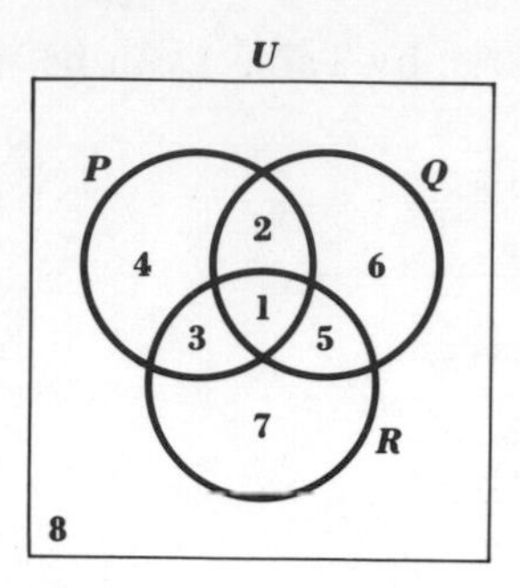

Diagram 5

Just as the preceding definition applies to the conjunction of three statements, a definition of the intersection of three sets can be formulated. Exercise 8 requires that this be done. We conclude with a diagram for the intersection of three sets and an example.

Diagrams have been used to illustrate the truth sets of two statements (and a set which represents the truth set for the conjunction of two statements). Similarly, a diagram which represents the truth set for three statements can be constructed. If three sets are involved, there are eight logical possibilities. The eight possibilities are represented by Diagram 5.

The numbers in the diagram refer to the logical possibilities in Truth Table 5.

The discussion has been without reference to specific sets, so an illustration using sets which are familiar is offered.

Example 5. Suppose that the president and the chairman of the board of CCC have decided that they need a small committee to study some of the policies of CCC. They have decided that they will select a three-man committee, to be chosen as follows: one member will be from the board of directors, the second member will be one of the vice-presidents, and the third member will be either Mr. Jett or Mr. Klam. First of all, consider in how many ways this three-man committee can be selected. There are four choices for the first position, three choices for the second position, and two choices for the third position. This means that there are $4 \cdot 3 \cdot 2 = 24$ different choices. The 24 different logical possibilities are listed in Table 6.

In order to illustrate the conjunction of three statements, let p, q, and r be as follows:

p: Mr. Dann is a member of the committee.
q: Mr. Hand is a member of the committee.
r: Mr. Klam is a member of the committee.

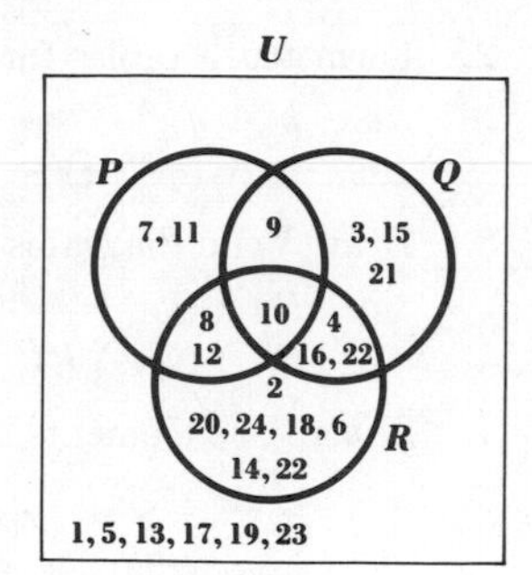

Diagram 6

The truth sets for each of these three statements follow. (The numbers in the sets refer to the logical possibilities in Table 6.)

$$P = \{7,8,9,10,11,12\}$$
$$Q = \{3,4,9,10,15,16,21,22\}$$
$$R = \{2,4,6,8,10,12,14,16,18,20,22,24\}$$

The universal set together with the eight subsets shown in Diagram 5 can now be given for this example, and this is done in Diagram 6. The numbers in Diagram 6 refer to the logical possibilities in Table 6.

Table 6

Logical possibilities	*Members*	*Logical possibilities*	*Members*	*Logical possibilities*	*Members*
1	*c,g,j*	9	*d,h,j*	17	*e,i,j*
2	*c,g,k*	10	*d,h,k*	18	*e,i,k*
3	*c,h,j*	11	*d,i,j*	19	*f,g,j*
4	*c,h,k*	12	*d,i,k*	20	*f,g,k*
5	*c,i,j*	13	*e,g,j*	21	*f,h,j*
6	*c,i,k*	14	*e,g,k*	22	*f,h,k*
7	*d,g,j*	15	*e,h,j*	23	*f,i,j*
8	*d,g,k*	16	*e,h,k*	24	*f,i,k*

If the set of logical possibilities in Table 6 is the one considered but statements p, q, and r are changed, then a resulting change is effected in Diagram 6. Exercises 12 and 13 are directed to this point.

Exercises

1. Given

$$S = \{a,b,c,d,e\}$$
$$T = \{a,e,i,o,u\}$$

Find $S \cap T$.

2. Form truth tables for:

   a. $p \wedge q$ *b.* $q \wedge p$
   c. $p \wedge (q \wedge r)$ *d.* $(p \wedge q) \wedge r$

3. Draw Venn diagrams for:

   a. $P \cap Q$ *b.* $Q \cap P$
   c. $P \cap (Q \cap R)$ *d.* $(P \cap Q) \cap R$

4. List all the elements of S if

$$S = \{x|x \text{ is a counting number and } x^2 \leq 40\}$$

5. Given the compound statement

   The cost of living increased and the national debt decreased in the decade 1940 *to* 1950.

   If the simple statement "Cost of living increased in the decade 1940 to 1950" is true and the simple statement "National debt decreased in the decade 1940 to 1950" is false, is the given statement true?

6. Given the compound statement

   Income per blue-collar employee has decreased, and their percentage of the U.S. labor force has also decreased.

   a. If the simple statement "Income per blue-collar employee has decreased" is false, is the given statement true? (Make no assumptions about the blue-collar percentage of the labor force.)
   b. If the simple statement "Income per blue-collar employee has decreased" is true, is the given statement true? (Make no assumptions about the blue-collar percentage of the labor force.)

7. Choose statements p and q such that each of the compound statements can be represented by "p and q."

   a. The waves are high, and the water is cold.
   b. Jack drove and arrived safely.
   c. Both Dee and Liliah came.
   d. Rain and sleet are both falling.
   e. Ralph and Richard went together.

8. Give a definition for the intersection of three sets P, Q, and R.

9. Determine which of the statements below are true. All statements refer to the set of officials of the CCC. Observe that the statements represent the four cases in the truth table for conjunction.

   a. Mr. Adams is vice-president in charge of advertisements, and Mr. Beam is president.
   b. Mr. Hand is a vice-president, and so is Mr. Beam.
   c. Mr. Ealy and Mr. Fain are members of the board of directors.
   d. Mr. Klam and Mr. Ivan are members of the board of directors.

10. Write four statements of the type "p and q" to represent each of the four logical possibilities in the truth table for conjunction (other than those in Exercise 9).

11. Give an example of a statement "p and q" such that $Q \subset P$.
12. For each of the situations below, refer to the set of logical possibilities in Table 6. Find the subsets of logical possibilities which are the truth sets for the statement p and for the statement q, and draw a diagram showing these subsets.
    a. p: Mr. Crow is a member of the committee.
    q: Mr. Jett is a member of the committee.
    b. p: Mr. Ealy is not a member of the committee.
    q: Mr. Dann is a member of the committee.
    c. p: Mr. Dann is a member of the committee.
    q: Mr. Fain is a member of the committee.
    d. p: Both Mr. Crow and Mr. Jett are members of the committee.
    q: Mr. Gumm is a member of the committee.
13. Do Exercise 12 for statements p, q, and r as listed below.
    a. p: Mr. Klam is not a member of the committee.
    q: Mr. Jett is a member of the committee.
    r: Mr. Gumm is a member of the committee.
    b. p: Mr. Jett is a member of the committee.
    q: Mr. Gumm is a member of the committee.
    r: Mr. Hand is a member of the committee.
    c. p: Mr. Klam is a member of the committee.
    q: Mr. Crow is a member of the committee.
    r: Mr. Hand is a member of the committee.
14. Refer to Exercises 6 to 8 of Sec. 1.2 and write the subset of logical possibilities if:
    a. The president of company B is correct.
    b. The president of company C is correct.
15. Make a truth table for $p \wedge (q \wedge r)$. Verify that it is the same as the one for $(p \wedge q) \wedge r$.

Answers to problems

A. Yes *B.* No *C.* No *D.* No

1.4 *Disjunction of statements and union of sets*

Problems

Given the following compound statement:

President Theodore Roosevelt was famed as a trustbuster, or he was a friend of big business.

A. If the simple statement "President Theodore Roosevelt was famed as a trustbuster" is true and the simple statement "He was a friend of big business" is true, is the given statement true?

B. If the simple statement "President Theodore Roosevelt was famed as a trustbuster" is true and the simple statement "He was a friend of big business" is false, is the given statement true?

C. If the simple statement "President Theodore Roosevelt was famed as a trustbuster" is false and the simple statement "He was a friend of big business" is true, is the given statement true?

D. If the simple statement "President Theodore Roosevelt was famed as a trustbuster" is false and the simple statement "He was a friend of big business" is false, is the given statement true?

Example 1. Miss Hardin, who is receptionist for a dentist, is busy at her desk when someone passes by to enter her employer's office. An insurance salesman is also waiting to call on her employer and asks Miss Hardin who it was that just entered the door. Since Miss Hardin was busy and only had a partial view of the visitor, she replies, "That was Mr. Jones or Mr. Smith." What Miss Hardin means by this remark is that the visitor is one or the other, but not both. In this example, the word *or* is used in the *exclusive* sense.

There is a different meaning for the word *or*, which is known as the *inclusive* sense. It is the inclusive meaning that is most frequently used in conversation involving mathematics. An illustration of such usage follows.

Example 2. When commenting on two of his salesmen, a sales manager remarks, "Mr. Persevere or Mr. Rover will meet his sales quota." What he means is that the quota will be achieved by one or the other or that *both* will meet the quota.

Study these examples to understand how the first excludes (the exclusive sense) the interpretation of *or* as both and the second includes (the inclusive sense) the meaning of *or* as both. The inclusive case will be discussed first, and toward the end of the section, the exclusive case will be treated.

The discussion in this section will be patterned rather closely after the discussion in Sec. 1.3, which dealt with the connective *and*. By the use of examples, "reasonable" meanings will be determined for each of the logical possibilities, but first for *inclusive or*.

The examples will be chosen from the set of officials of the CCC created for just such purposes.

The compound statement "p or q" is the *disjunction* of p and q. Consider the compound statement

Mr. Beam is president of CCC, or Mr. Fain is a member of the board of directors of CCC.

This compound statement is made up of two simple statements and the connective *or*. Since each of the simple statements is true (see page 9), it seems natural to agree that the compound statement is also true, and we so decide. (It is this case—true, true—that distinguishes the inclusive and exclusive usages of the word *or*, as will be shown later.)

1. *If p and q are true statements, then "p or q" is also true.*

Now consider the statement

Mr. Adams is chairman of the board of CCC, or Mr. Beam is a vice-president in charge of sales.

In this example, the first simple statement (which we designate by p) is true and the second simple statement (which we designate by q) is false. We must decide whether the compound statement "p or q" is true. But ordinary usage of the word *or* dictates that such statements are true if one or the other of the components is true. In conformity with everyday usage, it is agreed that in this case—true, false—"p or q" is true.

2. *If p is a true statement and q is a false statement, then "p or q" is true.*

Since the order in which statements p and q are written does not affect the meaning of the compound statement "p or q," the following agreement is made.

3. *If statement p is false and statement q is true, then statement "p or q" is true.*

The one remaining case is when both parts of the compound statement "p or q" are false. Such an example is

Mr. Crow is chairman of the board of directors of CCC, or Mr. Adams is in charge of customer response.

Each of the parts of the compound statement "p or q" is false—false, false—and it is agreed that the statement "p or q" is false.

4. *If p and q are false statements, then "p or q" is false.*

The commonly used symbol for the connective *or*, which will be in use throughout this book, is $\vee$. The agreements that have been reached in the discussion for the four logical possibilities are summarized in Table 7.

Table 7

p	q	$p \vee q$
T	T	T
T	F	T
F	T	T
F	F	F

Set off for easy reference is the definition that has been reached.

Definition: **(*Truth value of the disjunction of two statements*)** *If p and q are statements, then the compound statement (disjunction) "p or q" is false if and only if both p and q are false; otherwise, it is true.*

Reconsider Diagram 1 to see how the intersection of two sets is used to illustrate the connective *and* and compare it with Diagram 7, which illustrates similar circumstances for the connective *or*. If U is a set of logical possibilities (P denotes the subset of U which is the truth set for the statement p, and Q is the subset of U which is the truth set for the statement q), the truth set for the statement "p or q" is represented by the shaded area in Diagram 7.

Example 3. Another example will be instructive. Let p be the statement

p: x is a counting number greater than 3 and less than 10.

Let q be the statement

q: x is a counting number less than 7.

Then, in the usual notation for truth sets of statements,

$$P = \{4,5,6,7,8,9\}$$

and

$$Q = \{0,1,2,3,4,5,6\}$$

Diagram 8 is modeled after Diagram 7 except that the numbers in Diagram 8 refer to the elements of the sets P and Q in the example.

The truth set for the statement "p or q" is

$$\{0,1,2,3,4,5,6,7,8,9\}$$

The sets P and Q in the example just concluded were used to denote truth sets for given statements. Of course, sets can be considered independent of their use as illustrations for truth sets. However, the definition for the union of sets which follows does conform to the usage for sets when they are used as truth sets.

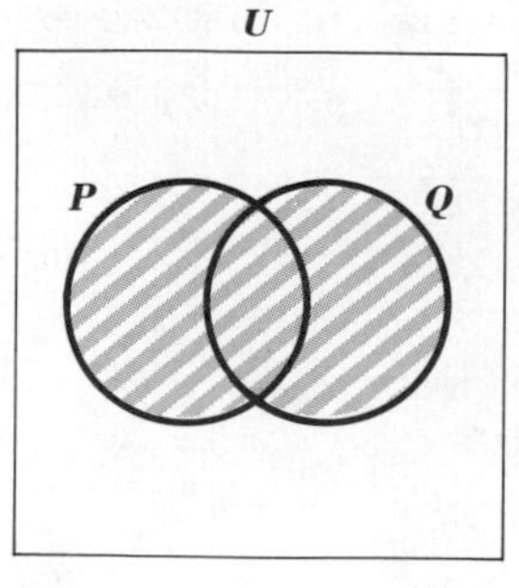

Diagram 7

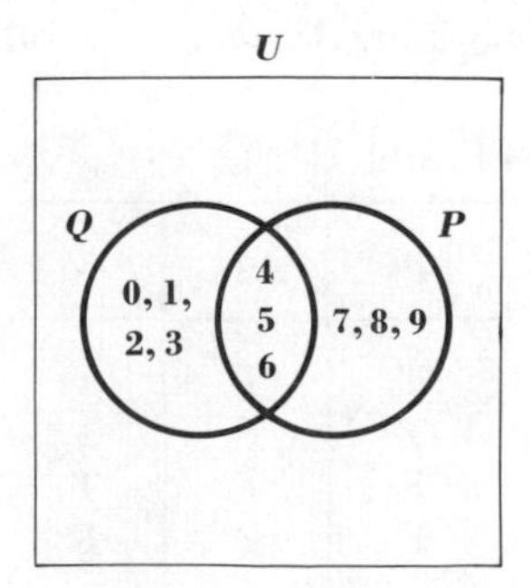

Diagram 8

Definition: **(*Union of sets*)** *If P and Q are subsets of the set U, then the union of P and Q is the set of elements in P or in Q (or in both). Symbolically, we write*

$$P \cup Q = \{x|x \in P \text{ or } x \in Q\}$$

Again the notation has been chosen to be suggestive. Note that the symbol chosen to represent the connective *or* as applied to statements is $\vee$ and the symbol chosen to represent the union of two sets is $\cup$. The similarity in appearance of the two symbols should aid in identifying the two concepts.

At this point in the discussion of the connective *and* in Sec. 1.3, it was pointed out that care must be taken not to be misled by the diagrams. Consequently, two other examples were considered to show that Diagrams 3 and 4 were different from Diagram 1. (Study those diagrams as a reminder of the difference.) Again, examples will illustrate how the union of two sets might differ in appearance from Diagrams 7 and 8.

Example 4. Consider the following situation. Let p be the statement

p: x is an executive in the promotion branch of CCC.

Let q be the statement

q: x is a vice-president of CCC.

Refer to the table of organization for CCC on page 9 and verify that the set of logical possibilities and the truth sets listed below are correct.

$$\begin{aligned} U &= \{a,b,c,d,e,f,g,h,i,j,k\} \\ P &= \{g,h,i,j,k\} \\ Q &= \{g,h,i\} \\ P \cup Q &= \{g,h,i,j,k\} \end{aligned}$$

The set Q is a subset (proper) of the set P, and $P \cup Q = P$. Diagram 9 illustrates this example.

It should be clear that if the set P is contained in the set Q, then the union of the sets P and Q is the set Q. Diagram 10 illustrates this case.

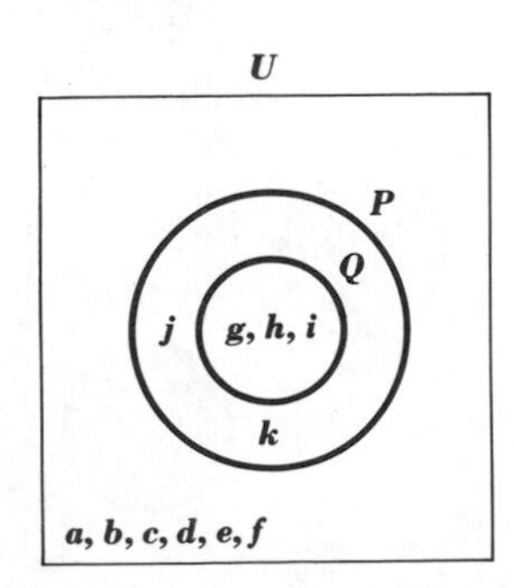

Diagram 9

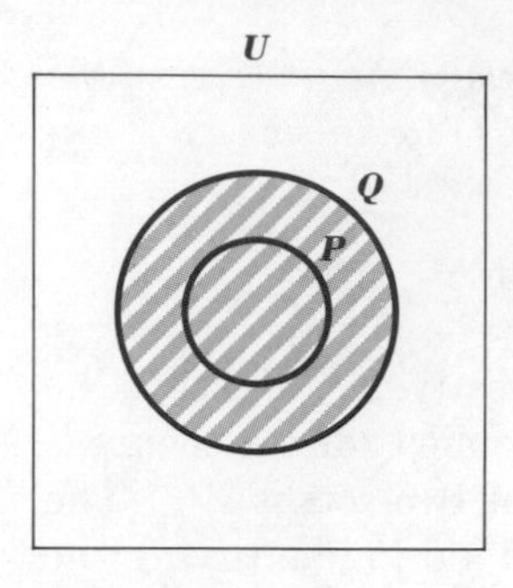

Diagram 10

Example 5. As another example, let p be the statement

p: x is an executive in the promotion branch of CCC.

Let q be the statement

q: x is the president of CCC.

Use of the list of executives of CCC indicates that the sets for this example (using the usual notation) are

$$U = \{a,b,c,d,e,f,g,h,i,j,k\}$$
$$P = \{g,h,i,j,k\}$$
$$Q = \{b\}$$

In this example, the sets, P and Q have no elements in common, and their union is

$$P \cup Q = \{b,g,h,i,j,k\}$$

This union is represented symbolically in Diagram 11.

Reconsider Diagrams 7 and 9 to 11 to see how the set $P \cup Q$ looks for each of the various relationships which may exist between the sets P and Q.

This concludes the discussion about the basic ideas concerned with *or* and two statements. But just as the idea of the connective *and* was extended to three (or more) statements, the concept of the connective *or* can be extended to three (or more) statements. The extension will be

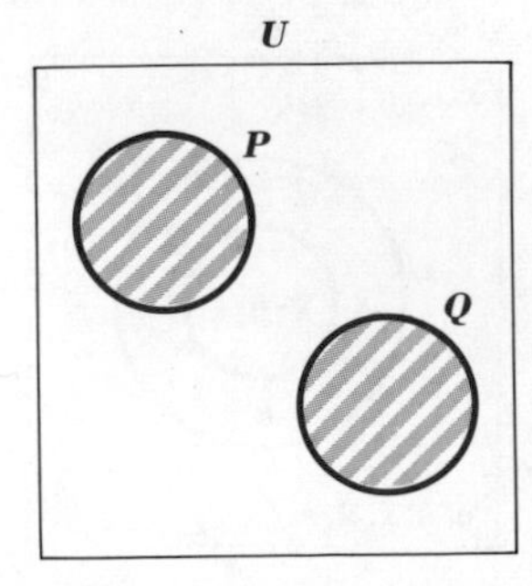

Diagram 11

made by the use of the truth tables. It may be helpful to reread how this was accomplished for the connective *and*. The sequence of truth tables which follow shows how to use the definition for disjunction to fill in the truth table for the connective *or* and three statements.

Table 8

Logical possibilities	p	q	$p \vee q$
1	T	T	T
2	T	T	T
3	T	F	T
4	T	F	T
5	F	T	T
6	F	T	T
7	F	F	F
8	F	F	F

Table 9

Logical possibilities	$p \vee q$	r	$(p \vee q) \vee r$
1	T	T	T
2	T	F	T
3	T	T	T
4	T	F	T
5	T	T	T
6	T	F	T
7	F	T	T
8	F	F	F

It is routine to verify that the truth table for $p \vee (q \vee r)$ yields the same result as Table 9. However, it is strongly suggested that these computations be made (and Exercise 5 is directed to this point) to justify the definition that follows, which is based on the common result.

Definition: (***Truth value for the disjunction of three statements***) *If p, q, and r are statements, the compound statement "p or q or r"* $[(p \vee q) \vee r = p \vee (q \vee r)]$ *is false if and only if p, q, and r are all false; otherwise, it is true.*

Just as the four logical possibilities for the connective *or* as applied to two statements give rise to a universal set with four subsets, the con-

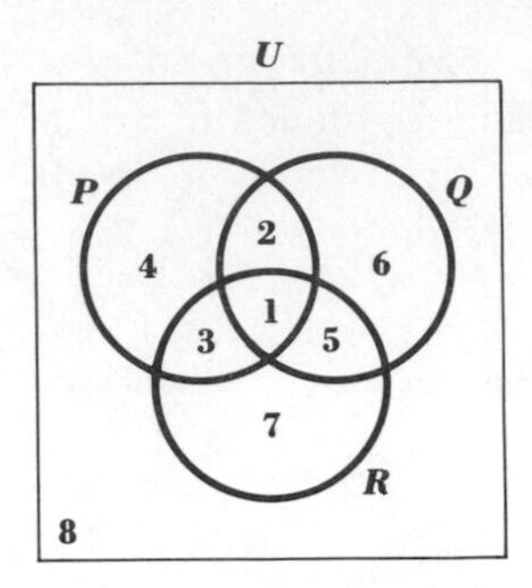

Diagram 12

nective *or* as applied to three statements partitions the universal set into eight subsets. Diagram 12 shows the eight subsets of the universal set of logical possibilities for statements p, q, and r and the associated truth sets P, Q, and R. The numbers in Diagram 12 refer to the logical possibilities in Table 9.

Example 6. An illustration of the use of the connective *or* for three statements will show how to apply the definition. A manager of an appliance store has just completed a questionnaire with regard to the three kinds of television sets that his store sells. For brevity in notation, we shall refer to these three makes of television sets as A, B, and C. The manager had his staff interview 100 persons with the following results:

1. 5 like all three products.
2. 9 like A and B.
3. 12 like A and C.
4. 37 like A.
5. 11 like B and C.
6. 42 like B.
7. 8 like none of the three products.

To display this information in a diagram, there must be statements, associated truth sets, and the number of elements in each subset of the universal set. Let p, q, and r be the statements

p: I like product A.

q: I like product B.

r: I like product C.

The diagram to be used in the analysis of these results is like Diagram 12, and counting will tell how many elements are in each of the eight subsets.

Now, to determine how many persons belong in each category, the results of the poll are used. The first fact, that five persons like all three products, enables us to enter the number 5 in area 1 (refer to Diagram 12). The second fact, that nine like A and B, enables us to enter the number 4

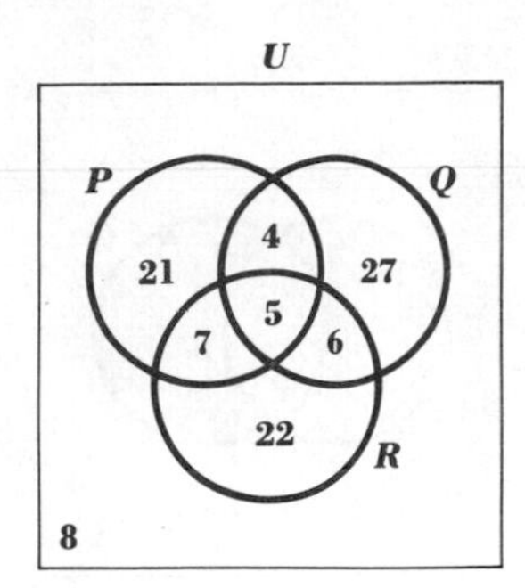

Diagram 13

(9 − 5) in area 2. From the third fact, we enter the number 7 (12 − 5) in area 3; from the fourth fact, the number 21 (37 − 5 − 4 − 7 = 21) in area 4; from the fifth fact, the number 6 (11 − 5) in area 5; and from the sixth fact, the number 27 (42 − 5 − 4 − 6) in area 6. The seventh fact enables us to enter 8 in area 8. This leaves area 7 unlabeled. However, 100 persons were interviewed, so

$$100 - (21 + 27 + 4 + 5 + 6 + 7 + 8) = 22$$

is entered in area 7. Diagram 13 is very similar to Diagram 12, but in Diagram 13 the numbers represent the results of the questionnaire.

The number of persons in the truth set of $p \vee q \vee r$, $P \cup Q \cup R$, is 92.

As we mentioned earlier in this section, it is possible to use the word *or* in two different ways. Up to this point, the discussion has centered around the inclusive meaning of *or*, which uses *or* to mean "one or the other or both." If the *exclusive* meaning is adopted, then *or* means "one or the other but not both." There are four logical possibilities, and three of them are the same as for the inclusive interpretation. Because of this fact, the truth table for the exclusive interpretation is the same as the truth table for the inclusive meaning except for the case when p and q are both true. For that case, the compound statement "p or q" is false. The notation $\underline{\vee}$ is used for the *exclusive or*, and the truth table is given in Table 10.

Table 10

p	q	$p \underline{\vee} q$
T	T	F
T	F	T
F	T	T
F	F	F

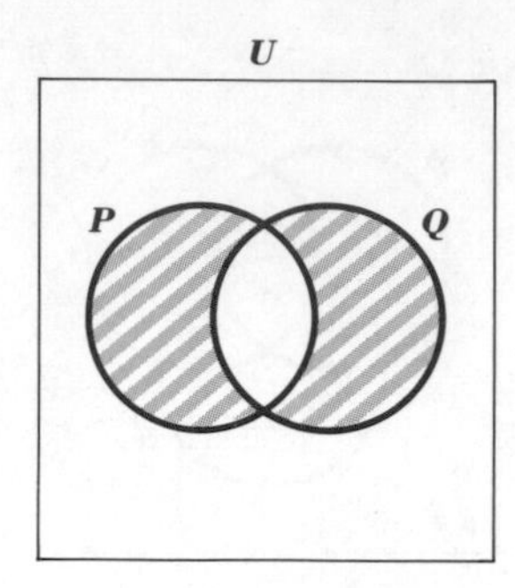

Diagram 14

Since the truth table, and hence the definition, has been changed, the diagrams that are used to illustrate truth sets must also be changed. Diagram 14 illustrates the truth sets for the exclusive meaning of *or* in the case where P and Q have some, but not all, elements in common.

Since some care has been exercised in the selection of symbols (examples are $\wedge$, $\cap$, and $\vee$, $\cup$), the choice for a symbol to represent the union of the truth sets for the disjunction using exclusive *or* is almost automatic. The shaded area of Diagram 14 is represented by $P \underline{\cup} Q$ and is motivated by the associated symbol $p \underline{\vee} q$. Exercise 9 requires you to draw diagrams to illustrate the cases where P is a subset of Q ($P \subseteq Q$), Q is a subset of P ($Q \subseteq P$), and P and Q have no elements in common ($P \cap Q = \{\quad\}$) for the exclusive *or*.

Example 7. This section is concluded with one example of the exclusive usage of *or*. Again we revert to the company created in Sec. 1.1 for a concrete application. Let p be the statement

p: x is a division chief of CCC.

Let q be the statement

q: x is an executive in the promotion branch of CCC.

Then using the set

$$O = \{a,b,c,d,e,f,g,h,i,j,k\}$$

as a universal set, it follows that $U = O$, and

$$U = \{a,b,c,d,e,f,g,h,i,j,k\}$$
$$P = \{j,k\}$$
$$Q = \{g,h,i,j,k\}$$

$p \underline{\vee} q$: x is a division chief of CCC or an executive in the promotion branch of CCC, but not both.

$$P \underline{\cup} Q = \{g,h,i\}$$

The shaded area of Diagram 15 represents the set $P \underline{\cup} Q$ for this last example. Explain why Diagrams 14 and 15 have different appearances.

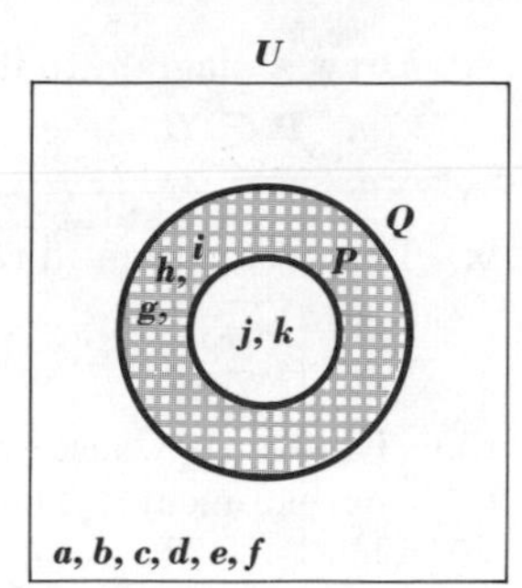

Diagram 15

Exercises

1. Determine which of the following uses of *or* are inclusive and which are exclusive:
   a. A large American automobile manufacturer makes Fords or Chevrolets or Chryslers.
   b. The building-construction industry hopes for lower interest rates or smaller down payments for home buyers.
   c. Inflation is caused by an excess of money or an excess of demand for goods and services.
   d. An Oregon plywood mill which is caught in a profit squeeze can either operate at a loss or close down the mill.
2. Give three statements that use *or* in the exclusive sense.
3. Give three statements that use *or* in the inclusive sense.
4. In each of the following, let $P = \{0,1,5\}$, $Q = \{1,2,6\}$, $R = \{0,5,6\}$, and $T = \{2,0,3\}$. Tabulate.
   a. $P \cup Q$ *b.* $P \cup R$ *c.* $P \cup T$
   d. $Q \cup R$ *e.* $Q \cup T$ *f.* $R \cup T$
5. Show by use of the truth tables that $(p \lor q) \lor r = p \lor (q \lor r)$.
6. Draw Venn diagrams for:
   a. $P \cup Q$ *b.* $P \cup Q \cup R$
   c. $R \cup (S \cap T)$ *d.* $R \cup (S \cap R)$
7. Express the following sentences in mathematical notation. For example,
   Problem: The new salesman is personable and has good references.
   Solution: Let p: The new salesman is personable.
   q: The new salesman has good references.
   The original sentence becomes $p \land q$.
   a. The union is demanding that either Washington's birthday or the day after Thanksgiving be a holiday.
   b. Television is a more expensive advertising medium than is radio, and it is also more effective and reaches a wider audience.
   c. Industry and the university must work together, or else the student is wasting his money going to college.

8. Draw a diagram to illustrate $p \vee q \vee r$ when:
   a. $P \subset Q$ *b.* $P = Q$ *c.* $P \cap Q = \{\ \}$
   d. $R \subset Q$ *e.* $P \cap Q \cap R = \{\ \}$
9. Draw a diagram to represent $P \cup Q$ when:
   a. $P \subseteq Q$ *b.* $Q \subseteq P$
   c. $P = Q$ *d.* $P \cap Q = \{\ \}$
10. The Sturdy Construction Company recently built 18 houses in a new housing development. The different materials used were brick, wood, and concrete blocks. The colors are red, yellow, and green, and the houses are either one-level or two-level. The possibilities are listed below.

House	*Material*	*Color*	*Level*
H_1	b	r	1
H_2	b	r	2
H_3	b	y	1
H_4	b	y	2
H_5	b	g	1
H_6	b	g	2
H_7	w	r	1
H_8	w	r	2
H_9	w	y	1
H_{10}	w	y	2
H_{11}	w	g	1
H_{12}	w	g	2
H_{13}	c	r	1
H_{14}	c	r	2
H_{15}	c	y	1
H_{16}	c	y	2
H_{17}	c	g	1
H_{18}	c	g	2

Use the set $\{H_1, H_2, \ldots, H_{18}\}$ as a universal set of logical possibilities. Determine P, Q, and $P \cup Q$ for each of the following:

a. p: x is a brick house.
   q: x is a two-level concrete-block house.

b. p: x is a wood house.
   q: x is a green wood house.

c. p: x is a yellow house.
   q: x is a one-level house.

d. p: x is a red house.
   q: x is a concrete-block house.

11. Use the information of Exercise 10 and the statements that follow to determine R and $P \cup Q \cup R$. (NOTE: Use part *a* of Exercise 10 with part *a* of this exercise, etc.)

a. r: x is a yellow house.
b. r: x is a one-level green wood house.
c. r: x is a brick house.
d. r: x is a yellow wood house.

12. Rework Exercise 10 using the definition of *exclusive or*.
13. Rework Exercise 11 using the definition of *exclusive or*.

Answers to problems

A. Yes *B.* Yes *C.* Yes *D.* No

1.5 Negation of statements and complements of sets

The following problems refer to the administration of the business organization described on page 9.

Problems

Determine which of the following statements are true:

A. It is false that Mr. Adams is president and Mr. Crow is a vice-president.
B. It is false that Mr. Adams is president and Mr. Gumm is a vice-president.
C. It is false that Mr. Beam is president and Mr. Dann is a vice-president.
D. It is false that Mr. Crow is president and Mr. Adams is a vice-president.
E. It is false that Mr. Hand is a director or Mr. Beam is chairman.
F. It is false that Mr. Jett is a director or Mr. Adams is chairman.
G. It is false that Mr. Ealy is a director or Mr. Beam is chairman.
H. It is false that Mr. Fain is a director or Mr. Klam is chairman.

The usual way to negate a statement is by use of the word *not*. To illustrate, recall the statement in Example 7 of Sec. 1.1:

The shipment did not arrive on time.

If p is the statement

p: The shipment did arrive on time.

then the statement

The shipment did not arrive on time.

is the negation of statement p.

The definitions for conjunction and disjunction of statements were given by the use of truth tables for logical possibilities, and this will be done for the negation of a statement. Observe how the following definition makes clear what the truth table must be.

Definition: (**Negation of a statement**) *If p is a statement, then q is the negation of p if and only if* (1) *q is true when p is false and* (2) *q is false when p is true.*

Adoption of this agreement makes the truth table for the negation obvious. (See Table 11.) The symbol $\sim$ represents the connective *not.*

Table 11

p	$\sim p$
T	F
F	T

Unlike previously considered connectives, *not* is concerned with only one statement, but it will be shown later in this section how this definition can be extended to include the negation of compound statements.

Just as diagrams have been utilized to indicate truth sets for compound statements, a diagram to show the truth set of a statement and the negation of the statement is readily constructed. If we adopt the notation that p is a statement, the associated truth set is P, $\sim p$ is the negation of statement p, and $\tilde{P}$ is the truth set of $\sim p$, then Diagram 16 illustrates the sets.

Notice that once again the notation chosen is helpful in relating statements to sets. The symbol $\tilde{P}$ represents the truth set of $\sim p$. As another aid, remember that the set $\tilde{P}$ is the subset of elements in U which are *not* in P.

A name for the set of elements in U but not in P, $\tilde{P}$, will be handy, and the name chosen is *complement.* The sets P and $\tilde{P}$ are complementary; each set complements the other. The idea of complementary sets does not depend on truth sets, as it has been developed here, but the definition that follows is consistent with the usage with regard to truth sets.

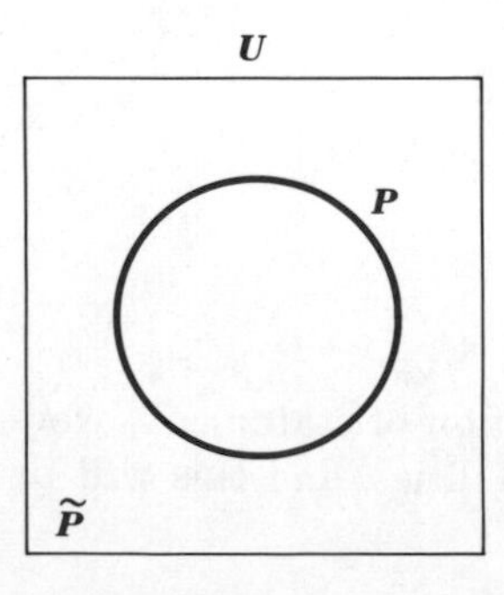

Diagram 16

Definition: **(*Complement of a set*)** *If P is a subset of a set U, then the complement of the set P is the set of those elements in the set U that are not in the set P. Symbolically, we write*

$$\tilde{P} = \{x | x \in U \text{ and } x \notin P\}$$

Listed below are some examples. The universal set is U; subsets P, Q, and R are listed with their complements.

$$U = \{0,1,2,3,4,5,6,7,8,9,10\}$$

Example 1. $P = \{0,1,2,4\}$ $\tilde{P} = \{3,5,6,7,8,9,10\}$

Example 2. $Q = \{0,2,4,6,8,10\}$ $\tilde{Q} = \{1,3,5,7,9\}$

Example 3. $R = \{0,1,4,6\}$ $\tilde{R} = \{2,3,5,7,8,9,10\}$

Now that agreement has been reached about what the negation of a statement means, the definition can be used to give meaning to the negation of compound statements where the connectives *and* and *or* are used. Since the negation of a statement t is a statement $\sim t$, which is false when statement t is true and true when t is false, it is simple to make a truth table for the negation of conjunctions and disjunctions. Applying the definition to the negation of statements connected by *and* yields Truth Tables 12 and 13.

Table 12

p	q	$p \wedge q$	$\sim(p \wedge q)$
T	T	T	F
T	F	F	T
F	T	F	T
F	F	F	T

Table 13

p	q	r	$p \wedge q \wedge r$	$\sim(p \wedge q \wedge r)$
T	T	T	T	F
T	T	F	F	T
T	F	T	F	T
T	F	F	F	T
F	T	T	F	T
F	T	F	F	T
F	F	T	F	T
F	F	F	F	T

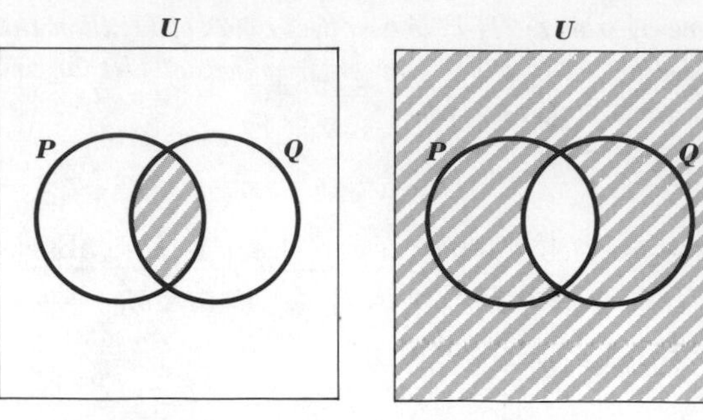

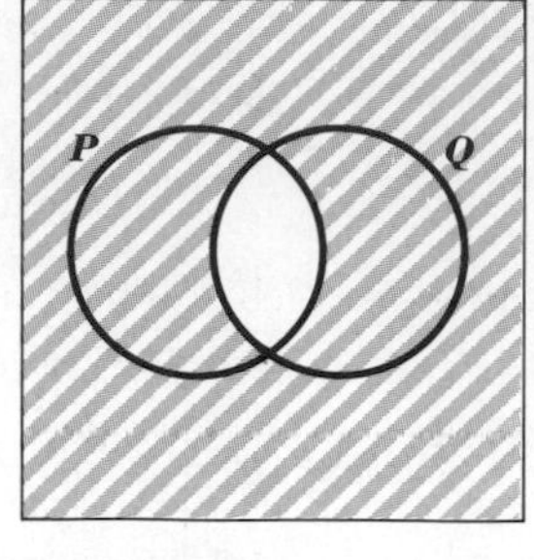

Diagram 17 *Diagram 18*

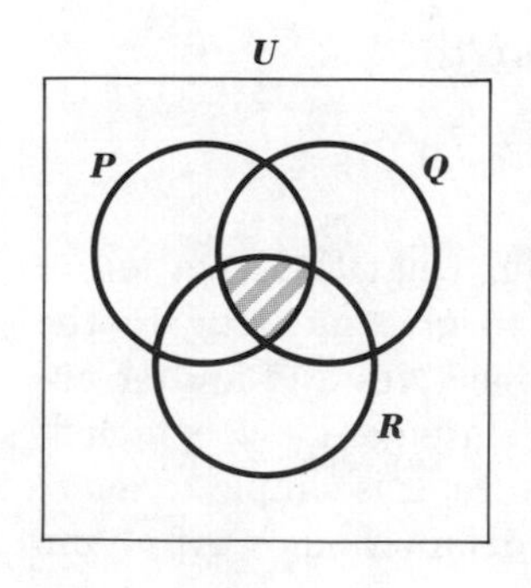

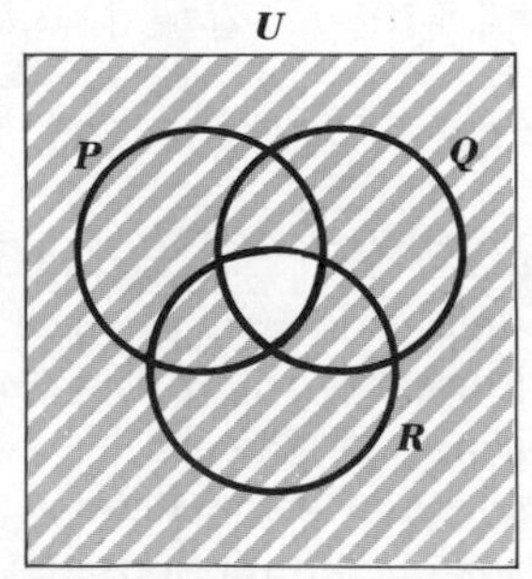

Diagram 19 *Diagram 20*

Exercise 14 requires that you formulate a statement about the truth value of the negation of a conjunction.

Diagrams have proved helpful in understanding truth sets of statements in earlier sections, and they will be helpful here. Diagrams 17 to 20 illustrate the truth sets for statements $p \wedge q$, $\sim(p \wedge q)$, $p \wedge q \wedge r$, $\sim(p \wedge q \wedge r)$, respectively.

Refer now to the sets U, P, Q, and R in Examples 1 to 3 and verify that the sets listed below are correct. Statement p is

x is a member of P.

and statements q and r are given similarly.

Example 4. $p \wedge q$: $P \cap Q = \{0,2,4\}$ $\quad \widetilde{P \cap Q} = \{1,3,5,6,7,8,9,10\}$

Example 5. $r \wedge q$: $R \cap Q = \{0,4,6\}$ $\quad \widetilde{R \cap Q} = \{1,2,3,5,7,8,9,10\}$

Example 6. $p \wedge r$: $P \cap R = \{0,1,4\}$ $\quad \widetilde{P \cap R} = \{2,3,5,6,7,8,9,10\}$

Example 7. $p \wedge q \wedge r$: $P \cap Q \cap R = \{0,4\}$

$\widetilde{P \cap Q \cap R} = \{1,2,3,5,6,7,8,9,10\}$

These tables, diagrams, and examples and the discussion make clear how to negate conjunctions. Now, in a fashion which closely parallels the treatment of negation of conjunctions, we turn to the negation of disjunctions—inclusive and exclusive, in that order.

Again using the definition for the negation of a statement ($\sim t$ is false when t is true and $\sim t$ is true when t is false), the truth table for the negation of disjunction (inclusive) is given in Table 14.

Table 14

p	q	$p \vee q$	$\sim(p \vee q)$
T	T	T	F
T	F	T	F
F	T	T	F
F	F	F	T

Exercise 14 requires the formulation of a statement about truth values for the negation of disjunctions. Exercise 17 requires the construction of the truth table for the negation of $p \vee q \vee r$.

Diagrams 21 to 24 illustrate the truth sets for the statements $p \vee q$, $\sim(p \vee q)$, $p \vee q \vee r$, $\sim(p \vee q \vee r)$ in the order indicated. The usual notation applies.

The sets U, P, Q, and R in Examples 1 to 3 are used to determine the sets below, which serve as examples of the current discussion.

Example 8. $p \vee q$: $P \cup Q = \{0,1,2,4,6,8,10\}$ $\widetilde{P \cup Q} = \{3,5,7,9\}$

Example 9. $r \vee q$: $R \cup Q = \{0,1,2,4,6,8,10\}$ $\widetilde{R \cup Q} = \{3,5,7,9\}$

Example 10. $p \vee r$: $P \cup R = \{0,1,2,4,6\}$

$\widetilde{P \cup R} = \{3,5,7,8,9,10\}$

Example 11. $p \vee q \vee r$: $P \cup Q \cup R = \{0,1,2,4,6,8,10\}$

$\widetilde{P \cup Q \cup R} = \{3,5,7,9\}$

Negation of statements, conjunctions, and disjunctions (inclusive) have been given by tables, diagrams, and examples. It is suggested that the verbal formalization of the negations be constructed. By now, the pattern is well established, and without much talk, the tables and diagrams for negation of disjunctions (exclusive) follow.

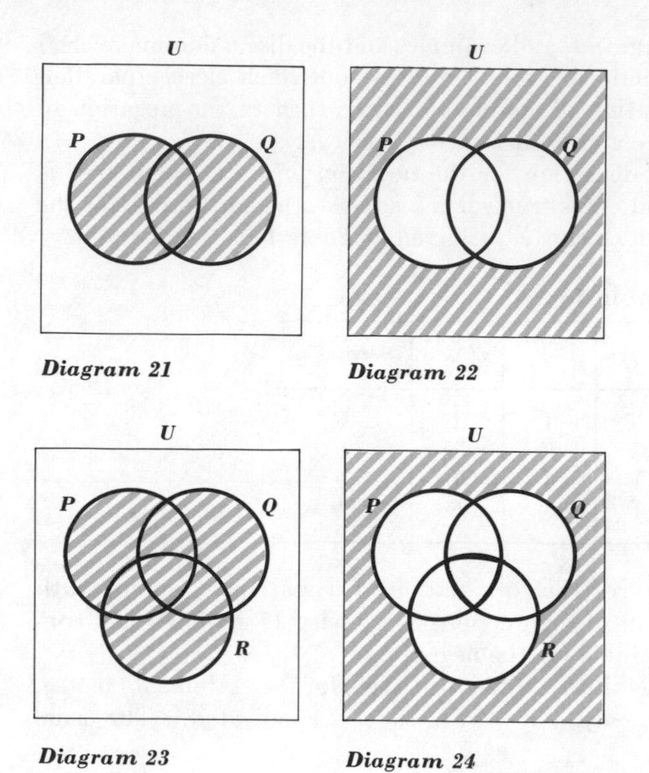

Diagram 21 *Diagram 22*

Diagram 23 *Diagram 24*

Applying the definition of negation to Table 10 for exclusive disjunction, Table 15 is the truth table for the negation of *exclusive or*.

Table 15

p	q	$p \veebar q$	$\sim(p \veebar q)$
T	T	F	T
T	F	T	F
F	T	T	F
F	F	F	T

Refer once again to the sets U, P, Q, and R in Examples 1 to 3 to see how the sets below are obtained.

Example 12. $p \veebar q$: $P \underline{\cup} Q = \{1,6,8,10\}$ $\widetilde{P \underline{\cup} Q} = \{0,2,3,4,5,7,9\}$

Example 13. $r \veebar q$: $R \underline{\cup} Q = \{1,2,8,10\}$ $\widetilde{R \underline{\cup} Q} = \{0,3,4,5,6,7,9\}$

The truth sets for $p \veebar q$ and $\sim(p \veebar q)$ are shown in Diagrams 25 and 26.

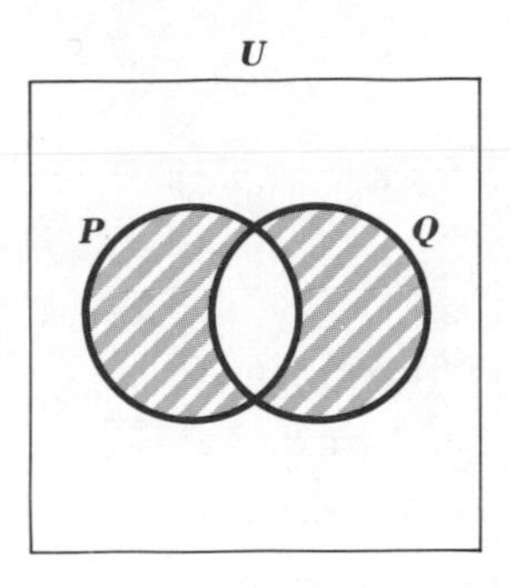

Diagram 25

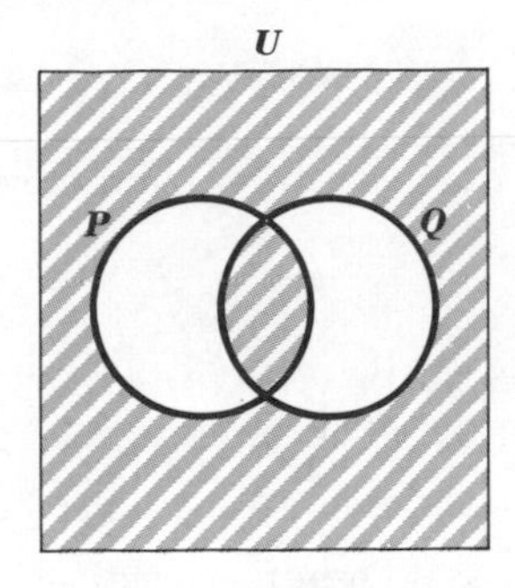

Diagram 26

Rules for negating statements, conjunctions, etc., have been developed, but what about the negation of a negation?

In everyday conversation, statements are often made which are called *double negatives*. An example of such a statement is

J. D. does not have no pants.

In ordinary usage, we interpret this to mean that the double negatives "cancel out" and that a statement free of negatives is meant. Perhaps the speaker of the statement above wishes to convey the meaning

J. D. does have pants.

Such an agreement of meaning is consistent with the definition that has been adopted here for negation of statements. It is easily seen that the negation of the negation of p has the same truth table as p. Truth Table 16 establishes this.

Table 16

p	$\sim p$	$\sim(\sim p)$
T	F	T
F	T	F

Frequently we write $\sim(\sim p) = p$, or $\tilde{\tilde{P}} = P$. (The entire matter of equivalent statements will be discussed later.)

The following list of symbols is offered as a partial summary of preceding considerations. Let p and q be statements and U a set of logical possibilities. Let P be the truth set for p and Q the truth set for q. Then

Set language	***Statement language***
$P = U$	p is a tautology
$p = \{\ \}$	p is self-contradictory
$P \cap Q$	p is true $\wedge q$ is true
$P \cup Q$	p is true $\vee q$ is true
$\tilde{P}$	$\sim p$ is true

Exercises

1. Construct truth tables for the following:
   a. $\sim p$ *b.* $\sim p \vee q$
   c. $p \wedge \sim q$ *d.* $\sim(p \wedge q)$
2. If $U = \{a,b,c,d,e\}$ and $S = \{a,b,c\}$, find $\tilde{S}$.
3. Show, using a Venn diagram, that $P \cap \tilde{P} = \{\ \}$.
4. Show, using a Venn diagram, that $P \cup \tilde{P} = U$.
5. Form the negation of "Profits are rising."
6. Which of the following are negations of "(Productivity goes up), and wages go up or the union will strike"?
   a. It is not true that [productivity goes up, and (wages go up or the union will strike)].
   b. Productivity does not go up, or [(it does not happen that wages go up) or (the union will strike)].
   c. Productivity does not go up, or [(wages do not go up), and (the union will not strike)].
7. In the text, the diagram which represents the truth set of $\sim(p \wedge q)$ is for the case when P and Q have some, but not all, elements in common. Draw a diagram to show the truth set of $\sim(p \wedge q)$ when
   a. $P \subset Q$
   b. Q is contained properly in P
   c. $P \cap Q = \{\ \}$
8. Do Exercise 7 for $\sim(p \vee q)$.
9. Do Exercise 7 for $\sim(p \veebar q)$.
10. Give $\widetilde{P \cup Q}$ for Exercise 4, Sec. 1.4.
11. Give $\widetilde{P \cup Q \cup R}$ for Exercise 4, Sec. 1.4.
12. Give $\widetilde{P \underline{\cup} Q}$ for Exercise 4, Sec. 1.4.
13. Give $\widetilde{P \underline{\cup} Q \underline{\cup} R}$ for Exercise 4, Sec. 1.4.
14. Formulate, in words, a definition of the truth value of the negation of:
    a. $p \wedge q$ *b.* $p \wedge q \wedge r$
    c. $p \vee q$ *d.* $p \vee q \vee r$
15. Show that the truth table for $\sim(p \wedge q)$ is the same as that for $\sim p \vee \sim q$.
16. Show that the truth table for $\sim(p \vee q)$ is the same as that for $\sim p \wedge \sim q$.
17. Construct the truth table for $\sim(p \vee q \vee r)$.

Answers to problems

A, B, C, D, and *E*

1.6 Other compound statements and decision making

Problems

A. In his monthly report, a salesman states

The promotional material did not arrive, and I cannot meet my sales quota this month.

If it is known that the promotional material did not arrive but that the quota could be met, did he lie?

B. The Internal Revenue Service makes an adverse ruling on the income tax report of a company. The ruling is based on the statement

(The company's reported costs were incorrect and the profits were misrepresented) or the sales were reported incorrectly.

The company knows and can produce evidence that the reported costs were correct. They also know that the profits were misrepresented and that the sales were reported correctly. Should the company appeal the ruling?

To place in proper perspective what has preceded so that the results can be easily used in this section, a short summary is given.

From the set of sentences, a subset of sentences was selected; these sentences were called *simple statements*. The connectives *and*, *or*, and *not* were used to form new statements (using simple statements as the building blocks). Agreements about the meaning of the newly constructed statements were based on consideration of several examples (and everyday usage of the connective words). These agreements were summarized in truth tables for compound statements. Three pertinent definitions were as follows: (1) compound statements of the form (symbolically) $p \wedge q$ are true if and only if p is true and q is true (and false otherwise); (2) a compound statement of the form $p \vee q$ is false if and only if p is false and q is false (and true otherwise); and (3) a statement of the form $\sim p$ is false when p is true and true when p is false. You are reminded once again that the usage we make of $p \vee q$ is in the inclusive sense.

Now that these definitions (truth tables) are at our disposal, they can be used to develop truth tables for other compound statements. As usual, the ideas will be developed through the use of examples.

Example 1. First, consider a compound statement of the form $p \wedge \sim q$. An example of such a statement is

r: I have met my sales quota, and I will not call on Burns Department Store today.

In this example, if p is the simple statement

p: I have met my sales quota.

and if q is the simple statement

q: I will call on Burns Department Store today.

then the statement r is (symbolically) $p \wedge \sim q$.

The definitions given earlier, and restated for handy reference above, make it a straightforward procedure to construct a truth table for $p \wedge \sim q$. This construction is performed, in steps, in Table 17.

Table 17

p	q
T	T
T	F
F	T
F	F

p	$\sim q$
T	F
T	T
F	F
F	T

p	$\sim q$	$p \wedge \sim q$
T	F	F
T	T	T
F	F	F
F	T	F

It has been helpful in previous considerations to draw a diagram which shows the truth sets for compound statements. Such a diagram can be drawn for $p \wedge \sim q$. The usual notation—P represents the truth set for statement p, Q represents the truth set for statement q, and $\tilde{Q}$ represents the truth set of the negation of statement q—is adopted so that the truth set for $p \wedge \sim q$ is $P \cap \tilde{Q}$. This notation (and assuming that the set P and Q have some, but not all, elements in common) yields Diagram 27.

Although Diagram 27 is for the case where the set P and Q have some, but not all, elements in common, you should now be sufficiently acquainted with such techniques to be able to draw the diagram for the cases where P is contained in Q, Q is contained in P, and P and Q have no elements in common. These are left as Exercise 5. The set illustrated

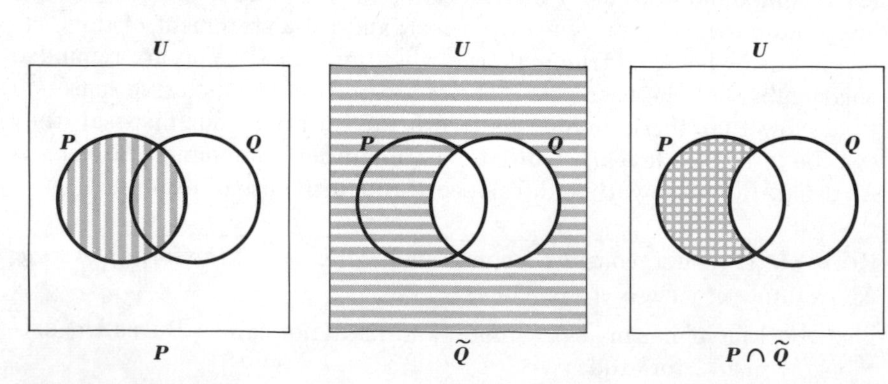

Diagram 27

in Diagram 27, $P \cap \tilde{Q}$, is sometimes represented by $P - Q$ and is referred to as the *difference* of the sets P and Q.

Example 2. Consider the compound statement which can be represented in the usual notation for statements by $\sim(p \vee \sim q)$. An example of such a compound statement is

It is not the case that [(*the counting number x is greater than* 10) *or* (*y is not greater than* 6)].

The parentheses are included to help punctuate the compound statement and are not usually (unfortunately?) used in writing expressions. As per our custom, a set of logical possibilities is made, and there are four elements in the set. The sequence of truth tables in Table 18 shows how to develop the truth table for the compound statement $\sim(p \vee \sim q)$. The first table gives the logical possibilities for p and q, the second table for p and $\sim q$, the third table for $p \vee \sim q$, and the final table for $\sim(p \vee \sim q)$.

Table 18

p	q
T	T
T	F
F	T
F	F

p	$\sim q$
T	F
T	T
F	F
F	T

p	$\sim q$	$p \vee \sim q$
T	F	T
T	T	T
F	F	F
F	T	T

$p \vee \sim q$	$\sim(p \vee \sim q)$
T	F
T	F
F	T
T	F

Again, a diagram can be used to show the truth set for the compound statement under consideration. So that the diagram will correctly reflect the truth table, some agreements must be made. First, let

p: The counting number x is greater than 10.

q: The counting number y is greater than 6.

The symbolic representation of the compound statement is $\sim(p \vee \sim q)$. Finally, note that we assume that $P \subset Q$. Diagram 28 is given in stages which correspond to the steps in Table 18. The purpose of a multistep diagram is to serve as a model to aid in the construction of similar diagrams.

Each of the two preceding examples deals with compound statements that use two simple statements as building blocks and connectives of a familiar type. This section is concluded with one other example in which three simple statements are used to build the compound statement.

Example 3. Consider the compound statement which can be represented by $(p \wedge q) \vee \sim r$. There are eight logical possibilities. The truth table for the compound statement is Table 19, but for this illustra-

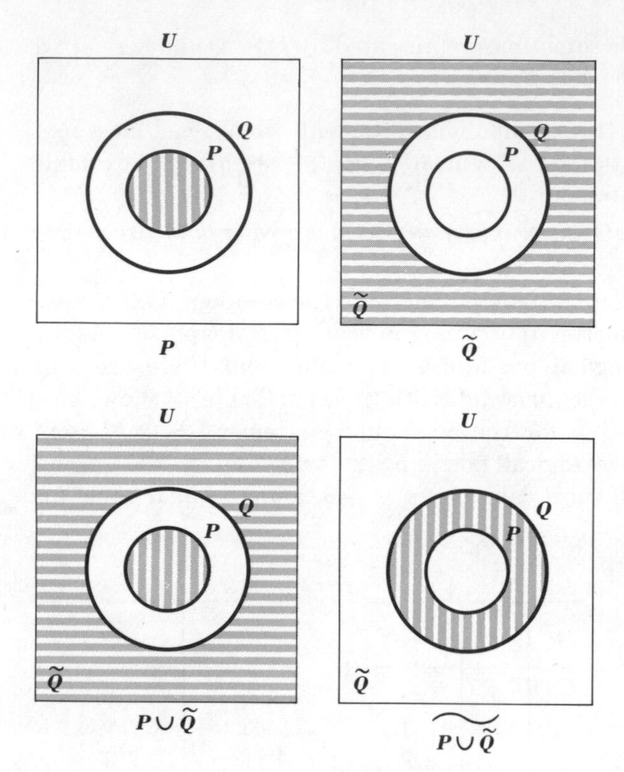

Diagram 28

tion, some of the intermediate steps are combined and the final result is entered in a column which has double vertical lines. The intermediate steps—the truth tables of $p \wedge q$ and of $\sim r$—are also given, but they appear just to the left and to the right, respectively, of the final result. The numbers under the columns show the order of steps.

Table 19

p	q	r	$p \wedge q$	$(p \wedge q) \vee \sim r$	$\sim r$
T	T	T	T	T	F
T	T	F	T	T	T
T	F	T	F	F	F
T	F	F	F	T	T
F	T	T	F	F	F
F	T	F	F	T	T
F	F	T	F	F	F
F	F	F	F	T	T
			1	3	2

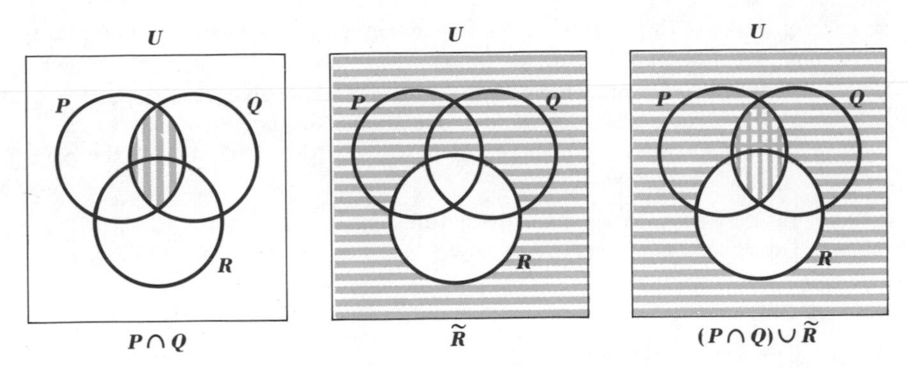

Diagram 29

Diagram 29 illustrates the compound statement whose truth values are in Table 19. It is given in three stages; the first diagram indicates the set $P \cap Q$, the second diagram the set $\tilde{R}$, and the third diagram the union of the two sets $(P \cap Q)$ and $\tilde{R}$. The diagram is given for the special case when P, Q, and R have some, but not all, elements in common.

This concludes the examples of compound statements that will be given in this section, but no doubt it is clear that the list of such examples could be extended indefinitely. The techniques that have been used both in forming the truth tables and in drawing thc diagrams can be applied to any other compound statement that is built using these connectives. The Exercises afford an opportunity to develop skills in making truth tables and in drawing diagrams for such compound statements. Since such compound statements are usual in ordinary conversation and since the job of assigning a truth value to such statements is not easy (unless some procedures have been established), the Exercises should develop an appreciation of the mathematical machinery that has been constructed. In the Exercises, we have included some compound statements that might occur in business applications in the hopes of demonstrating the "applied" aspect of the material.

Exercises

1. Given:

   p: Profits are rising.
   q: Dividends are rising.
   r: Stockholders are satisfied.

   Using p, q, and r as defined above, translate the following symbols into English:

   a. $p \wedge q$ *b.* $\sim(p \wedge q) \wedge \sim r$
   c. $(\sim p \vee \sim q) \wedge \sim r$ *d.* $(p \wedge q) \wedge (r \wedge \sim r)$

In each of the next three exercises, statements are given. Assume that P and Q have some, but not all, elements in common.

a. Make statements which represent the symbols given.
b. Make truth tables for the compound statements.
c. Draw a diagram which represents the truth set of each of the compound statements.

2. p: Interest payments are tax deductible.
 q: Interest rates are very high now.
 a. $p \wedge \sim q$ *b.* $\sim p \wedge q$
 c. $\sim p \wedge \sim q$ *d.* $\sim(\sim p \vee \sim q)$

3. p: Tax accountants are well trained.
 q: All businesses file income tax returns.
 a. $p \wedge \sim q$ *b.* $\sim p \wedge q$
 c. $\sim p \wedge \sim q$ *d.* $\sim(\sim p \vee \sim q)$

4. p: Sales taxes are paid on medicines.
 q: Medicare provides free medicine.
 a. $p \wedge \sim q$ *b.* $\sim p \wedge q$
 c. $\sim p \wedge \sim q$ *d.* $\sim(\sim p \vee \sim q)$

5. Draw diagrams to represent $p \wedge \sim q$ when:
 a. $P \subset Q$ *b.* $Q \subset P$ *c.* $P \cap Q = \{\ \}$

6. Draw diagrams to represent $(p \wedge q) \vee \sim r$ when:
 a. $P \subset Q$ *b.* $R \subset P \cap Q$ *c.* $R \cap P \cap Q = \{\ \}$

7. Let

$$U = \{1,2,3, \ldots ,10\}$$
$$A = \{1,5,9\}$$
$$B = \{2,5,8\}$$
$$C = \{1,3,5,7,9\}$$

 Write the following by the tabular method:
 a. $A \cup B$ *b.* $A \cap B$ *c.* $\tilde{C}$
 d. $(A \cup B) \cap C$ *e.* $A \cap B \cap C$

8. Draw a diagram illustrating each of the following when P, Q, and R have some, but not all, elements in common.
 a. $(P \cap \tilde{Q}) \cup R$ *b.* $(P \cap Q) \cap \tilde{R}$

9. The Midwest Belt Company has classified its customers according to the method of payment used, the number of products purchased, and the geographic location of the customer. This classification results in $2 \cdot 3 \cdot 4 = 24$ categories, as given in the table below. Select from the set of 24 categories those which satisfy the following compound statements:
 a. x is a cash customer from the Eastern area or a credit customer who buys two products.
 b. x is a cash customer who does not buy two products and is from the Western area, or x is a customer who is not from the Eastern area.

c. x is a Rocky Mountain customer who buys one product, and x is not a customer who buys one product.

d. x is not a Central customer, a Western customer, or a credit customer, and x is a three-product customer.

Customer	***Pays cash***	***Buys one, two, or three products***	***Geographic location***
1	Yes	1	Eastern
2	Yes	1	Central
3	Yes	1	Rocky Mountain
4	Yes	1	Western
5	Yes	2	Eastern
6	Yes	2	Central
7	Yes	2	Rocky Mountain
8	Yes	2	Western
9	Yes	3	Eastern
10	Yes	3	Central
11	Yes	3	Rocky Mountain
12	Yes	3	Western
13	No	1	Eastern
14	No	1	Central
15	No	1	Rocky Mountain
16	No	1	Western
17	No	2	Eastern
18	No	2	Central
19	No	2	Rocky Mountain
20	No	2	Western
21	No	3	Eastern
22	No	3	Central
23	No	3	Rocky Mountain
24	No	3	Western

Answers to problems

A. Yes. If

p: The promotional material did not arrive.
q: I can meet my sales quota this month.

then the compound statement is $p \wedge \sim q$. It is given that p is true and q is true. Therefore, $p \wedge \sim q$ (see Table 17) is false. Hence, the salesman lied.

B. Yes. If

p: The reported costs were incorrect.
q: The profits were misrepresented.
r: The reported sales were correct.

then the Internal Revenue Service's statement becomes $(p \wedge q) \vee \sim r$. It is given that p is false, q is true, and r is true. Therefore, the compound statement is false (see Table 19). The case should be appealed.

1.7 Conditional statements and equivalent statements

Problems

A. A foreman on a production line reports

If the rate of production does not increase, then our sales requests cannot be met.

If the rate of production increases, is his statement true?

B. Is the statement

If production goes up, then prices go down.

equivalent to the statement

If prices do not go down, then production does not go up.

In everyday conversation, many expressions are used which are examples of the type of statement to be discussed in this section: the *conditional* statement. One example of a conditional statement is

If the sun shines, then I shall go to the game.

This type of statement is characterized by the presence of the words *if* (which precedes the remainder of the statement) and *then* (which occurs somewhere within the statement). These words divide a conditional statement into two parts. Using the usual notation and denoting the first part of the statement by p and the second part by q, a conditional statement is expressed as

If p, then q.

The two portions of conditional statements are given special names, and this terminology is rather well established. Conforming with the usual practice, we call the first part the *hypothesis* and the second part the *conclusion*. It will be one of the main purposes of this section to decide what should be the truth table for conditional statements.

Since p and q are statements, it follows that a set of logical possibilities can be determined in the usual fashion. It is easy to see that there are four cases to consider. Before deciding on the truth-table entries, an example will assist in the decision.

Example 1. Suppose that a friend says to you

If the sun shines, then I shall go to the game.

The hypothesis is "The sun shines." The conclusion is "I shall go to the game." If the sun shines, you expect to see your friend at the game. So, if the sun shines and the friend attends the game, you regard his statement as true. However, if the sun shines and the friend does not attend, then you regard his statement as false. Perhaps on the basis of this example, you would agree that if both the hypothesis and the conclusion are true, then the conditional is true. It should also seem reasonable that if the hypothesis is true and the conclusion is false, the conditional is false.

But what about the other two logical possibilities? Assume that the sun does not shine. What behavior do you expect of your friend? More importantly, what can he do and not be a liar? Of most importance, what truth value should be assigned to the conditional statement for the two cases when the hypothesis is false? One solution is to avoid the issue. It can be reasoned that since the sun did not shine, it can never be determined if the speaker told the truth or lied. Some books adopt this attitude and dismiss the matter. There are some good reasons for doing so. However, in this book we shall persist and shall assign a truth value when the hypothesis is false. Of course, we are at liberty to assign the values in any agreeable way since this is a definition. But an attempt will be made to explain why the particular values that are chosen are the ones selected.

If the sun does not shine and if you attend the game and your friend does not, then you might decide that this is what his statement meant. So to give him the "benefit of the doubt," you decide he did not lie. (If the hypothesis is false and the conclusion is false, then the conditional is true.)

Now suppose, as you are standing in the rain at the game, your friend appears. You are indignant. You tell him he lied because he said that if the sun shone, he would attend. But he is offended and argues: "Of course, I said if the sun shines, then I will attend the game; and I would have. But I did not say what I would do if it rained. I did not have any intention of missing this game. Regardless of the weather, I expected to be here. And anyway, you cannot call me a liar, because the sun is not shining and you do not know what I would have done if the sun had shone." You feel the logical ground slipping from under your feet and give in. (If the hypothesis is false and the conclusion is true, then the conditional is true.)

The four cases have been discussed with reference to one example, and the conclusions reached in the discussion agree with those that are usually taken for the truth table of a conditional statement. In Table 20, these agreements are summarized, with the symbol $\rightarrow$ used to represent "if . . . then . . . ." The symbol $p \rightarrow q$ is read "If p, then q."

The information given in Table 20 is expressed in the definition which follows.

Table 20

p	q	$p \to q$
T	T	T
T	F	F
F	T	T
F	F	T

Definition: **(*Truth values for conditional statements*)** *Let p and q be statements. The conditional statement "If p, then q" is false if and only if p is true and q is false; otherwise, it is true.*

It has been our practice to follow a definition of a truth value with a diagram illustrating the truth set for the type of statement in the definition. There is already a method for illustrating truth sets if it is possible to express the statement using some combination of the connectives *and*, *or*, and *not*. To take advantage of this method, we shall construct a compound statement (expressed using the connectives mentioned) which is equivalent to the conditional. Of course, no meaning has yet been given to the term *equivalent statement*, but such a meaning will be established with the aid of an example.

Example 2. Consider a new compound statement that is related to the one in Example 1 of this section:

The sun will not shine, or I shall go to the game.

This example is of a familiar type, and it can be assigned truth values by methods already discussed. Designating by p the statement

p: The sun will shine.

and by q the statement

q: I shall go to the game.

the example is then of the form $\sim p \vee q$. The truth table for $\sim p \vee q$ is constructed (in two steps) in Table 21. The final result is in the column with double lines.

Table 21

p	q	$\sim p$	$\sim p \vee q$
T	T	F	T
T	F	F	F
F	T	T	T
F	F	T	T

Table 21 should look familiar. The entries in the final column are the same as those in Table 20 for conditional statements. Consider the four cases for the statement in Example 1 to see that it expresses exactly what is expressed by the conditional statement in Example 2. Because the two statements express, in different ways, exactly the same idea, the two statements are *equivalent.*

Definition: (***Equivalent statements***) *If t and s are statements which have as components the same simple statements, then t and s are equivalent if and only if they have the same truth table. Symbolically, we write*

$$t \equiv s$$

Now that a compound statement which is equivalent to the conditional statement (and which uses the connectives *not* and *or*) has been discovered, a truth-set diagram using techniques developed earlier can be constructed. The truth set for $\sim p \lor q$, hence for $p \rightarrow q$, is designated by $\tilde{P} \cup Q$. Diagram 30 represents, in stages, the truth set for the statement $\sim p \lor q$ in the case where the sets P and Q have some, but not all, elements in common.

You should draw a diagram for the case where P is contained in Q, Q is contained in P, and P and Q have no elements in common; Exercise 7 is directed to this point.

Consider again the truth set $\tilde{P} \cup Q$ that is shaded in Diagram 30. Observe that the truth set is the complement of the set $P \cap \tilde{Q}$. This indicates that there is another equivalent statement which could be used just as well for the conditional (or the statement $\sim p \lor q$). It is a simple matter to verify that this is the case. First note that the statement that corresponds to the set $\widetilde{P \cap \tilde{Q}}$ is $\sim(p \land \sim q)$. The truth table given in Table 22, with the final result recorded in the column with double lines, verifies that $p \rightarrow q \equiv \sim p \lor q \equiv \sim(p \land \sim q)$.

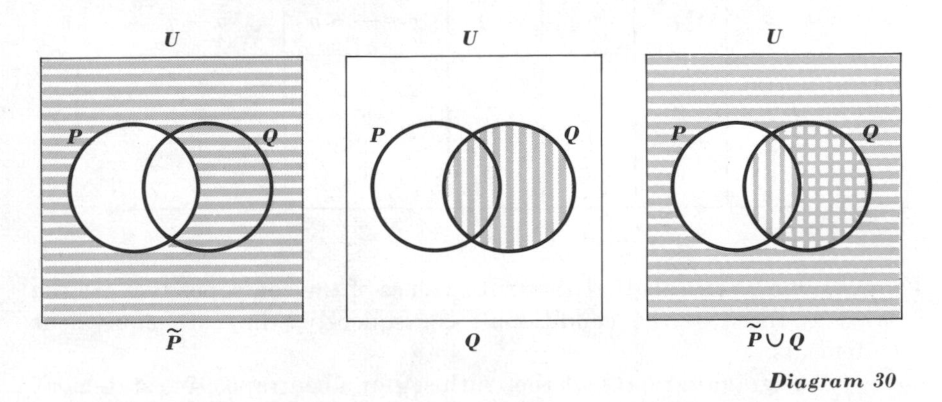

Diagram 30

Table 22

p	q	$\sim q$	$p \wedge \sim q$	$\sim(p \wedge \sim q)$
T	T	F	F	T
T	F	T	T	F
F	T	F	F	T
F	F	T	F	T

Table 22 demonstrates that the statement $\sim(p \wedge \sim q)$ can be substituted for either of the two that have been given previously in this section.

One of the reasons that the conditional is of prime importance in mathematics is that a vast majority of all theorems are stated in the "if . . . then . . ." form. Proof of theorems will not constitute a major portion of this book; in fact, very few formal proofs will be given. However, logical arguments to support many theorems will be given. There is one type of proof referred to in mathematics as the *contrapositive method*. You have undoubtedly already encountered such proofs, probably in geometry, although they may have been referred to as *indirect proofs*. This method of establishing theorems uses another statement equivalent to the conditional, which is "If not q, then not p." If statements p and q are as they have been for the examples of this section, then

If I do not go to the game, then the sun will not shine.

is the expression for "If not q, then not p." Carefully consider the four logical possibilities for this statement, and for the conditional simultaneously, to verify that (in this example) $p \rightarrow q$ and $\sim q \rightarrow \sim p$ are equivalent. That they are equivalent for any statements is established by Truth Table 23.

Table 23

p	q	$\sim q$	$\sim p$	$\sim q \rightarrow \sim p$	$p \rightarrow q$
T	T	F	F	T	T
T	F	T	F	F	F
F	T	F	T	T	T
F	F	T	T	T	T
		1	2	3	4

Table 23 reveals that the truth values of the contrapositive are the same as those of the conditional; consequently, they are equivalent statements.

A diagram to represent the truth set for a contrapositive statement

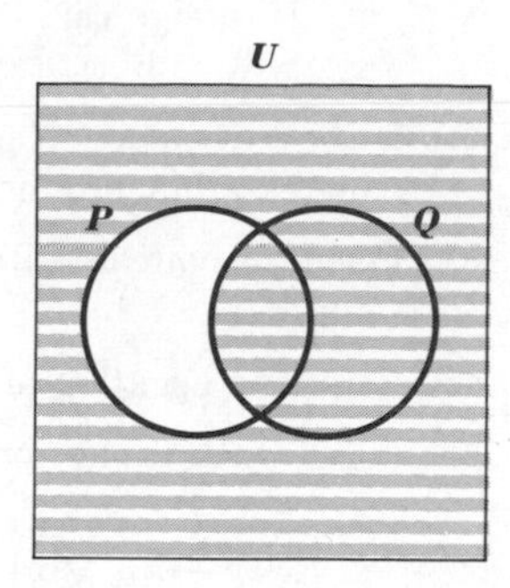

Diagram 31

is given next. It is convenient to first convert the statement $\sim q \rightarrow \sim p$ to the form $\sim(\sim q) \vee \sim p$, or equivalently, $q \vee \sim p$. To see how these equivalent forms are derived, recall that in this section it is established that $t \rightarrow s$ is equivalent to $\sim t \vee s$. If this equivalence is applied to $\sim q \rightarrow \sim p$, then $\sim q \rightarrow \sim p$ can be written as $\sim(\sim q) \vee p$. The equivalence $\sim(\sim q) \equiv q$ is used to write $\sim(\sim q) \vee \sim p$ as $q \vee \sim p$.

Diagram 31 represents the truth set for $\sim q \rightarrow \sim p \equiv q \vee \sim p$ for the case where P and Q have some, but not all, elements in common.

Previously in this section, simple statements have been used for both the hypothesis and the conclusion in the conditional. Now that a definition for the conditional statement (in terms of a truth table) is known, compound statements can be used for either the hypothesis or the conclusion. One example worth noting is $p \rightarrow (p \vee q)$. This conditional has a simple hypothesis and a compound conclusion. Truth Table 24 is for $p \rightarrow (p \vee q)$.

Table 24

p	q	$p \vee q$	$p \rightarrow (p \vee q)$
T	T	T	T
T	F	T	T
F	T	T	T
F	F	F	T

Since the results in Table 24 are all T, this conditional is an example of a type of statement that was earlier called a *tautology*. Other examples of conditionals with compound components are included in the exercises.

Exercises

1. In the following sentences, write the hypotheses and conclusions:
   a. If the sales forecast is accurate, then we will make a million dollars.
   b. If we are to succeed, then we must have a plan of action.

c. If foreign cars become popular, then the American automobile industry will be hurt.

2. Is the statement "If productivity increases, wages will increase" true if wages increase but productivity does not increase?
3. Form the contrapositive of "If prices rise, then people with fixed incomes will suffer."
4. Form the negation of "If costs rise, then profits will decrease."
5. Form truth tables for the following:

   a. $p \to q$ *b.* $\sim q \to \sim p$ *c.* $p \vee (q \to \sim p)$
   d. $\sim p \to (p \vee q)$ *e.* $(q \to p) \wedge p$
6. Translate the following statement into symbols:

   If our high prices cause customers to buy from our competitors, then the stockholders will be unhappy and the general manager will be fired.
7. Draw a diagram to represent the truth set for the conditional for each of the cases $P \subset Q$, $Q \subset P$, and $P \cap Q = \{ \ \}$.
8. Change each of the statements in Exercise 1 to the contrapositive form.
9. Do the following sentences have exactly the same meaning?

   a. It is not true that sales will fall or profits will fall.
   b. Sales will not fall, and profits will not fall.

Answers to problems

A. Yes. The hypothesis is false; hence, in any case the conditional is true.

B. Yes. If

p: Prices go up.
q: Production goes down.

then the two statements are $\sim q \to \sim p$ and $p \to q$, respectively. But $p \to q \equiv \sim q \to \sim p$, so the two statements are equivalent.

1.8 The converse of a conditional statement and the biconditional

Problem

Are the following statements equivalent?

s: The shipment will be accepted if and only if no defects are found in the sample.

t: If there are no defects in the sample, the shipment will be accepted; and if the shipment is accepted, the sample yielded no defects.

The fifth, and last, connective to be considered in this book is called the *biconditional.* Just as with the terms *conditional* and *conditional statement,* the terms *biconditional* and *biconditional statement* will be used interchangeably. One way to define biconditionals is by a truth table, and of course this will ultimately be done. But in the meantime, there is a related topic that offers an alternate approach which will be investigated.

Conditional statements have associated statements called *converses*, and it is this concept that will be explored first. In order to achieve an understanding of what the converse of a conditional statement is, recall that conditional statements have two parts: hypothesis and conclusion. In the expression "If p, then q," the letter p represents a statement which is the hypothesis and q represents the conclusion. If these roles are interchanged, then the resulting statement is the converse. This is summarized in the following.

Definition: **(*Converse of a conditional statement*)** *If p and q are statements, then the converse of the conditional statement "If p, then q" is the conditional statement "If q, then p." Symbolically, the converse of*

$$p \rightarrow q \text{ is } q \rightarrow p$$

Some examples will aid in understanding the concept of converse. As a starting point, consider the five conditional statements which follow.

Example 1. If $x + 2 = 5$, then $x = 3$.

Example 2. If $x(x - 1) = 0$, then $x = 0$.

Example 3. If x is a square, then x is a rectangle.

Example 4. If x is a rectangle, then x is a square.

Example 5. If $x + x = 4$, then $x = 2$.

The converse of each of these conditional statements is easily constructed by use of the definition of a converse.

Example 6. If $x = 3$, then $x + 2 = 5$.

Example 7. If $x = 0$, then $x(x - 1) = 0$.

Example 8. If x is a rectangle, then x is a square.

Example 9. If x is a square, then x is a rectangle.

Example 10. If $x = 2$, then $x + x = 4$.

Any replacement of x which causes Example 1 to be true also causes Example 6 to be true. A similar remark holds for the conditional in Example 5 and its converse in Example 10. This means that there are conditional statements which have the same truth set as their converses. However, Example 2 is false for $x = 1$ (hypothesis true, conclusion false), but Example 7 is true for $x = 1$ (hypothesis false, conclusion true). This shows that not *all* conditionals have the same truth set as their converses. It will be instructive to consider Example 3 and its converse in Example 8 to see if there are replacements which cause Example 3 to be true and Example 8 to be false (or Example 8 true and Example 3 false). Do the same for the statement in Example 4 and its converse in Example 9.

This discussion of converses, although far from complete, is adequate for the definition of biconditional statements.

Definition: (**Biconditional statements**) *If p and q are statements, then the biconditional statement "p if and only if q" is equivalent to the conjunction of the conditional "If p, then q" and its converse "If q, then p." Symbolically, we write*

$$p \leftrightarrow q \equiv (p \rightarrow q) \wedge (q \rightarrow p)$$

The symbol $\leftrightarrow$ will be used to represent the biconditional. This definition permits the formation of five biconditionals from the previous examples.

Example 11. $x + 2 = 5$ if and only if $x = 3$.

Example 12. $x(x - 1) = 0$ if and only if $x = 0$.

Example 13. x is a square if and only if x is a rectangle.

Example 14. x is a rectangle if and only if x is a square.

Example 15. $x + x = 4$ if and only if $x = 2$.

Since, by definition, $p \leftrightarrow q$ is equivalent to $(p \rightarrow q) \wedge (q \rightarrow p)$, the truth values for the biconditional cannot be assigned arbitrarily because there are already rules for the construction of the table for conditionals and conjunctions. But although some freedom is lost, it is not difficult to use previous tables to construct Table 25. The construction is given in three steps: the first gives the truth table for the conditional "If p, then q"; the second, the truth table for the conditional "If q, then p"; and the third, the truth table for the biconditional "If p, then q" and "If q, then p."

Table 25

p	q	$p \to q$
T	T	T
T	F	F
F	T	T
F	F	T

q	p	$q \to p$
T	T	T
F	T	T
T	F	F
F	F	T

p	q	$p \to q$	$\wedge$	$q \to p$
T	T	T	T	T
T	F	F	F	T
F	T	T	F	F
F	F	T	T	T

The techniques that have been employed in previous sections to make truth tables for compound statements can be employed to make a diagram which illustrates the compound statement $p \leftrightarrow q$. The truth set for the conditional $p \to q$ can be expressed as $\tilde{P} \cup Q$, and for the conditional $q \to p$ as $\tilde{Q} \cup P$. Consequently, the truth set for the biconditional can be expressed as $(\tilde{P} \cup Q) \cap (\tilde{Q} \cup P)$. Diagram 32 represents the truth set for the biconditional in the special case where the truth sets P and Q have some, but not all, elements in common. Again, a sequence of diagrams is given to show how the final result is obtained.

The diagrams for the cases when $P \subset Q$, $Q \subset P$, and P and Q have no elements in common are left to the reader (Exercise 7).

The diagram (and the truth table) illustrates that the biconditional is true for elements common to P and Q and for elements common to their complements. These sets can be represented by $(P \cap Q) \cup (\tilde{P} \cap \tilde{Q})$. Translated to the notation for statements, this becomes

$$(p \wedge q) \vee (\sim p \wedge \sim q)$$

This suggests that there is an equivalent statement for the biconditional, and even more, it makes clear what the equivalent statement is. Table 26, written in three steps for easier comprehension, verifies that this is the case.

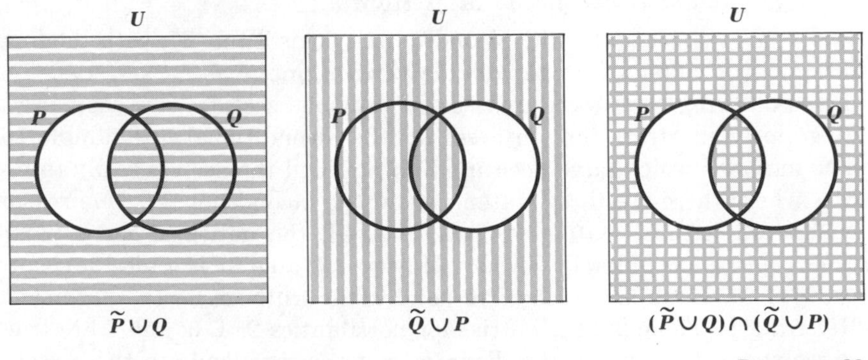

Diagram 32

Table 26

p	q	$p \wedge q$
T	T	T
T	F	F
F	T	F
F	F	F

$\sim p$	$\sim q$	$\sim p \wedge \sim q$
F	F	F
F	T	F
T	F	F
T	T	T

$p \wedge q$	$\sim p \wedge \sim q$	$(p \wedge q) \vee (\sim p \wedge \sim q)$
T	F	T
F	F	F
F	F	F
F	T	T

Example 16. Now that a method for determining the truth table of a biconditional has been established (Table 25), it will serve to form biconditionals which have compound statements for each of the parts. To extend the generalization one more step, observe that in the previous consideration only two statements were used. There are many such examples that could be given, but space limits the illustrations to one: $(p \vee q) \leftrightarrow \sim r$. So that there may be something concrete before us, let simple statements p, q, and r be given by

p: Miss Hardin arrived late.

q: Mr. Adams's ulcer hurts.

r: Mr. Adams is nice to Miss Hardin.

A biconditional of the form $(p \vee q) \leftrightarrow \sim r$ using these statements is expressed in words by

Miss Hardin arrived late or Mr. Adams's ulcer hurts if and only if Mr. Adams is not nice to Miss Hardin.

Use of Table 25 for the biconditional yields Table 27, with the final result in the column with the double vertical lines.

Following the procedure used in previous sections, we shall use a diagram to illustrate the truth set for the biconditional in Example 16. The methods which have been used do not lend themselves easily to the case where there are three statements in the biconditional. One reason is that when three statements are involved, the universal set is partitioned into eight different subsets. However, Table 27 is a nice device to use to shade the areas which correspond to the truth set for the statement. Referring to Table 27, the entries for possibilities 2, 4, 6, and 7 are true, so we shade the area in the diagram which corresponds to these cases.

Table 27

Logical possibilities	p	q	r	$p \vee q$	$\leftrightarrow$	$\sim r$
1	T	T	T	T	F	F
2	T	T	F	T	T	T
3	T	F	T	T	F	F
4	T	F	F	T	T	T
5	F	T	T	T	F	F
6	F	T	F	T	T	T
7	F	F	T	F	T	F
8	F	F	F	F	F	T

That portion of Table 27 that has T in the final column is reproduced in Table 28.

Table 28

Logical possibilities	p	q	r
2	T	T	F
4	T	F	F
6	F	T	F
7	F	F	T

The subsets that correspond to these cases are 2, 4, 6, and 7 in Diagram 33. The numbers in Diagram 33 are in the same areas as in all previous examples, but it may be helpful to study Diagram 5 again.

Although it was necessary to consider each of the eight logical possibilities in Example 16, it is sometimes possible to eliminate some of the possibilities due to information available. So another variation of the same example is offered.

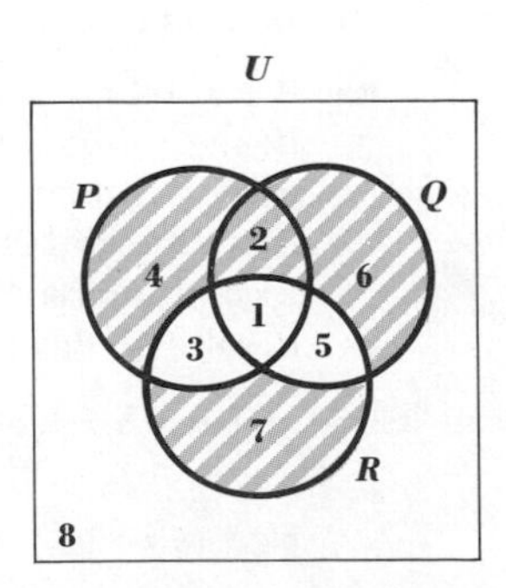

Diagram 33

Example 17. Suppose it is known that Mr. Adams is always nice to Miss Hardin. We need now consider only those cases where the statement r is true. We can eliminate cases 2, 4, 6, and 8 and consider only cases 1, 3, 5, and 7. On the basis of this additional information, the truth table for this particular biconditional statement reduces to Table 29.

Table 29

Logical possibilities	p	q	r	$p \vee q$	$\leftrightarrow$	$\sim r$
1	T	T	T	T	F	F
3	T	F	T	T	F	F
5	F	T	T	T	F	F
7	F	F	T	F	T	F

The table reveals that if Mr. Adams is always nice to Miss Hardin, then the statement

Miss Hardin arrived late or Mr. Adams's ulcers hurt if and only if Mr. Adams is not nice to Miss Hardin.

is true when Miss Hardin is on time and Mr. Adams's ulcers do not hurt—a reasonable result.

Although only a few examples have been given, the important aspects of biconditional statements have been covered. The Exercises offer ample opportunity to develop facility in the construction of both tables and diagrams for other particular cases of biconditional statements.

Exercises

1. Give the converse for each of the following conditional statements:

   a. If a good replacement is not found, then the organization will suffer.
   b. If the sales forecast is too low, then actual profits will be less than potential profits.
   c. If we are to borrow money from a bank, then we must have accurate accounting records.
   d. If sales go down, then costs should go down.

2. Construct truth tables for the following biconditionals:

   a. $p \leftrightarrow q$
   b. $(p \wedge q) \leftrightarrow (p \vee q)$
   c. $p \vee (q \leftrightarrow p)$
   d. $\sim(p \vee q) \leftrightarrow (p \wedge q)$

3. A member of the advertising agency of a company has made a survey of 100 customers who use the products of the company. For convenience, these products are referred to as A, B, and C. The results of his survey are as follows:

   5 like all three products.
   9 like A and B.
   12 like A and C.
   37 like A.
   11 like B and C.
   42 like B.
   8 like none of the three products.

   a. Draw a diagram to illustrate the results of the survey showing the number of persons in each of the eight categories.
   b. For each of the biconditional statements which follow, shade the area for which the biconditional is true.
      (1) x likes A if and only if x likes B.
      (2) y likes A and C if and only if y likes C.
      (3) u likes all three if and only if u likes none of the three.
      (4) v likes A if and only if v likes B.
      (5) w likes A and B if and only if w likes B and C.

4. The Health Food Company sells pickled eel, kumquats, and pumpkin seeds. In a survey of 100 people, the following results were obtained:

   61 like kumquats.
   73 like pickled eel.
   80 like pumpkin seeds.
   40 like kumquats and pickled eel.
   46 like kumquats and pumpkin seeds.
   60 like pumpkin seeds and pickled eel.
   30 like all three.

   a. Draw a diagram with the proper number in each of the eight areas.
   b. Use the diagram drawn in part *a* to answer the following questions:
      (1) How many liked none of the three?
      (2) How many liked kumquats only?
      (3) How many liked pumpkin seeds only?
      (4) How many liked kumquats if and only if they liked pumpkin seeds?

5. Let p be the statement "The quality of product x is good." Let q be the statement "The market for product x is poor." Write each of the following in logic symbols and make a truth table for each statement.

   a. The quality of product x is good, and the market for product x is poor.
   b. If the quality of product x is good, then the market for product x is poor.
   c. Either the quality of product x is good or the market for product x is poor, but not both.
   d. The market for product x is poor if and only if the quality of product x is good.

6. In Examples 3 and 8, are there replacements for x which make:
   a. 3 true and 8 false?
   b. 8 true and 3 false?

7. Draw a diagram for $p \leftrightarrow q$ for the case when:
   a. $P \subset Q$ *b.* $Q \subset P$ *c.* $P \cap Q = \Phi$

Answer to problem

They are equivalent. Let

p: The shipments will be accepted.
q: No defects are found in the sample.

Statement A is $p \leftrightarrow q$. Statement B is $(p \rightarrow q) \wedge (q \rightarrow p)$.

p	q	$p \leftrightarrow q$	$p \rightarrow q$	$q \rightarrow p$	$(p \rightarrow q) \wedge (q \rightarrow p)$
T	T	T	T	T	T
T	F	F	F	T	F
F	T	F	T	F	F
F	F	T	T	T	T

The statements have identical truth tables.

1.9 *Logical implication and consistency*

Problems

A. Mr. Crow says
If we make fewer blue sheets, then our sales will increase.
Mr. Dann says
If our sales do not increase, then production costs will decrease.
Mr. Fain says
We will not make fewer blue sheets, and our production costs will decrease.
Mr. Ealy says
I can believe two of you, but not all three since your statements are inconsistent.
Is Mr. Ealy right?

B. The plant manager of a factory says
If we package our product more attractively, then it will sell better.
One of his colleagues says
Our product will sell better if and only if we package it more attractively.

A third person says that the statement of the manager implies the statement of his colleague. Still another party says that the second statement implies the first statement. Is the analysis of the third party correct? Of the fourth party?

Many questions with answers which can be determined by applying mathematics involve such concepts as counting, estimation, measurement, and arithmetical computations. Another class of questions have solutions which depend on a study of relationships rather than on quantitative analysis. Hence, the concept of a *relation* is an important mathematical one. This idea is not a new one, in this book, because one relation has already been introduced, that is, the *relation of equivalence* of a set of statements. To review, recall that statements are equivalent if and only if the truth tables of the statements are the same. (An additional requirement is that the statements must involve the same simple statements.)

In this section, two more relations on sets of statements will be introduced. The relations to be considered are *logical implication* and *consistency*. Again, examples will be used to lead to definitions of these relations.

Example 1. One day in the cafeteria of CCC, the board members, Mr. Crow, Mr. Dann, and Mr. Ealy, overhear what appears to be a heated argument between Mr. Adams and Mr. Beam. The argument involves proposed changes in the colors of the blankets that CCC manufactures and the effect of such changes on sales. After considerable conversation, it is clear that the position taken by Mr. Adams can be summarized by the statement

Our sales will increase, and we shall not change the colors of the blankets.

The position taken by Mr. Beam can be summarized by the statement

If we change the color of the blankets, then our sales will increase.

After they leave, Mr. Crow, Mr. Dann, and Mr. Ealy discuss the positions taken by Mr. Adams and Mr. Beam. Mr. Crow says he believes that the two men just do not understand one another and that their positions are the same; that is, Mr. Crow believes that the statements which summarize the positions taken by Mr. Adams and Mr. Beam are equivalent. Mr. Dann does not agree with Mr. Crow. Mr. Dann says that Mr. Adams and Mr. Beam are in disagreement, but that the statement of Mr. Adams implies the statement of Mr. Beam. Mr. Ealy takes a different viewpoint from Mr. Dann and Mr. Crow and asserts that the statement of Mr. Beam implies the statement of Mr. Adams.

The statement of the problem is somewhat involved; there are five persons, each of whom has a position, and two different arguments.

Nevertheless, enough mathematical machinery has been constructed in previous sections so that an analysis can be made. It will be instructive to try to decide which (if any) of the men—Mr. Crow, Mr. Dann, or Mr. Ealy—you agree with before reading what follows. The first task is to isolate the statements which are crucial to the argument and label them with the usual notation. One way to do this is

p: Sales will increase.

q: The color of the blankets will be changed.

With this agreement, the position of Mr. Adams is represented by $p \wedge \sim q$. Reread the statement that summarized his position to see that this is true. Similarly, the position of Mr. Beam is expressed by $q \rightarrow p$. The truth tables developed in earlier sections can be used to make tables for each of these statements. It may be helpful to reread Tables 1, 11, and 20 to see how Tables 30 and 31 are constructed. Table 30 represents the position of Mr. Adams, and Table 31 the position of Mr. Beam.

Table 30

p	q	$\sim q$	$p \wedge \sim q$
T	T	F	F
T	F	T	T
F	T	F	F
F	F	T	F

Table 31

p	q	$q \rightarrow p$
T	T	T
T	F	T
F	T	F
F	F	T

Now the positions taken by Mr. Crow, Mr. Dann, and Mr. Ealy can be evaluated by use of Tables 30 and 31.

First, Mr. Crow had asserted that the statements of Mr. Adams and Mr. Beam were *equivalent*. But two statements are equivalent if and only if they have the same truth tables. A glance at Tables 30 and 31 shows that the tables are not the same. Mr. Crow is wrong!

Second, Mr. Dann had asserted that the statement of Mr. Adams *implies* the statement of Mr. Beam. The men agree that what this means is that *in every case for which Mr. Adams's statement is true, Mr. Beam's statement is also true.* (You may now wish to reassess your position in view of this agreement.) Look at the tables. There is only one case when $p \wedge \sim q$ is true (the second), and $q \rightarrow p$ is true for this case also. Mr. Dann is correct!

Third, Mr. Ealy sees that he is incorrect. If the word *implies* is to mean that when $q \rightarrow p$ (Mr. Beam's statement) is true, then $p \wedge \sim q$ (Mr. Adams's statement) is true also, then $q \rightarrow p$ does *not* imply $p \wedge \sim q$. In the tables, $q \rightarrow p$ is true in the first case but $p \wedge \sim q$ is false! (This situation also prevails in the fourth case, but one such case is enough to illustrate that $q \rightarrow p$ does not imply $p \wedge \sim q$.)

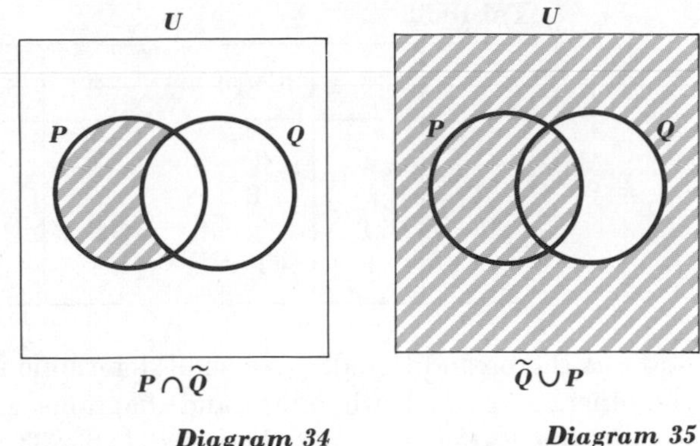

$P \cap \tilde{Q}$ $\tilde{Q} \cup P$

Diagram 34 ***Diagram 35***

Diagrams 34 and 35 illustrate the truth sets for the compound statements and will serve as a visual aid in demonstrating the relation *implies*. The diagrams reveal that the sets $P \cap \tilde{Q}$ and $\tilde{Q} \cup P$ are not equal (the statements are not equivalent), $(P \cap \tilde{Q}) \subseteq (\tilde{Q} \cup P)$ (the statement $p \wedge \sim q$ implies $q \rightarrow p$), and $(\tilde{Q} \cup P) \nsubseteq (P \cap \tilde{Q})$ ($q \rightarrow p$ does not imply $p \wedge \sim q$).

During the discussion of this example, a definition of *implies* has been reached; this relation is called *implication*.

Definition: (***Implication***) *If t and s are statements, then t implies s if and only if, when statement t is a true statement, s is true also. Symbolically, we write*

$t \Rightarrow s$ if and only if $T \subseteq S$.

As for the relation of equivalence, the assumption is made that statements t and s involve the same simple statements.

Example 2. Before leaving the matter of implication, a comparison of two different pairs of statements will be made because these pairs are frequently of use in particular examples. First, a pair of statements is chosen which use connectives *and* and *or*. Let t be the compound statement "p or q," and let s be the compound statement "p and q." The definition will be used to determine if "t implies s" and if "s implies t." The truth tables for the statements t and s are given in Table 32, and Diagrams 36 and 37 illustrate the truth sets for each statement.

From the truth tables and the diagrams, we see that t does not imply s but that s does imply t.

Example 3. As another example, consider t given by "If p, then q" and s given by "p if and only if q." You will recognize t as the conditional

Table 32

p	q	$p \vee q$
T	T	T
T	F	T
F	T	T
F	F	F

p	q	$p \wedge q$
T	T	T
T	F	F
F	T	F
F	F	F

and s as the biconditional. We shall determine if either of these implies the other. Again, truth tables and diagrams will be used to illustrate each of the statements, and these are Table 33 and Diagram 38.

Table 33

p	q	$p \rightarrow q$
T	T	T
T	F	F
F	T	T
F	F	T

p	q	$p \leftrightarrow q$
T	T	T
T	F	F
F	T	F
F	F	T

From the truth tables and the diagrams, it is clear that the biconditional implies the conditional but the conditional does not imply the biconditional. This should not come as a surprise, because another way to express this is: A theorem and its converse imply the theorem, but a theorem does not imply the theorem and its converse.

The second relation between statements that will be considered in this section is *consistency*. Again an example will be offered to point the

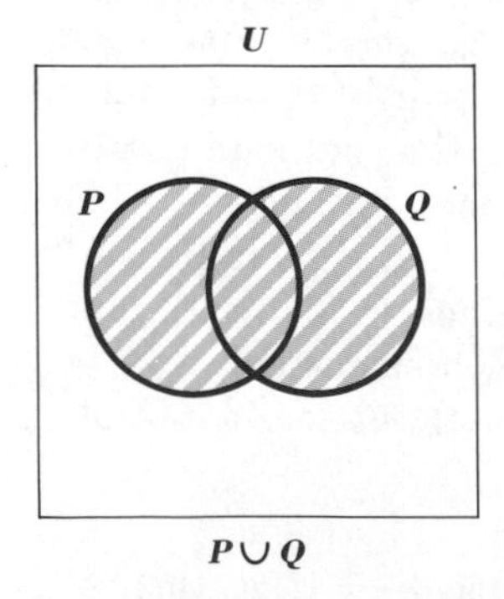

$P \cup Q$

Diagram 36

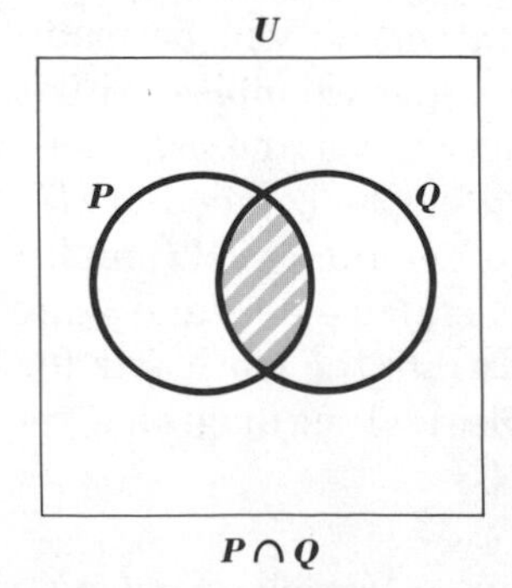

$P \cap Q$

Diagram 37

way to a definition. Keep in mind as you read the example that the goal is to determine when statements are consistent.

Example 4. The sales manager of a paper-supply firm is upset over shipping schedules and is discussing the matter with Mr. Lane of the shipping department. He says

> *If shipments leave the factory on time, then the salesmen meet their sales quotas.*

Mr. Lane, the head of the shipping department, replies

> *Shipments leave the factory on time, and the salesmen do not meet their sales quotas.*

One of the vice-presidents, who has brought the two men together to air their differences, interjects that their statements are *inconsistent.* At that point, the antagonists demand that the vice-president explain his remark. He explains that shipments may or may not go out on time and that the salesmen may or may not meet their quotas but under no circumstances are the statements of the two men both true. All three men agree that if both statements are *never* true simultaneously, then they are *inconsistent,* and they decide to put the statements of the sales manager and Mr. Lane to this test. (Is the vice-president right?)

In what follows, names are given to the statements involved, and the statements are written in symbolic form. The next step in the analysis is to construct truth tables and diagrams. Read Tables 34 and 35 to see that under no circumstances are the statements ever both true.

p: Shipments leave the factory on time.
q: The salesmen meet their sales quotas.

First: $p \rightarrow q$
Second: $p \wedge \sim q$

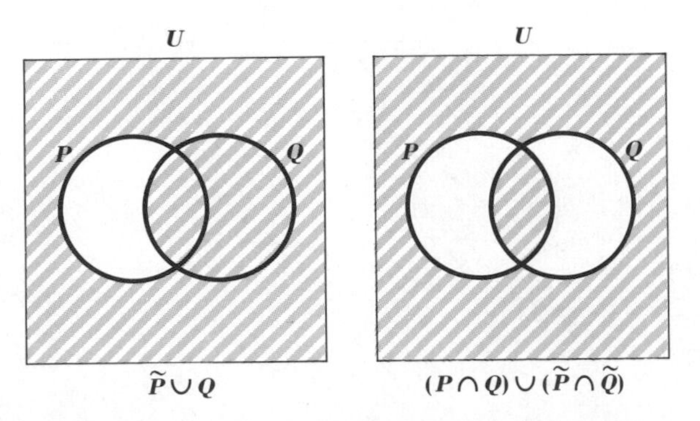

Diagram 38 $\tilde{P} \cup Q$ $(P \cap Q) \cup (\tilde{P} \cap \tilde{Q})$

Table 34

p	q	$p \rightarrow q$
T	T	T
T	F	F
F	T	T
F	F	T

Table 35

p	q	$\sim q$	$p \wedge \sim q$
T	T	F	F
T	F	T	T
F	T	F	F
F	F	T	F

Diagram 39 shows that the truth sets for these statements have no element in common.

What does the example indicate for the definition of consistency? Actually, the dispute was about inconsistency, with an agreement that inconsistency means that for no case are both statements true. But if *inconsistent* means that for *no* cases are both statements true, then *consistent* should mean that for *at least one case* both statements are true, and this is the definition adopted.

Definition: **(*Consistency*)** *If t and s are statements, then t and s are consistent if and only if there is at least one logical possibility for which both statements are true. Equivalently, t and s are inconsistent if and only if for no logical possibility are both statements true. As usual, it is understood that t and s involve the same simple statements.*

The definition for consistency is given for pairs of statements. It is clear that the definition can be extended. For example, statements t, s, and r are consistent if and only if there is a logical possibility such that all three are true; otherwise, they are inconsistent.

The definition of consistency can also be translated into the language of sets in an obvious way. If T and S denote the truth sets for statements t and s and if t and s are consistent, then $T \cap S \neq \{\quad\}$. If

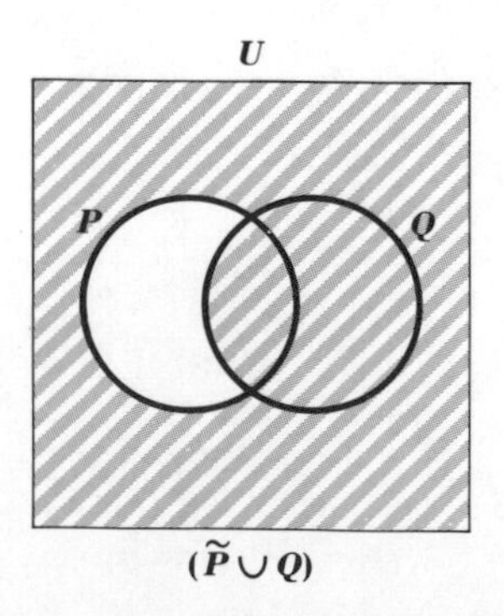

$(\widetilde{P} \cup Q)$

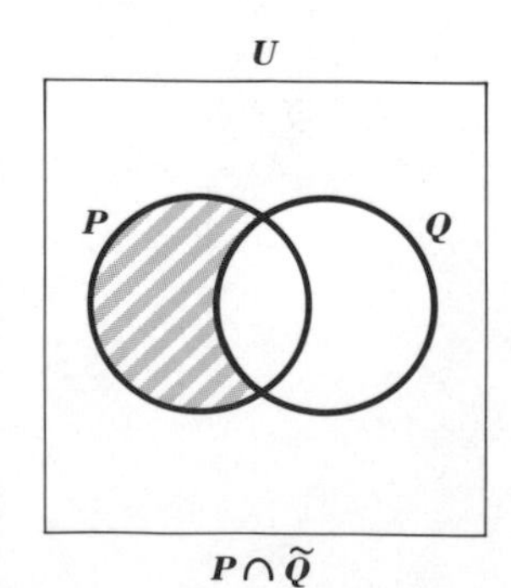

$P \cap \widetilde{Q}$

Diagram 39

t and s are inconsistent, then $T \cap S = \{\ \}$. Exercise 15 requires you to determine the condition of the truth sets when three statements are consistent.

Example 5. Example 4 led to the definition of consistency, and now the definition can be applied to such pairs of statements as

$$t: \ p \vee q$$
$$s: \ p \wedge q$$

Are these statements consistent? Either truth tables or diagrams can be used to answer the question.

Example 6. Try the definition on the pair of statements

$$t: \ p \rightarrow q$$
$$s: \ p \leftrightarrow q$$

Are these statements consistent?

Two more examples will be given. The first of these involves only two statements, but each statement is made up of three simple statements.

Example 7. First, the statements

t: If Mr. Jones resigns, then (Mr. Smith will become president of the company or sales will not increase).

s: Sales will increase if and only if Mr. Jones resigns and Mr. Smith becomes president of the company.

will be tested for consistency. The analysis is started by assigning names to the simple statement components of t and s. Let

p: Mr. Jones resigns.

q: Mr. Smith becomes president of the company.

r: Sales will increase.

Then, in symbolic language, we can write

$$t: \ p \rightarrow (q \vee \sim r)$$
$$s: \ r \leftrightarrow (p \wedge q)$$

The final step is to prepare truth tables; they are given in Table 36.

Inspection of the tables shows that in the first case statements t and s are both true, so the two statements are consistent. You can verify that there are three other cases for which both statements are true.

Table 36

p	q	r	p	$\rightarrow$	q	$\vee$	$\sim r$
T	T	T	T	T	T	T	F
T	T	F	T	T	T	T	T
T	F	T	T	F	F	F	F
T	F	F	T	T	F	T	T
F	T	T	F	T	T	T	F
F	T	F	F	T	T	T	T
F	F	T	F	T	F	F	F
F	F	F	F	T	F	T	T

p	q	r	$\leftrightarrow$	$p \wedge q$
T	T	T	T	T
T	T	F	F	T
T	F	T	F	F
T	F	F	T	F
F	T	T	F	F
F	T	F	T	F
F	F	T	F	F
F	F	F	T	F

Example 8. The final example deals with three statements. Again, it is desired to test for consistency the statements

t: Miss Hardin looks sophisticated if and only if her hair is blond.

s: If Miss Hardin does not look sophisticated, then her hair is blond.

r: Miss Hardin looks sophisticated, or her hair is blond, but not both.

The choice for simple statement components is

p: Miss Hardin looks sophisticated.

q: Miss Hardin has blond hair.

Then, stated in symbols, the three statements become

$$t:\quad p \leftrightarrow q$$
$$s:\quad \sim p \rightarrow q$$
$$r:\quad p \underline{\vee} q$$

It may be helpful to review the meaning of *exclusive or* as given on page 37. The truth table for these statements is given in Table 37.

Table 37

p	q	$p \leftrightarrow q$	$\sim p \rightarrow q$	$p \underline{\vee} q$
T	T	T	T	F
T	F	F	T	T
F	T	F	T	T
F	F	T	F	F

The entries in the table show that these three statements are inconsistent since for *no* case are all three statements true.

Exercises

1. Listed below are some pairs of statements. For each pair, determine by the use of truth tables whether the first implies the second.

   a. $p \to q,\ p \lor q$ *b.* $p \lor q,\ p \land q$
   c. $p \leftrightarrow q,\ \sim p$ *d.* $\sim(p \lor q),\ \sim p \land \sim p$
   e. $p \to q,\ \sim q \to p$ *f.* $p \lor (q \to r),\ \sim r \lor (p \leftrightarrow q)$

2. Work Exercise 1 again using diagrams for truth sets instead of truth tables.
3. Test each of the pairs of statements in Exercise 1 for consistency by the use of truth tables.
4. Work Exercise 3 again using diagrams for truth sets instead of truth tables.
5. What is the largest number of the 12 statements listed in Exercise 1 that a person can believe at any time?
6. Determine the largest number of the three statements in each exercise below that can be believed at any time.

   a. $p \to q,\ p \lor q,\ p \land q$
   b. $p \lor q,\ p \leftrightarrow q,\ \sim p$
   c. $\sim(p \lor q),\ \sim(p \land q),\ \sim p \land \sim q$
   d. $\sim q \lor \sim p,\ p \to q,\ \sim q \to p$
   e. $(p \lor q) \to r,\ \sim r \lor (p \leftrightarrow q),\ \sim r$

7. The board of directors of a toy-manufacturing plant is meeting to discuss the colors of the toy wagons to be manufactured and the sales of the wagons. The first director asserts that

   Our wagons should not be red, or our sales will not increase.

   The second director asserts that

   Our wagons should be red, and our sales will increase.

   The president of the firm asserts that the directors are in agreement. Let

   p: The wagons should be red.
   q: The sales will increase.

   a. Write the statements of the directors in logic notation.
   b. Draw a diagram for the statements of the directors and determine if the truth sets are the same.

8. A third director (see Exercise 7) asserts that

   If our wagons are red, then the sales will increase.

   A fourth director asserts that

   If our sales increase, then the wagons are red.

   The president asserts that the statement of the third director implies the statement of the fourth director.

   a. Write the statements of the directors in logic notation.
   b. Draw a diagram for the statements of the directors and determine if the president is correct.

9. A fifth director (see Exercise 7) asserts that

   It is false that (if the wagons should not be red, then the sales will increase), or the wagons should not be red.

   The president asserts that this statement is true for all cases.

   a. Write the statement of the director in logic notation.

   b. Draw a diagram for the statement of the director and determine if the president is right.

10. Make a truth table for the statements in Exercise 8 and verify your answer.

11. Make a truth table for the statement in Exercise 9 and verify your answer.

12. Determine if the following are consistent:

$$\begin{array}{c} p \wedge q \\ p \veebar q \\ p \rightarrow q \\ p \leftrightarrow q \end{array}$$

13. The statements above the solid line in each of the exercises below are the hypothesis, and the statement below the line is the conclusion in an *argument*. An argument is said to be valid if the conjunction of the hypotheses implies the conclusion. Determine which of the arguments are valid.

a. $$\begin{array}{c} p \leftrightarrow q \\ p \\ \hline q \end{array}$$

b. $$\begin{array}{c} p \vee q \\ \sim p \\ \hline q \end{array}$$

c. $$\begin{array}{c} p \wedge q \\ \sim p \rightarrow q \\ \hline \sim q \end{array}$$

d. $$\begin{array}{c} p \rightarrow q \\ \sim q \rightarrow \sim p \\ \hline p \rightarrow p \end{array}$$

e. $$\begin{array}{c} p \rightarrow q \\ \sim p \rightarrow \sim q \\ \hline \sim p \rightarrow \sim p \end{array}$$

f. $$\begin{array}{c} \sim q \rightarrow \sim p \\ \hline p \rightarrow q \end{array}$$

g. $$\begin{array}{c} p \veebar q \\ \sim q \rightarrow p \\ p \vee \sim p \\ \hline \sim p \end{array}$$

h. $$\begin{array}{c} p \rightarrow q \\ \sim p \rightarrow \sim q \\ p \wedge \sim p \\ \hline p \wedge q \end{array}$$

14. Choose simple statements and write each of the following in logic notation. Determine whether the final statement is implied by the conjunction of the other statements; i.e., determine which are valid arguments.

   a. If this is a good course, then it is worth taking. Either the grading is lenient, or the course is not worth taking. But the grading is not lenient. Therefore, this is not a good course.

   b. This is a good course if and only if the grading is lenient. Either the grading is lenient, or this course is not worth taking. This course is worth taking. Therefore, this is not a good course.

   c. If the United States is a democracy, then its citizens have the right to vote. Its citizens have the right to vote. Therefore, the United States is a democracy.

15. If t, s, and r are consistent statements, what can be said of their truth sets T, S, and R?

Answers to problems

A. Mr. Ealy is wrong. For the case "The company does not make fewer blue sheets and the sales increase and the production costs decrease," all three statements are true.

B. The third party is wrong. The fourth party is correct.

1.10 Tree diagrams

Problems

A. A job shop has received an order for a part. This part can be made from either of two grades of steel, and the required machining can be done on any of three machines. How many combinations of materials and machines does the production manager have to choose from? Draw a tree diagram to represent this.

B. The State of Complacency is going to have a new highway built. The highway can be cement or asphalt, two-lane or four-lane, and restricted-access or unrestricted-access. Draw a tree diagram showing the logical possibilities for the highway.

It has been the procedure to use visual aids, such as truth-set diagrams, whenever possible to assist in analyzing business situations. Such procedures are extremely helpful—provided, of course, that the diagrams do not mislead. In this section, the main topic will be still another type of diagram, one which is helpful in analyzing certain situations that do not easily lend themselves to truth tables or truth sets.

Example 1. A sales manager is talking to his assistant about the unusual success that one of their salesmen has in selling. They cannot be unhappy with the salesman since his sales quotas are always met, and in fact, he is one of their very best salesmen in terms of results. However, his habits are also well known. They know that on any given day, he will not call on more than six customers. Furthermore, if on any day he makes sales to three customers, then he quits work for that day. Since his results are good and since they know they cannot afford to reprimand him too severely, they are not unhappy. On the other hand, other salesmen for the company also know of his habits, and they know that frequently he does not work all day. Since this is bad for morale, the manager is convinced that something should be done about the salesman. His assistant believes that as long as the salesman's results are good, they should not interfere. The discussion becomes heated, and finally the manager says that something must be done about the situation since on at least half of the workdays, the salesman does not see six customers. The assistant retorts that perhaps this is true, but on 90 percent of the workdays, the salesman sees at least five customers. We shall assume that the salesman's chances of selling each customer are equal to his chances of rejection. (Precise meaning will be given to such statements in Chap. 3.) If we accept the habits of the salesman as they are stated, is the manager correct? Is the assistant correct? Are they both correct?

The diagram to be used to illustrate this example is called a *tree diagram*. Before examining this tree diagram, it will be helpful to notice how some previously considered situations also lend themselves to this new kind of diagram.

Example 2. Recall that if p and q are statements, then the set of four logical possibilities may be written as {TT,TF,FT,FF}. This display is still another way to represent the set of logical possibilities, which until now have been written in tables. The tree diagram in Diagram 40 also illustrates the four logical possibilities. This tree has four *branches*, which correspond to the four elements listed in the set

$$\{TT,TF,FT,FF\}$$

Example 3. To illustrate again, if p, q, and r are statements, then the set of eight logical possibilities can be written as

$$\{TTT,TTF,TFT,TFF,FTT,FTF,FFT,FFF\}$$

You will recognize these as the eight entries in a truth table expressed in a different manner. A tree diagram, somewhat like the one in Diagram 40 for two statements, is shown in Diagram 41 for the three statements p, q, and r. By tracing down the eight branches of this tree, it can be seen that they correspond exactly to the elements in the set whose tabulation is

$$\{TTT,TTF,TFT,TFF,FTT,FTF,FFT,FFF\}$$

(and also to the eight entries in the truth table).

If all eight of the logical possibilities for the three simple statements can actually occur, the statements are *independent;* otherwise, they constitute a *dependent* set of statements.

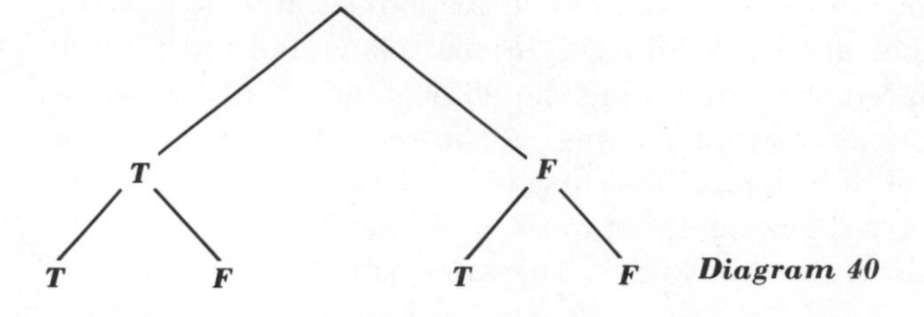

Diagram 40

Example 1. (Continued) Now that two familiar examples have been illustrated by tree diagrams, we return to the example given at the beginning of this section. A tree diagram to represent the logical possibilities for the calls that the salesman may make will be drawn. If the salesman were to make calls on six customers, with the condition that each of them will buy or will not buy, then the tree diagram would have $2^6 = 2 \cdot 2 \cdot 2 \cdot 2 \cdot 2 \cdot 2 = 64$ branches. However, as an added condition it is given that he quits any time he has three sales in one day. This means that not all the 64 branches on the tree will be required to represent the logical possibilities. Diagram 42 shows the tree of logical possibilities for the salesman's calls in one day. A count of the number of logical possibilities (branches of the tree) shows that there are 42. Furthermore, a count of the branches of the tree that terminate with a visit to the sixth customer shows that there are 32. The manager's statement about the salesman not seeing six customers on half of his workdays can now be evaluated in terms of this information. Similarly, his assistant's statement that on 90 percent of the days, the salesman sees five customers can also be analyzed by use of this diagram, although it will require a new count of a different type of branch. Is the manager correct? Is his assistant correct? If you need help with these questions, see Exercise 6.

One application of tree diagrams is as an aid in making business decisions. An example of such usage follows.

Example 4. A grain elevator is located in a rural area which produces large quantities of wheat. The elevator is small, and often the operator finds it profitable to sell all the grain in storage and restock from the current harvest. The operator wishes to consider various possibilities and their financial results. He decides to look over the records for previous years and determine the price usually paid for grain and the price usually obtainable if the grain is shipped to a metropolitan market. From the records, he obtains the data given in Table 38, applied to the next four weeks of his operation. Prices are in dollars per bushel.

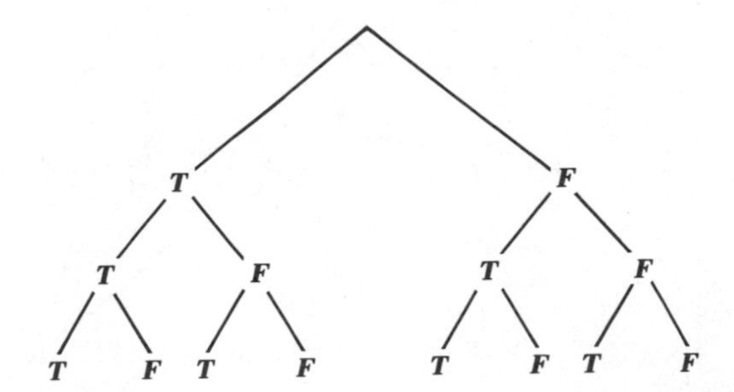

Diagram 41

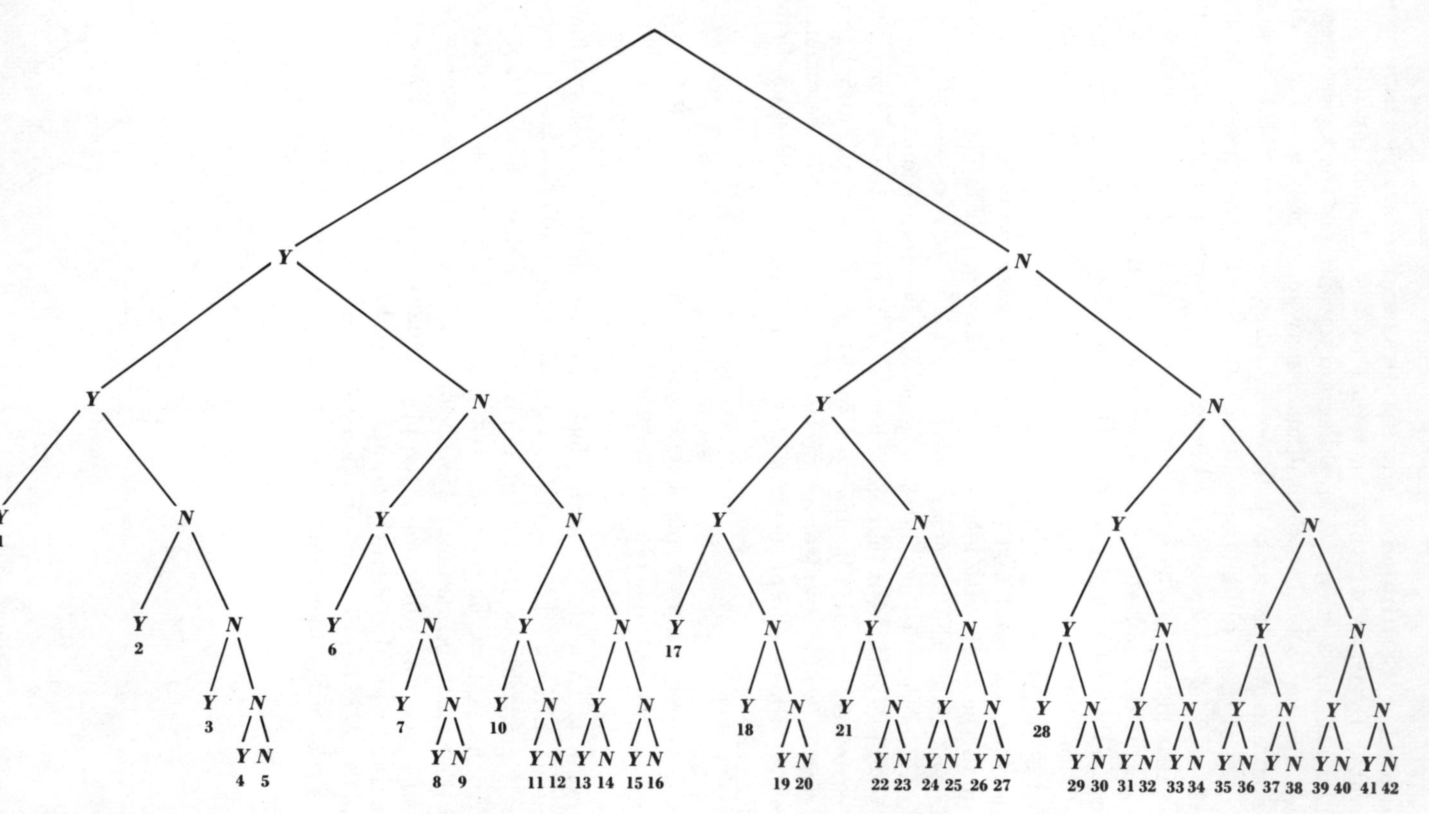
Y N
Y N Y N
Y N Y N Y N Y N
1 Y N Y 6 N Y N Y 17 N Y N Y N Y N
2 Y N Y N Y N Y N Y N Y N Y N Y N Y N Y N
3 Y N 7 Y N 10 Y N Y N Y N 18 Y N 21 Y N Y N Y N 28 Y N Y N Y N Y N Y N Y N Y N
4 5 8 9 11 12 13 14 15 16 19 20 22 23 24 25 26 27 29 30 31 32 33 34 35 36 37 38 39 40 41 42

Diagram 42

Table 38

Week	*Cost*	*Selling price*
1	$1.80	
2	$1.60	$1.70
3	$1.80	$1.60
4	$2.10	$2.40

He decides that at the end of each week he will do one of the following:

1. Sell the entire elevator of grain.
2. Restock.
3. Do neither.

If these acts are abbreviated as sell, buy, and hold (or even more concisely, as S, B, and H), respectively, then a tree diagram can be drawn to represent the possibilities. The elevator is empty as the harvest starts, so the first week the operator must buy or hold. If he buys (the first week), then the second week he must sell or hold. If he holds (the first week), then the second week he must buy or hold. And so it goes. What is the best sequence of decisions? At any point in the diagram, only two of the three acts are possible. The tree is illustrated in Diagram 43.

There are 16 possible choices, and they are numbered in the diagram. Of these choices, 1, 4, 6, 7, 10, 11, 13, and 16 lead to an empty elevator with profits of $0.50, −0.10, −0.20, 0.60, 0, 0.80, 0.60, and 0 per bushel,

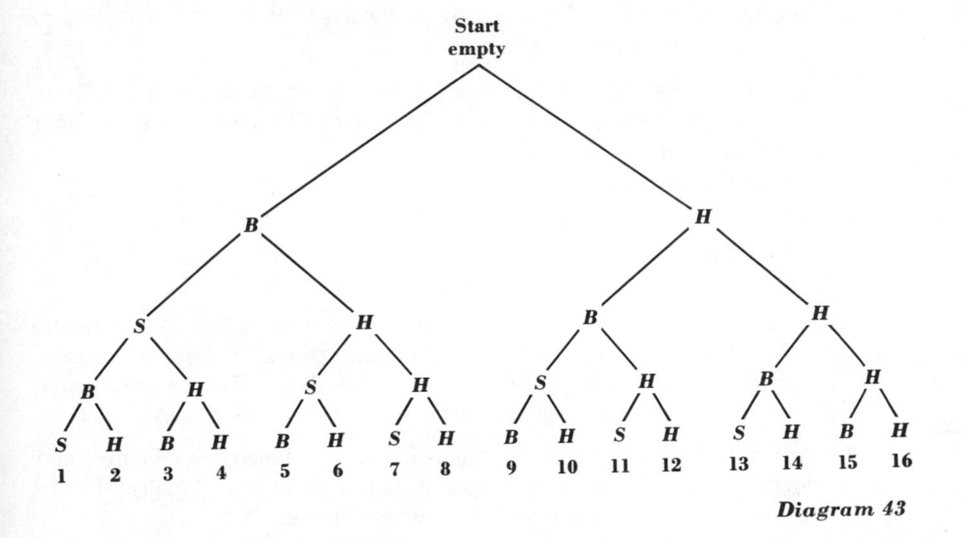

Diagram 43

respectively. The diagram, with the options as given, is sufficient to show that these choices lead to an empty elevator. To see how the profits are computed, we give the arithmetic for choices 1 and 10. Choice 1: Buy at $1.80 followed by sell at $1.70 yields loss of $0.10; followed by buy at $1.80 yields cost at $1.90; followed by sell at $2.40 yields profit of $0.50. Choice 10: Hold, no profit; followed by buy at $1.60 yields cost of $1.60; followed by sell at $1.60 yields no profit; followed by hold yields no grain and no profit.

These two computations can be used as a pattern for the other cases.

Other choices 2, 3, 5, 8, 9, 12, 14, and 15 lead to a full elevator with profits of $-0.10, −0.10, −0.20, 0, 0, 0, 0, 0, respectively. Again, the tree diagram can be used to see that these choices lead to a full elevator, and profit or loss is computed as for choices 1 and 10. If the operator wishes to end the season with an empty elevator, then the choices are clearly 7 or 11. If he wishes to end the season with a full elevator, then the choice is not so clear. For example, if the elevator holds 10,000 bushels, then choice 2 results in a loss of $1,000 and an $18,000 investment. For comparative purposes, choice 3, under the same circumstances, is not so favorable a choice. Why? Choice 9 yields no loss, but the elevator of wheat costs $21,000. You can determine that choice 12 is the best if the elevator is to be full at the end of the season.

Exercises

1. Acme Linen Company makes sheets in the colors white, pink, and blue. They are made with and without fitted corners. Make a tree diagram to show the possibilities.

2. If the same number of sheets are made in each of the three colors (see Exercise 1) and the same number are made with fitted corners as without fitted corners, then the statement

   This sheet is pink with fitted corners.

   is true for one-sixth of the sheets manufactured. The statement

   This sheet is pink.

   is true for one-third of the sheets made.

   a. Make a statement which is true for one-half of the sheets made.
   b. Make a statement which is true for two-thirds of the sheets made.
   c. Make a statement which is true for five-sixths of the sheets made.
   d. Make a statement which is true for one-third of the sheets made.

3. Classify the persons enrolled for this course as freshmen, sophomores, and upperclassmen. Then classify them as men and women. Finally, classify them as those who will fail and those who will pass.

   a. Construct a tree for the class enrollment.

b. Choose for statements

p: x is a sophomore.
q: x will pass the course.
r: x is a man.

Write in logic notation

x will pass this course if and only if x is not a sophomore man.

c. Construct a truth table for *b*.

d. List one logical possibility which makes part *b* true and one which makes part *b* false.

4. In selecting its personnel, a company considers the overall grade-point average of the applicant, the grade-point average in the major department, and the recommendations given the applicant. The overall grade-point average is divided into those over or equal to 2.5 and those under 2.5. The major-department grade-point average is divided into those over or equal to 2.8 and those under 2.8. Recommendations are classified as high, average, and low.

a. Make a table of logical possibilities.

b. Construct a tree diagram for the table.

c. Choose for statements

p: x has an overall grade-point average under 2.5.
q: x has a major-department grade-point average over or equal to 2.8.
r: x receives an average recommendation.

Write in logic notation the statement

x receives an average recommendation if and only if x has an overall grade-point average over or equal to 2.5 and a major-department grade-point average over or equal to 2.8.

d. Make a truth table for the statement in part *c*.

e. Write in logic notation

If x has an overall grade-point average over or equal to 2.5 and a major-department average over or equal to 2.8, then x receives an average recommendation.

f. Make a truth table for the statement in part *e*.

5. Company A classifies salesmen as good or bad, the supply of a product as high or low, and the quality of a product as excellent, average, or poor. x is a salesman for company A with product y.

a. Make a table of the logical possibilities for x's ability, the supply of product y, and the quality of product y.

b. Construct a tree diagram illustrating this table.

c. Choose for statements

p: x is a good salesman.
q: The supply of product y is high.
r: The quality of product y is excellent.

Write in logic notation

If x is a good salesman and the quality of product y is excellent, then the supply of product y is low.

d. Make a truth table for the statement in part *c*.

e. Write in logic notation

x is a bad salesman, and if the quality of product y is poor, then the supply of product y is high.

f. Make a truth table for the statement in part *e*.

g. If it is known that if the quality is poor, then the supply is high, which of the logical possibilities of part *a* are eliminated? Illustrate on the tree diagram.

6. Refer to Diagram 42 and verify that of the 42 visits to customers, 6 end on the fifth visit. Do the same to see that 32 of the visits end with 6 visits. Is the number $6 + 32 = 38$, 90 percent of 42? Is the assistant's statement true?

7. Verify that each of the results listed in the text for the 16 possibilities of Example 4 is correct.

Answers to problems

A. He has six combinations to choose from.

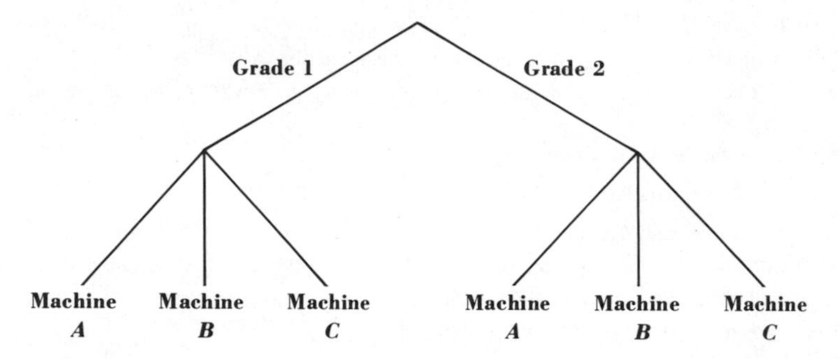

B.

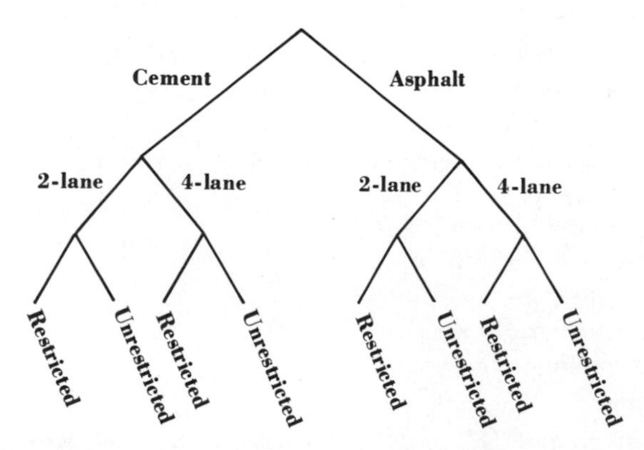

2

FORMULAS FOR COUNTING

2.0 Introduction

The primary purpose of the introductory section of each chapter, as we explained in Sec. 1.0, is to present a preview of what is to come in the chapter. But before that task is begun for this chapter, think about the concepts studied in Chap. 1. Some of these are statement, set, connective, truth table, relation, and tree diagram. These terms all have special mathematical meanings, and hopefully those meanings have been made clear. One important aspect of any job in business (or any job, for that matter) is decision making. In what precedes, an attempt was made to show how business circumstances can be translated into statements which, in turn, permit analysis by the use of truth tables. Sets and diagrams are useful in illustrating the analysis. A tree diagram is another visual aid in the analysis.

These procedures may have seemed somewhat nonmathematical to you—especially if your prior training in mathematics dealt largely with its computational aspects. But this chapter will offer quite a change of pace. Although mathematics as an aid in decision making will continue to be an integral part of the book, we shall devote this chapter to a different kind of mathematics, one that is familiar to all: *counting*. This change in emphasis will result in more of the computational aspect of mathematics and hence will result in the kind of mathematical activity which may more closely parallel the kind of exercise required of you in your previous mathematical training.

You are reminded that the process of counting as you have experienced it in earlier studies is fairly easy. Now some questions will be posed for consideration. They all involve counting, but the answers, though fairly easily obtained, are not easily computed by counting in the straightforward "1, 2, 3, . . ." manner.

1. How many different committees of five people each can be chosen from the United States Senate? (Two committees are the same if and only if they are made up of the same group of five people.)
2. In a certain state, automobile license plates have either two or three letters followed by three or four digits. How many different license plates are possible in that state?
3. In the American League baseball standings, how many outcomes are possible, i.e., with one team in first place, one in second place, one in third place, and so forth on to tenth place (there being 10 teams in all)?

The answers to the three questions can be determined by counting, but each is more than a million. Clearly, you would not want to count all possible elements in the sets to arrive at the answer! By the time you finish this chapter, you will be able to answer these (and even harder) questions.

Thus we are led to the necessity of some structured means of computation. The end result will invariably be what is called—in this book—a *formula*. The emphasis will continue to be nonrigorous. Formulas will be motivated by examples, and the formulas will then be used on other examples. There will be some discussions of a mathematical nature to justify the formulas, but the emphasis will be on understanding and usage—not proof.

Some specific formulas are the number of elements in the union of sets, permutations, combinations, binomial formulas, and multinomial formulas. But these are technical terms best understood through a more detailed study than that intended in a preliminary section like this.

The important thing to remember as you proceed is that you will be provided with some important mathematical tools that will assist you in answering questions which arise quite naturally out of everyday situations.

2.1 Counting the number of elements in sets

Problems

A. An adding-machine company makes a basic right-column hand-operated machine which costs \$120. Additional features are available, as follows:

Wide carriage—add \$30
Electric drive—add \$60
Grand-total register—add \$50
Two additional columns—add \$40

If each agency carries one machine of each possible combination of features:

1. How many machines would each agency have to stock?
2. If there are 300 agencies, what dollar deduction in inventory would result from dropping the wide-carriage and grand-total features on the hand-operated machines?
3. One agency reported the following sales:

   16 wide-carriage machines
   18 hand-operated machines
   11 grand-total machines
   42 eight-column machines
   94 electrically driven machines
   2 hand-operated wide-carriage machines with grand-total registers
   3 ten-column electric machines with grand-total registers
   3 electric machines with grand-total registers and wide carriages
   2 hand-operated machines with grand-total registers and narrow carriages
   10 ten-column hand-operated machines

   All wide-carriage machines had 10 columns, and all grand-total machines had 10 columns. How many machines were sold, and how many of each type were sold?
4. How much would the inventory of the agency in part 3 be reduced if the models which had zero sales were not carried in stock?

B. The sales manager of a company is to divide his sales force into two groups. One group is to concentrate on selling to wholesale organizations, and the other to retail outlets. He wishes to have seven members in the first group and eight in the second, with three salesmen to work in both categories. How many salesmen must he employ?

In this section, we shall begin an intensive study of problems whose solutions are arrived at by counting. Such problems are certainly not new to the reader since most persons normally face circumstances calling for some sort of counting process at a very early age; nor are such problems new in this book. For example, several questions were answered by counting branches of the trees in Sec. 1.10.

Example 1. Often a question can be answered by counting the union of two sets. As an example, John has seven books and Mary has five books. How many books do they possess? One way to think of the answer is to put the seven books with the five books and count the elements in the union of the sets. After a bit of practice (usually in the first grade), the sets are ignored and a mental process yields "seven and five is twelve." In symbols, $7 + 5 = 12$.

The process is easily extended to three, four, five, or more sets. So that the next example will be in terms of something familiar (in this book), recall that if statements p and q are given, a universal set U of logical possibilities for these two statements is determined. If P and Q denote truth sets of each of these statements, respectively, then a diagram can be used to illustrate (see Diagram 1).

In Diagram 1, the numbers 1 to 4 have been inserted to indicate the four subsets of U that correspond to the entries in the truth table for p and q. If the number of elements in each of the four subsets of U is known, then the sum of the numbers of elements in the sets yields the total number of elements in the universal set U.

Example 2. To illustrate, if the number of elements in the subsets is 3, 5, 4, and 7, then the number of elements in U is the sum $3 + 4 + 5 + 7 = 19$. This example, and the previous one, clearly shows that the number of elements in a set is arrived at in a trivial way if the set has been partitioned into subsets and if each of the subsets has already been counted (or can be counted).

There is a temptation to write a formula for counting the number of elements in the union of two (or more) sets on the basis of these examples, and such will soon be done. But a little caution must be used. In the examples, the sets used to form the union to be counted were selected so that (1) no element was counted twice and (2) every element was counted. Certainly, these two principles must prevail if the answer determined by forming a sum is to be meaningful.

With this word of warning in mind and with a new word and a new notation, the first formula can be presented. If sets have the property that they have no elements in common, they are *disjoint*. For two sets A and B to be disjoint means that $A \cap B = \{\ \ \}$. The notation for

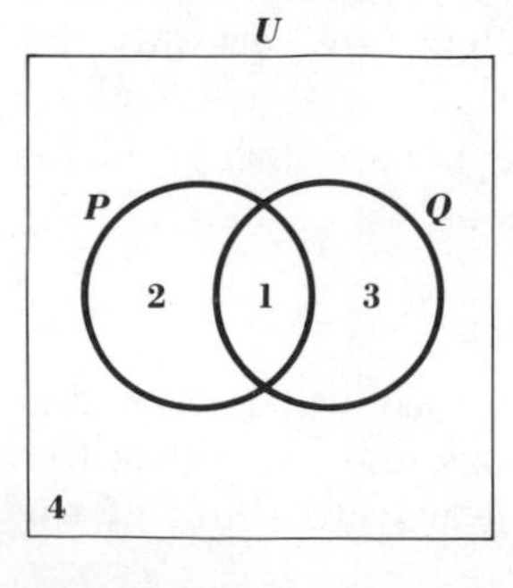

Diagram 1

the number of elements in set S is $N(S)$. Read the symbol $N(S)$ as "the number of elements in set S." Now the formula.

Formula: (**Number of elements in the union of disjoint sets**) *If the sets $S_1, S_2, \ldots, S_n$ are disjoint and if $N(S_1), N(S_2), \ldots, N(S_n)$ are the number of elements in the sets, then the number of elements in the set S formed by the union of $S_1, S_2, \ldots, S_n$ is the sum of $N(S_1), N(S_2), \ldots, N(S_n)$. Symbolically, we write*

$$S = S_1 \cup S_2 \cup \cdots \cup S_n$$

and also $$N(S) = N(S_1) + N(S_2) + \cdots + N(S_n)$$

The breaking down of a set S into subsets $S_1, S_2, \ldots, S_n$ such that no two of the subsets have any elements in common (the sets are said to be *pairwise disjoint*) and such that S is the union of the subsets is described by the phrase "a partition of S." The use of the word *partition* in this context agrees with its usage in everyday circumstances, such as a partition of a building into rooms and a partition of the chairs in a room into rows. Some of the examples of Chap. 1 are easily interpreted as partitions, and some additional examples will illustrate the concept.

Example 3. The next example of a partition of a set is a subset and its complement. If R is a subset of U, then the pair of sets R and $\tilde{R}$, written $(R,\tilde{R})$, is a partition of U. Diagram 2 should remind you that R and $\tilde{R}$ have no elements in common and that $U = R \cup \tilde{R}$.

Example 4. Reconsider the subsets $P \cap Q$, $P \cap \tilde{Q}$, $\tilde{P} \cap Q$, and $\tilde{P} \cap \tilde{Q}$ as given in Diagram 1. They are pairwise disjoint, and their union is U. Be sure it is clear why the set labeled 1 in Diagram 1 is $P \cap Q$, the set labeled 2 is $P \cap \tilde{Q}$, the set labeled 3 is $\tilde{P} \cap Q$, and the set labeled 4 is $\tilde{P} \cap \tilde{Q}$. No two of the sets $P \cap Q$, $P \cap \tilde{Q}$, $\tilde{P} \cap Q$, $\tilde{P} \cap \tilde{Q}$ have any elements in common. For example, the intersection of $P \cap Q$ and $P \cap \tilde{Q}$, denoted $(P \cap Q) \cap (P \cap \tilde{Q})$, is empty since any elements in the intersection must be in both Q and $\tilde{Q}$, which is an impossibility. Test other pairs of these four subsets to verify that their intersections are empty. The fact that the union of the four sets is the set U has already been demonstrated.

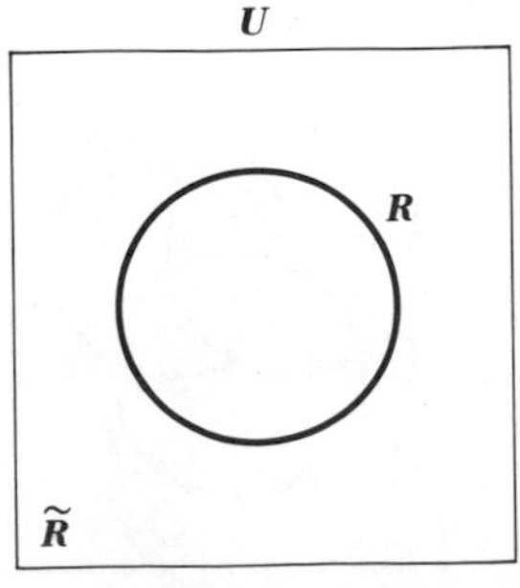

Diagram 2

A notation for indicating a partition will be handy, and grouping by parentheses seems natural. For a set R and its complement, we write $(R,\tilde{R})$ is a partition of U. Similarly, $(P \cap Q, P \cap \tilde{Q}, \tilde{P} \cap Q, \tilde{P} \cap \tilde{Q})$ is a partition of U. More generally, if $S_1, S_2, \ldots, S_n$ are subsets that are *pairwise disjoint* and $U = S_1 \cup S_2 \cup \cdots \cup S_n$, then the symbol $(S_1, S_2, \ldots, S_n)$ is used to indicate that the sets form a partition of U.

Example 5. For the next example, statements p, q, and r are given, and the associated truth sets are P, Q, and R, respectively. The partition of the set of logical possibilities U into subsets is illustrated in Diagram 3. The numbers 1 to 8 have been inserted in Diagram 3 to denote the subsets which partition the universal set of logical possibilities into eight disjoint subsets. These eight subsets can be (and have been) thought of as arising from the eight logical possibilities in the corresponding truth table.

But perhaps a more reasonable way to view these eight subsets, at least for present purposes, is to consider the partition

$$(P \cap Q, \tilde{P} \cap Q, P \cap \tilde{Q}, \tilde{P} \cap \tilde{Q})$$

and a second partition

$$(R,\tilde{R})$$

If each of the four elements of the first partition is intersected with each element of the second partition, the result is the eight sets indicated in Diagram 3. If the usual set notation is used, the eight sets in Diagram 3 are written

(1) $P \cap Q \cap R$ (3) $P \cap \tilde{Q} \cap R$ (5) $\tilde{P} \cap Q \cap R$ (7) $\tilde{P} \cap \tilde{Q} \cap R$
(2) $P \cap Q \cap \tilde{R}$ (4) $P \cap \tilde{Q} \cap \tilde{R}$ (6) $\tilde{P} \cap Q \cap \tilde{R}$ (8) $\tilde{P} \cap \tilde{Q} \cap \tilde{R}$

The eight subsets of U are such that they are pairwise disjoint (that is, no pair of sets has any elements in common), and the union of the eight subsets is the universal set U. As before, if the number of elements in each of the eight subsets is known, then the number of elements in the universal set is the sum of the numbers of elements in the eight subsets.

The partition which results from the two partitions

$$(P \cap Q, P \cap \tilde{Q}, \tilde{P} \cap Q, \tilde{P} \cap \tilde{Q}) \qquad \text{and} \qquad (R,\tilde{R})$$

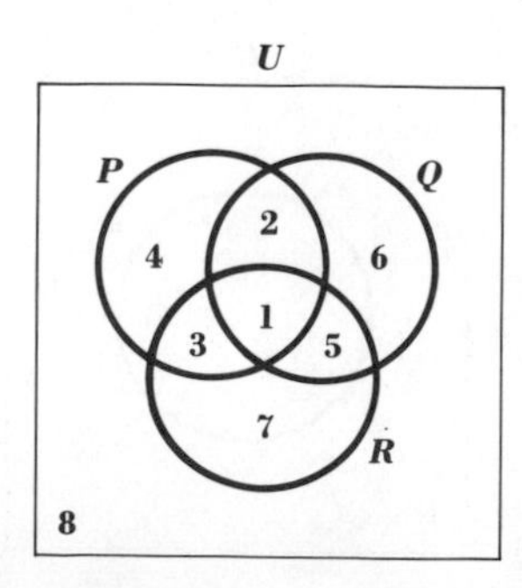

Diagram 3

by intersecting each member of the first partition with each member of the second partition is called the *cross partition* of the two partitions. The partition with eight subsets is called a *refinement* of each of the other two partitions.

Everything said thus far in this section is directed to the question: What is the number of elements in a set S if S has been partitioned into subsets such that the number of elements in each subset is known? The answer is determined by following the direction in the formula: *add*. The remainder of this section is directed to a second question, which is only a slight variation of the first question. For the first question, care was taken that no element was counted twice (the subsets are pairwise disjoint) and that each element was counted (the union of the subsets is the set). If a set is to be counted, it must *always* be the case that every element is counted, so the second condition will continue to be imposed on the subsets. But what about cases where elements might be counted more than once? The second question then is

What is the number of elements in S if the subsets have elements in common?

An example will illustrate.

Example 6. A sales analyst, Mr. Sanchez, has just completed a survey showing that 65 persons who were interviewed purchased pink candles and 48 persons purchased blue candles. Answer the questions:

How many persons were interviewed?

How many persons purchased candles of one of the colors but did not purchase both types of candles?

A hasty reply might be that there were $65 + 48 = 113$ persons interviewed. This result assumes that the two sets have no elements in common; but there might be persons who purchased both pink and blue candles. Mr. Sanchez is consulted, and he reports that there were 100 persons interviewed and that every person made a purchase. With this new information, a diagram can be drawn to illustrate the results of the survey.

Consider the two statements

p: x is a purchaser of pink candles.

b: x is a purchaser of blue candles.

If U is the universal set, P the truth set of p, and B the truth set of b, then we already have the information

$$N(U) = 100$$

$$N(P \cap B) = 13 \qquad N(P \cap \tilde{B}) = 52$$

$$N(\tilde{P} \cap B) = 35 \qquad N(\tilde{P} \cap \tilde{B}) = 0$$

This is portrayed in Diagram 4.

This example shows that the formula for finding the number of elements in the union of sets when the sets are disjoint fails if the sets are not disjoint. Expressed in symbols, this becomes

$$N(P) + N(B) = 65 + 48 = 113$$
$$N(P \cup B) = 100$$
$$N(P) + N(B) \neq N(P \cup B)$$

The discrepancy between the numbers 113 and 100 arises because some elements (13) are counted twice: once as elements of P and again as elements of B. The example also makes clear that the number of elements in $P \cup B$ is obtained by adding the number of elements in P and B and subtracting the number of elements in $P \cap B$.

Diagram 4 is also helpful in determining the answer to the second question of this example. There are 52 persons who purchased pink candles but not blue, and 35 persons purchased blue candles but not pink. These sets are disjoint, so the previous formula applies; $52 + 35 = 87$ persons are in the stated category. Symbolically,

$$N(P \cap \tilde{B}) + N(\tilde{P} \cap B) = 52 + 35$$

This special case points the way to the formula developed in the next paragraphs.

Let S and T be sets, disjoint or not, with $N(S)$ and $N(T)$ elements in S and T, respectively. The number of elements in $S \cup T$, $N(S \cup T)$, will be determined by the use of the three sets indicated in Diagram 5. The number of elements in S, $N(S)$, is the sum of the number of elements of S in T, $N(S \cap T)$, and the number of elements of S not in T, $N(S \cap \tilde{T})$. (These are disjoint sets, so the previous formula can be applied to them.) Therefore,

$$(1) \quad N(S \cap T) + N(S \cap \tilde{T}) = N(S)$$

Similarly, the set T has a partition $(S \cap T, \tilde{S} \cap T)$. Applying the formula for counting the elements in a partition, we have

$$(2) \quad N(S \cap T) + N(\tilde{S} \cap T) = N(T)$$

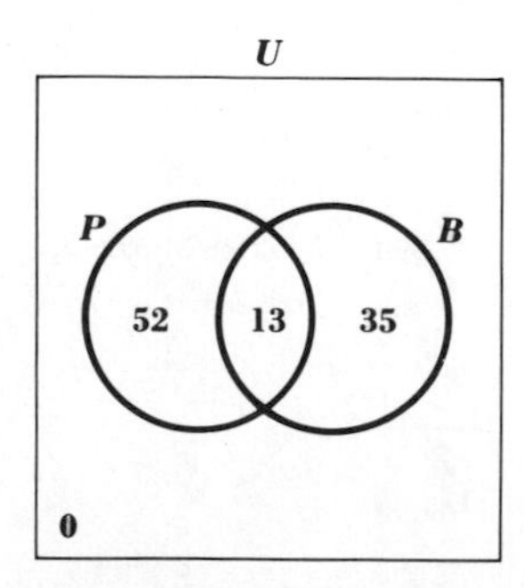

Diagram 4

Addition of Eqs. (1) and (2) yields

$$(3)\quad 2[N(S \cap T)] + N(S \cap \tilde{T}) + N(\tilde{S} \cap T) = N(S) + N(T)$$

If $N(S \cap T)$ is subtracted from both sides of Eq. (3), then

$$(4)\quad N(S \cap T) + N(S \cap \tilde{T}) + N(\tilde{S} \cap T) = N(S) + N(T) - N(S \cap T)$$

Now consider $(S \cap T,\ S \cap \tilde{T},\ \tilde{S} \cap T)$. These sets have been written as if they were a partition of the set $S \cup T$, and they are. The three sets are pairwise disjoint (no two of them have elements in common)—a fact that should be verified—and the union of the three sets is $S \cup T$. One way to see that these statements are true is to study Diagram 5.

The formula on page 93 offers a way to count the elements in the set $S \cup T$ when the number of elements in the members of a partition is given. So applying the formula, there is another result:

$$(5)\quad N(S \cap T) + N(S \cap \tilde{T}) + N(\tilde{S} \cap T) = N(S \cup T)$$

The left-hand sides of Eqs. (4) and (5) are equal, so the right-hand sides of these expressions are also equal. Therefore,

$$N(S \cup T) = N(S) + N(T) - N(S \cap T)$$

This is the formula that has been sought.

Formula: (***Number of elements in the union of two sets***) *If S and T are sets with N(S) and N(T) elements, respectively, then the number of elements in $S \cup T$ is determined by adding the number of elements in S and the number of elements in T and subtracting the number of elements in $S \cap T$. Symbolically, we write*

$$N(S \cup T) = N(S) + N(T) - N(S \cap T)$$

Certainly, this result should not come as a surprise. The question dealt with how to count when some elements might be counted twice. The solution is to subtract those elements that have been counted twice.

In a similar fashion, a formula for determining the number of elements in the union of three sets can be derived. Since the method employed is just an extension of the computational techniques in the paragraphs above, the derivation will not be given here in detail. The result is summarized in the formula that follows.

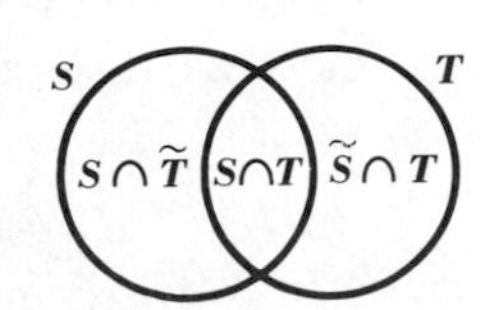

Diagram 5

Formula: **(*Number of elements in the union of three sets*)** *If S, T, and R are sets with $N(S)$, $N(T)$, and $N(R)$ elements, then the number of elements in $S \cup T \cup R$ is*

$$N(S \cup T \cup R) = N(S) + N(T) + N(R) - N(S \cap T) - N(S \cap R) - N(T \cap R) + N(S \cap T \cap R)$$

The section will conclude with three examples which use the formulas for the number of elements in the union of sets.

Example 7. In a sample of defective blankets, 63 have defects in the cloth, 82 have defects caused by manufacturing, and 21 have defects of both types. How many blankets are in the sample?

Using suggestive notation,

$$N(C) = 63 \qquad N(M) = 82 \qquad N(C \cap M) = 21$$
$$N(S) = N(C) + N(M) - N(C \cap M) = 63 + 82 - 21 = 124$$

There are 124 blankets in the sample.

Example 8. In a sample of 100 defective sheets, 82 have tears and 32 are improperly dyed. How many have both defects?

Using suggestive notation,

$$\begin{aligned} N(S) &= 100 \qquad N(T) = 82 \qquad N(D) = 32 \\ N(S) &= N(T) + N(D) - N(T \cap D) \\ 100 &= 82 + 32 - N(T \cap D) \\ 100 &= 114 - N(T \cap D) \\ N(T \cap D) &= 14 \end{aligned}$$

Fourteen sheets have both defects.

Example 9. An employee makes a survey of 300 purchases of sheets to analyze the most popular color. The survey shows:

190 purchased pink sheets.
170 purchased blue sheets.
110 purchased white sheets.
120 purchased both pink and blue sheets.
70 purchased both pink and white sheets.
50 purchased both blue and white sheets.

However, the employer disbelieves the results of the survey. Why?

The answer is obtained as an application of the formula for counting the union of three sets. Using suggestive notation,

$$N(C) = 300 \qquad N(P) = 190 \qquad N(B) = 170 \qquad N(W) = 110$$
$$N(P \cap B) = 120 \qquad N(P \cap W) = 70 \qquad N(B \cap W) = 50$$
$$N(C) = N(P) + N(B) + N(W) - N(P \cap B) - N(P \cap W) - N(B \cap W) + N(P \cap B \cap W)$$

$$300 = 190 + 170 + 110 - 120 - 70 - 50 + N(P \cap B \cap W)$$
$$300 = 470 - 240 + N(P \cap B \cap W)$$
$$300 = 230 + N(P \cap B \cap W)$$
$$70 = N(P \cap B \cap W)$$

The number purchasing all three products is 70; that is, 70 customers purchased both blue and white sheets, as well as pink sheets. But the survey had reported only 50 such customers. Therefore, the data submitted in the survey are incorrect.

Exercises

1. A survey of purchasers of gasoline reveals that 43 bought regular and 68 bought premium. How many persons were surveyed if:
   a. No purchaser bought gasoline of both qualities?
   b. 12 purchasers bought gasoline of both qualities?
   c. 29 purchasers bought gasoline of both qualities?
   d. 51 purchasers bought gasoline of both qualities?

2. Statement p has eight elements in truth set P; statement q has 12 elements in truth set Q. How many elements are in the truth set of $p \vee q$, $(P \cup Q)$, if:
   a. p and q are inconsistent?
   b. p implies q?
   c. p and q are both true for exactly two cases?

3. What can be said about $N(P)$ and $N(Q)$ if p and q are equivalent?

4. A survey of 580 housewives reveals that:
   70 purchase pink, blue, and striped towels.
   110 purchase pink and blue towels.
   100 purchase pink and striped towels.
   120 purchase blue and striped towels.
   200 purchase pink towels.
   260 purchase blue towels.
   150 purchase none of the three.
   How many purchase striped towels?

5. A salesman has six prospects: a_1, a_2, a_3, a_4, a_5, a_6. He partitions the set into those who have purchased previously and those who have not, and he obtains

$$(P,\tilde{P}) = (\{a_1,a_3\}, \{a_2,a_4,a_5,a_6\})$$

   He partitions the set into those expected to purchase in the future and those not expected to purchase in the future, and he obtains

$$(Q,\tilde{Q}) = (\{a_1,a_2,a_4\}, \{a_3,a_5,a_6\})$$

   a. Draw a diagram illustrating the cross partition. In each of the four sets of the cross partition, list the elements.

b. Use the formula for the number of elements in the union of two sets to obtain $N(P \cup \tilde{Q})$.
c. Write in words a description of the set $P \cap Q$.
d. Write in words a description of the set $\tilde{P} \cup Q$.

6. Let

p: The quality of Klar detergent is good.
q: The supply of Klar is large.
r: The sales of Klar are good.

a. Write the partition naturally induced on the set of all logical possibilities by statement p. Do the same for statement q and for statement r.
b. Write the cross partition of the partitions in part *a*.

What can you say about the cross partition if:

c. It is inconsistent that quality is good, supply is large, and sales are good?
d. The sales of Klar are good implies that (supply is large and quality is high)?

7. If (A_1,A_2) and (B_1,B_2) are partitions of S, then show that the cross partition $(A_1 \cap B_1, A_1 \cap B_2, A_2 \cap B_1, A_2 \cap B_2)$ satisfies the two conditions of a partition.

8. A partition is a refinement of another partition if every subset of the first is a subset of some element of the second. Show:

a. $(A_1 \cap B_1, A_1 \cap B_2, A_2 \cap B_1, A_2 \cap B_2)$ is a refinement of (A_1,A_2).
b. $(A_1 \cap B_1, A_1 \cap B_2, A_2 \cap B_1, A_2 \cap B_2)$ is a refinement of (B_1,B_2).

9. An oil firm which has a large credit-card business classifies its customers according to (1) whether the card is used zero, one, or two or more times per month; (2) whether the monthly total purchase is less than $7 or is $7 or more; and (3) whether or not the bill is paid when due.

a. Develop a notation and write the three partitions indicated.
b. Write the cross partition of partitions 1 and 2.
c. Write the cross partition of part *b* and partition 3.
d. Draw a tree diagram for part *c*.

The company assumes from past experience that the conditions imposed to effect the partitions are such that of their 120,000 credit-card holders, each subset of partition 1 has 40,000 members, each subset of the cross partition of partitions 1 and 2 contains 20,000 members, and the cross partition of partitions 1, 2, and 3 is such that each subset has 10,000 members. Use these numbers for the following questions:

e. If a card of thanks is sent to those customers who use the card two or more times per month or spend more than $7, how many cards will be sent?
f. If a calendar for a wallet is sent to those who pay promptly, how many are sent?
g. How many receive both a card of thanks and a wallet calendar?

10. Derive the formula for the number of elements in the union of three sets. [HINT: In the formula $N(A \cup B) = N(A) + N(B) - N(A \cap B)$, let $A = S$, $B = T \cup R$. Show that $S \cap (T \cup R) = (S \cap T) \cup (S \cap R)$ and that $S \cap T \cap S \cap R = S \cap T \cap R$.]

Answers to problems

A. 1. 16
2. $348,000
3. 112 sold

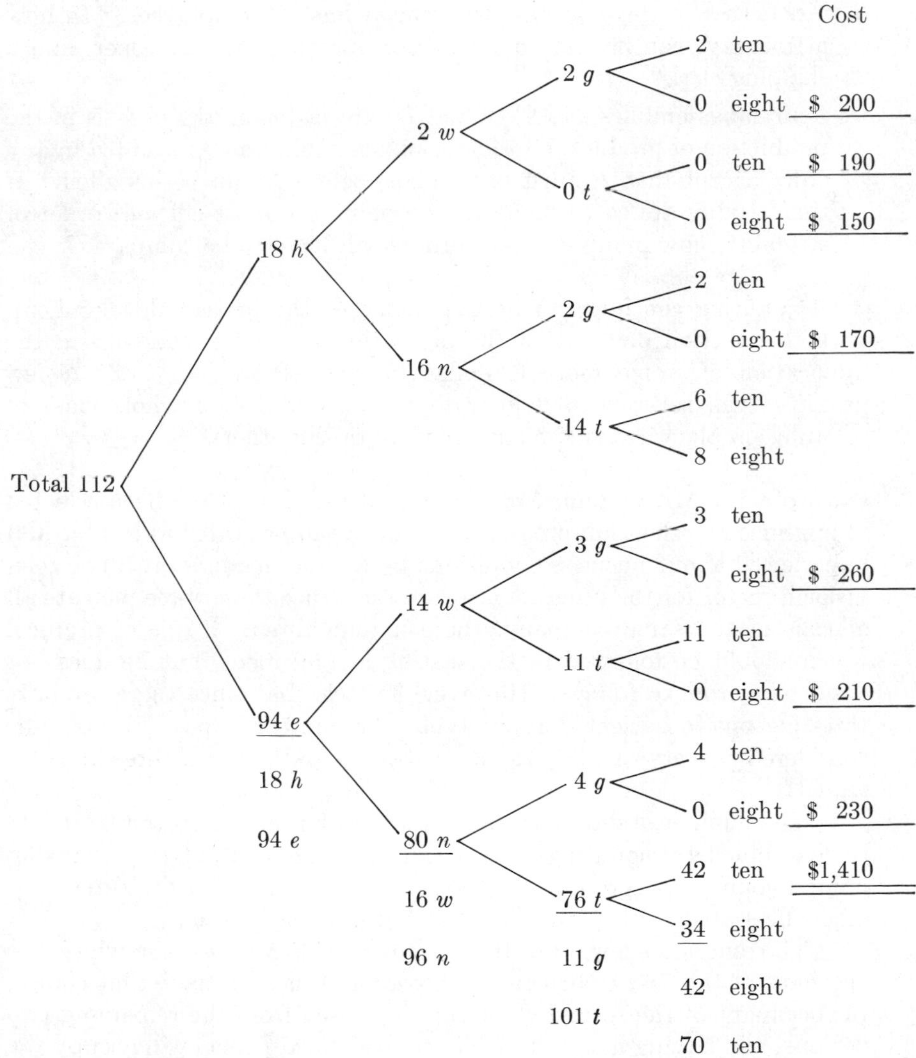

4. $1,410

B. 12

2.2 Permutations

Problems

A. A sales manager is taking over a new geographic region. In this new region, there are to be three different sales routes, with one salesman assigned to each route. The manager has narrowed his choice of the three salesmen to assign to five of his employees. Under these conditions, in how many ways can he make the assignment?

B. Mr. Lerner in the shipping department has 128 employees. In how many ways can he choose a platform foreman, a timekeeper, and a shipping clerk?

C. Four subassemblies (A, B, C, and D) are fastened to a chassis in the production of product X. These subassemblies may be added in any order except that B must be in place before D can be installed. If time studies are to be made to determine the most efficient order of assembly, how many different orders will have to be compared?

Counting continues to be the theme of the present development, with this section devoted to the derivation of, and instruction in the application of, some more formulas to be used in counting. A few examples will serve to demonstrate the existence of a whole class of counting problems, each similar in nature to the others.

Example 1. Mr. Gumm, Mr. Hand, and Mr. Ivan have been selected to appear on a television program. They are to be seated at a table and interviewed about business conditions by a panel moderator. The television director for the program is aware that since these three men are all officials of the same company, there is undoubtedly a rule of protocol which should be followed in the seating arrangement, but he does not know what rule to follow. However, he feels that since there are only three persons to be seated at the table, the number of possibilities to be considered is very small. In how many ways can the three men be seated?

The number of different ways in which the three men can be seated is determined by counting, and a tree diagram will aid in visualizing how to count. Diagram 6 illustrates the ways in which the three men might be seated, with each man's initial representing him.

The diagram shows that there are three different ways in which the occupant of the first chair can be chosen; and having made that choice, the occupant of the second chair can be chosen from the remaining two persons; and having made those choices, the third person will occupy the third chair. So the number of different seating arrangements (number of branches of the tree diagram) is 6, which arises from the product $3 \cdot 2 \cdot 1$.

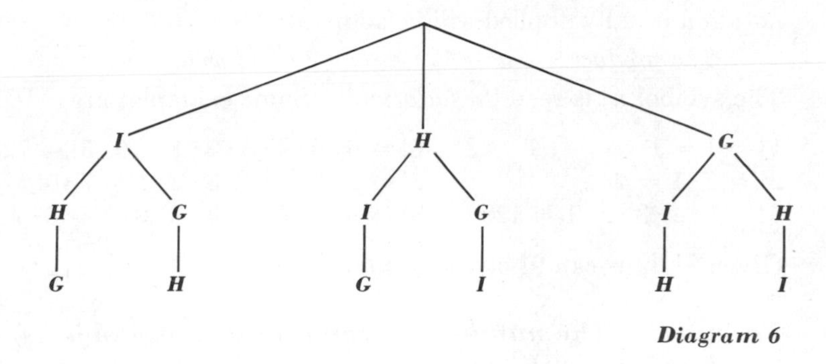

Diagram 6

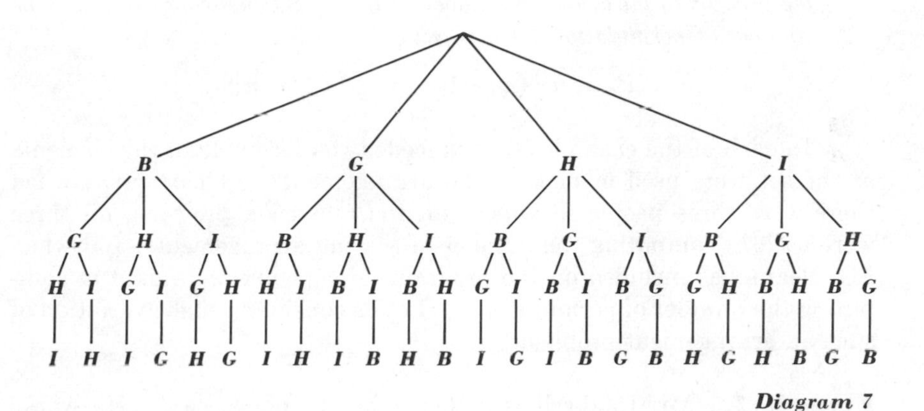

Diagram 7

Example 2. For a second example, suppose that at the last minute it is decided that the president of the company, Mr. Beam, will also be invited to appear on the program. The program director reasons (correctly) that now there are $4 \cdot 3 \cdot 2 \cdot 1$ different seating arrangements. This type of arrangement is called a *permutation*. The tree diagram that illustrates these 24 permutations is Diagram 7.

These two examples are special cases of the more general problem of determining the number of permutations of n objects (all n are used in each arrangement). The word *arrangement* is suggestive of the nature of the problem. Observe that in the second example, one permutation is B, G, H, I, while another (distinguishable) permutation is B, G, I, H. Both of these arrangements are counted in determining the total number of permutations.

These examples show that the number of permutations of n objects (all n are used in each arrangement) is given by the product of the counting numbers n to 1. Since this particular product will occur throughout much of the remainder of this chapter (and also in later chapters), the notation usually applied will be adopted:

The product $n \cdot (n - 1) \cdot \cdots \cdot 2 \cdot 1$ *is written* $n!$.

The symbol $n!$ is read "*n factorial.*" Some examples are

$1! = 1 = 1$ | $6! = 6 \cdot 5 \cdot 4 \cdot 3 \cdot 2 \cdot 1 = 6 \cdot 5! = 720$
$2! = 2 \cdot 1 = 2$ | $7! = 7 \cdot 6 \cdot 5 \cdot 4 \cdot 3 \cdot 2 \cdot 1 = 7 \cdot 6! = 5{,}040$
$5! = 5 \cdot 4 \cdot 3 \cdot 2 \cdot 1 = 120$ | $8! = 8 \cdot 7 \cdot 6 \cdot 5 \cdot 4 \cdot 3 \cdot 2 \cdot 1 = 8 \cdot 7! = 40{,}320$

Given 8!, how can 9! be easily computed?

Formula: **(*The number of permutations of* n *objects*)** *The number of permutations of* n *objects (all* n *are used in the arrangement) is the product of the counting numbers* n *to* 1. *Symbolically, if* ${}_nP_n$ *is the number of permutations of* n *objects, then*

$${}_nP_n = n \cdot (n - 1) \cdot \cdots \cdot 2 \cdot 1 = n!$$

In each of the examples that preceded the formula, all the elements of the set were used in making the arrangement. In particular, when there were three people to appear on the television program, all three were used in computing the number of seating arrangements; and when Mr. Beam was included on the program, four people were used in computing the number of permutations. Let us consider a slight variation of such an arrangement problem.

Example 3. Mr. Caldwell is a dispatcher of police cars. He works the night shift, and there are 11 police cars on duty during his shift. During one evening, there are four different calls to which he must dispatch police cars, and he decides not to use any car more than once. In how many different ways can he dispatch the cars?

In this example, the dispatcher has 11 objects from which to choose, but not all the police cars will be used in the assignments. It is the feature "not all the objects are to be used in the arrangement" that distinguishes this example from the previous ones in this section. To see that this is a problem in arrangement, observe that if A, B, C, and D designate police cars, then the choices A, B, C, D and A, B, D, C (as well as other arrangements such as A, C, B, D; A, D, C, B) must be counted. Mr. Caldwell has a choice of 11 different cars for the first call, 10 cars for the second call, 9 cars for the third call, and 8 cars for the fourth call. A tree diagram to illustrate this example is not given since the tree has $11 \cdot 10 \cdot 9 \cdot 8 = 7{,}920$ branches (possible assignments of cars).

The factorial notation adopted earlier can be used to express the result. Another way to write the product $11 \cdot 10 \cdot 9 \cdot 8$ is

$$\frac{11 \cdot 10 \cdot 9 \cdot 8 \cdot 7 \cdot 6 \cdot 5 \cdot 4 \cdot 3 \cdot 2 \cdot 1}{7 \cdot 6 \cdot 5 \cdot 4 \cdot 3 \cdot 2 \cdot 1} = \frac{11!}{7!}$$

since the denominator of the fraction "cancels out" all the factors in the numerator except those that are used in the computation.

This example illustrates another type of arrangement problem and shows how a variation of the first permutation formula can be applied when not all the objects are to be used in the arrangement.

Formula: ***(The number of permutations of*** n ***objects when*** r ***objects are used in each arrangement)*** *The number of arrangements possible, using the objects from a set with n elements, such that only r* ($r < n$) *objects are used in each arrangement is the product of the counting numbers n to* $(n - r + 1)$. *Symbolically, if* ${}_nP_r$ *represents the number of permutations, then*

$${}_nP_r = n \cdot (n - 1) \cdot \cdots \cdot (n - r + 1) = \frac{n!}{(n - r)!}$$

This formula is to be used only if r is less than n, since the previous formula covers those cases when all n objects are to be used in each arrangement.

Example 4. One more example of the last formula will be given. Miss Richman, who is a buyer for Superior Department Store, and Miss Trojan, who is a buyer for Gallant Department Store, will be at the same manufacturing plant at the same time on Tuesday. The president of the firm decides to choose a guide to show each of the buyers around the plant from among five of his assistants. If no man is to serve as a guide twice, in how many ways can the assignments be made?

In this example, the set from which elements are to be chosen has five elements; hence, in the formula, $n = 5$. The number of guides to be chosen is two, so that $r = 2$. Now the second permutation formula can be applied, and the number of choices is

$${}_5P_2 = \frac{5!}{(5 - 2)!} = \frac{5!}{3!} = \frac{5 \cdot 4 \cdot 3 \cdot 2 \cdot 1}{3 \cdot 2 \cdot 1} = 5 \cdot 4 = 20$$

Example 5. In Examples 1 to 4, the choices, after the first, were influenced by previous choices. In forming an arrangement from a set of elements, it was assumed that once an element was used, it could not, or would not, be reused in the arrangement. Recall, for example, that in Example 3, the assumption was that no police car was to be used more than once. Of course, if the calls were sufficiently well spaced so that any of the 11 cars could be used for any of the four calls, then the number of different ways that Mr. Caldwell could dispatch the four cars would be $11 \cdot 11 \cdot 11 \cdot 11$, which is 14,641. This result does not agree with the result of Example 3 for the obvious reason that in this second case, each choice of police car to be dispatched is *independent* of any previous choice.

Example 6. Another example will illustrate the same principle. Miss Hardin can drive between her residence and the office by three different routes. In how many ways can she drive to work and return home?

The return route is independent of the route chosen to go to work, so there are $3 \cdot 3 = 9$ different routes.

In the last two examples, the number of choices possible at each stage of the process was the same as for each other choice (11 for the fifth example, 3 for the sixth). This need not be the case, as another example will show.

Example 7. Oscar's Clothing Company manufactures skirts, blouses, and jackets. Suppose that skirts are made in seven different colors, blouses in five different colors, and jackets in three different colors. If a woman is going to purchase one of each, how many different purchases can she make that result in a different color arrangement for the skirt, blouse, and jacket? (The arrangement "blue skirt, aqua blouse, green jacket" is different from "blue skirt, green blouse, aqua jacket.")

There are seven choices of skirt colors; for each of these choices, there are five choices for the color of the blouse. Considering these two purchases alone, the customer has 35 choices. But now, for each of these choices, she has three choices for the color of the jacket, or a total of 105 different choices. It is clear that a complete tree diagram cannot be drawn here because of space limitation, but it will be instructive to construct at least a partial diagram.

Each time the customer makes a choice, it is independent of previous choices, and the answer is obtained by forming the product of 7, 5, and 3. This example illustrates the same principle about arrangements as that in Examples 5 and 6, and these properties are summarized in the following formula.

Formula: **(*The number of ways in which a sequence of independent events can occur*)** *If events $E_1, E_2, \ldots, E_n$ can occur in $w_1, w_2, \ldots, w_n$ different ways, respectively, and if the events are independent, then the sequence of events $E_1, E_2, \ldots, E_n$ can occur in $w_1 \cdot w_2 \cdot \cdots \cdot w_n$ different ways.*

This concludes the formulas of this section, but one other special case is given in Exercise 10. Be sure that it is clear why the formula ${}_{15}P_3$ cannot be used for Example 7.

Exercises

1. A telephone company is going to use seven digits in its telephone numbers.
   a. How many different telephone numbers are possible?
   b. How many are possible if the company decides not to use zero in the first position?
   c. How many are possible if the company decides not to use zero in the first or fourth position?

2. For display purposes, a clerk is to drape one each of seven different-colored kinds of draperies across a wall. In how many ways can she arrange the draperies?

3. In Exercise 2, if it is possible to display only four of the draperies on the wall, how many arrangements are possible?

4. The clerk in Exercise 2 decides that she can repeat some or all of the colors.
   a. Work Exercise 2 again.
   b. Work Exercise 3 again.

5. The clerk in Exercise 2 finally decides that there is room for five draperies and that she will repeat no colors.
   a. How many arrangements are possible?
   b. How many arrangements are possible if she places a purple one first?

6. A salesman will call on the owners of five stores during his two-day stay in a certain city.
   a. In how many ways can he make the five calls?
   b. In how many ways can he make the five calls if he makes three the first day and two the second day?
   c. In how many ways can he make the five calls if he calls on the subset $P = \{S_1,S_3,S_4\}$ the first day and the remainder $\{S_2,S_5\}$ the second day?
   d. In how many ways can he make the five calls if he calls on the subset $Q = \{S_2,S_5\}$ the first day and the remainder $\{S_1,S_3,S_4\}$ the second day?

7. In a bicycle assembly plant, there are 14 components that can be assembled in any order. The efficiency expert decides to time each possible order of assembly. If it takes an average of 45 min for each assembly, how long will it take to time all possible orders of assembly?

8. A men's clothing firm expects to bring out the fall models with three-button or two-button coats, each of which is available in eight colors and with or without two pairs of trousers. How many distinct choices can a customer make?

9. On an automobile there are 14 accessories that a customer can buy or reject. How many distinct choices can he make?

10. There is one special case of the formula

$$_nP_r = \frac{n!}{(n-r)!} \qquad r < n$$

that is worthy of note. Consider again the director of the television show in Example 1 and his problems. For one of the programs for which he is the director, in the warm-up before the program is to go on the air, a set of six people is selected from the audience as possible participants in the show. Only three of these candidates are actually to be used in the show. Under these circumstances, the number of different seating arrangements at the table is $_6P_3 = 6 \cdot 5 \cdot 4 = 120$. However, just before the show goes on the air, it is observed that one of the persons selected in the preliminary set is a vice-president of a firm which does a great deal of advertising on the network. He must not be neglected, so he will be one of the persons chosen. Furthermore, the director decides to allow him to occupy the first seat in the arrangement behind the table. How many arrangements are now possible? The remaining two selections are made from five persons; therefore the number of seating arrangements is $_5P_2 = 5 \cdot 4 = 20$.

This is an example of the formula:

The number of arrangements of $r - 1$ objects from the set of $n - 1$ objects is

$$_{n-1}P_{r-1} = \frac{_nP_r}{n} \qquad n > r > 0$$

a. What is the formula for $_{n-2}P_{r-2}$ $(n > r > 2)$?
b. What is the formula for $_{n-k}P_{r-k}$ $(n > r > k)$?

Answers to problems

A. $_5P_3 = 5 \cdot 4 \cdot 3 = 60$

B. $_{128}P_3 = 128 \cdot 127 \cdot 126 = 2{,}048{,}256$

C.

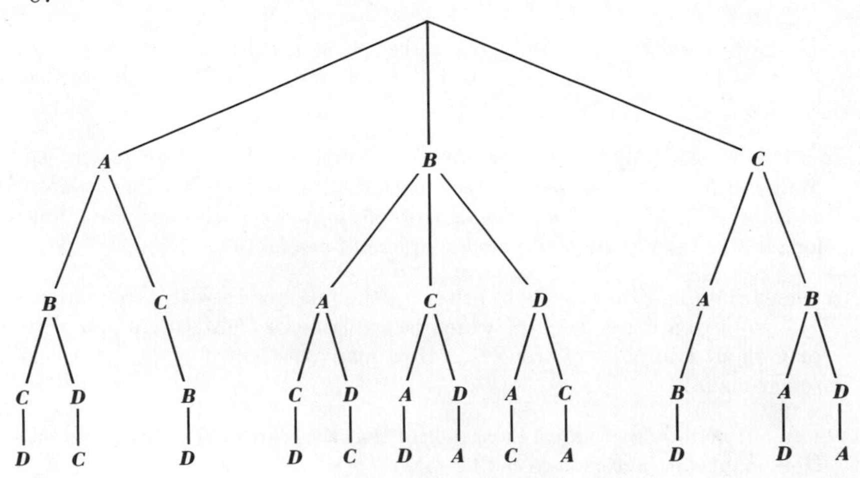

2.3 *Combinations*

Problems

A. Mr. Thorp has given Miss Hardin 10 letters to write. She knows that she can type only two of them before mail time. How many different choices of the two letters to type can she make? (The answer is *not* ${}_{10}P_2 = 90$.)

B. If there are 540 sweaters coming off an assembly line per day, how many different samples of five sweaters can be selected? (The answer is *not* ${}_{540}P_5$, $540!$, or $540 \cdot 5$.)

The formulas developed in Sec. 2.2 serve very useful purposes for counting the number of ways that certain kinds of events may occur. However, there are other kinds of problems for which the formulas are not suited, and it is the main purpose of this section to exhibit some of these new kinds of problems together with formulas for solving them. To illustrate, three examples of these kinds of problems are offered.

Example 1. The president of a food company is to select a committee of two from the firm's promotion department to conduct a study of the problem of purchasing land for a new location. How many different committees of two persons can be chosen from the five members of the promotion team?

This question sounds like the type of question considered in the previous section, so one of the formulas for arrangements which was derived in the previous section will be used (temporarily). The number of permutations possible using two objects chosen from a set with five objects is

$${}_5P_2 = \frac{5!}{(5-2)!} = \frac{5!}{3!} = \frac{5 \cdot 4 \cdot 3 \cdot 2 \cdot 1}{3 \cdot 2 \cdot 1} = 5 \cdot 4 = 20$$

The result of applying this formula indicates that there are 20 different ways that a committee of two can be chosen from the given candidates. But read the question again. The number of *different* committees is to be determined. A listing of these 20 arrangements will be helpful. Suppose that the initials of the names of the members are G, H, I, J, and K. They are used for brevity in the display.

The symbol [ ], called *brackets*, is used to enclose the elements of the *arrangement*. This notation is important since the symbol { }, called *braces*, has been reserved for the enclosure of elements in a *set*. Recall that *the order in which the elements of a set are tabulated does not change the set.*

[G,H]	[H,G]	[I,G]	[J,G]	[K,G]
[G,I]	[H,I]	[I,H]	[J,H]	[K,H]
[G,J]	[H,J]	[I,J]	[J,I]	[K,I]
[G,K]	[H,K]	[I,K]	[J,K]	[K,J]

But the committee [G,H] consists of the same members as the committee [H,G], the committee [H,I] consists of the same members as the committee [I,H], and there are clearly other pairs of arrangements that result in the same committee. That is, although the order of the elements in [G,H] and the order of the elements in [H,G] are different, the two committees are the same since the order in which the two men are chosen to be on the committee does not affect the membership of the committee. To count the number of *different* committees, all such duplications must be eliminated. Some duplications have been listed, but there are others. How many others? To answer this question, observe that each committee is listed twice because the positions of the members in each committee can be interchanged to yield the same committee. So the number of *distinct* committees is 10, and this answers the question that was originally asked. Also observe that although the formula ${}_nP_r$ does not work for this example, the correct answer can be written with the aid of the notation for permutations and factorials as

$$10 = \frac{20}{2} = \frac{20}{2 \cdot 1} = \frac{5 \cdot 4 \cdot (3 \cdot 2 \cdot 1)}{2 \cdot 1 \cdot (3 \cdot 2 \cdot 1)} = \frac{5!}{2!3!} = \frac{n!}{r!(n-r)!} = \frac{{}_nP_r}{r!}$$

Example 2. As a second example, suppose that the same president has decided to use three members on the committee instead of two. (Reread the first example.) How many *different* committees can be selected? The formula for the number of arrangements of three objects chosen from a set of five objects yields

$${}_5P_3 = \frac{5!}{(5-3)!} = \frac{5!}{2!} = \frac{5 \cdot 4 \cdot 3 \cdot 2 \cdot 1}{2 \cdot 1} = 5 \cdot 4 \cdot 3 = 60$$

The result of applying the formula ${}_nP_r$ is 60, which is the number of arrangements. However, this is not the number of *distinct* committees. A total list of the 60 different arrangements will not be given here (because of space limitation, for one reason), but consider a proper subset of the arrangements that follow:

[G,H,I]	[H,G,I]	[I,G,H]
[G,I,H]	[H,I,G]	[I,H,G]

Each of the six arrangements is distinct, but all result in the same committee. It is clear (is it not?) that each different choice of three letters can be expressed in $3 \cdot 2 \cdot 1 = 6$ different arrangements, and hence among the 60 different arrangements, there are only $^{60}/_{6} = 10$ different committees. If it is not clear why the number 60 must be divided by 6, then all 60 arrangements should be listed and sorted into sets such that each set of six arrangements results in the same committee.

Again, the correct answer to the question posed will be written, using notation adopted earlier, so that a comparison can be made between the formula ${}_nP_r$ and the formula to be adopted for this type of problem.

$$10 = \frac{60}{6} = \frac{60}{3 \cdot 2 \cdot 1} = \frac{5 \cdot 4 \cdot 3}{3 \cdot 2 \cdot 1} = \frac{5 \cdot 4 \cdot 3 \cdot (2 \cdot 1)}{3 \cdot 2 \cdot 1 \cdot (2 \cdot 1)} = \frac{5!}{3!2!} = \frac{n!}{r!(n-r)!}$$

$$= \frac{{}_nP_r}{r!}$$

Compare the result in this example with that of the preceding example to see that the same formula results.

Example 3. The numbers in the preceding examples have been kept small for a good reason. To illustrate the convenience of such a decision, suppose that the problem is to choose a committee of five from a set of 11 executives. An intermediate step in the determination of how many different committees can be chosen is the writing out of all the arrangements (${}_{11}P_5 = 55{,}440$), and the listing of all the distinct committees ($55{,}440/120 = 462$) becomes prohibitive. (Decide why the number of arrangements was divided by 120 to determine the number of distinct committees.)

These three examples show that there is another type of counting problem, which is closely associated with the counting of arrangements but for which the formulas for the number of arrangements will not work. These committees are examples of *combinations*. As an aid in distinguishing between permutations and combinations, note that for permutations (arrangements), the order of the elements is of importance, while for combinations the order of the elements is *not* important.

In each of the three examples, the number of combinations was determined by dividing the number of arrangements by the number of ways each arrangement can be expressed by rearranging the elements. These are but particular examples of the formula below.

Formula: ***(The number of combinations of*** n ***objects when*** r ***objects are in each combination)*** *The number of combinations possible, using the elements from a set with* n *elements, such that only* r *objects are used in each combination is the number of permutations divided by the number of permutations of* r *objects (using all* r *in each permutation). Symbolically, we write the number of combinations,* ${}_nC_r$*, as*

$${}_nC_r = \frac{{}_nP_r}{{}_rP_r} = \frac{{}_nP_r}{r!} = \frac{n!}{r!(n-r)!}$$

To illustrate the formula, two examples will be given which concern a set and the number of certain kinds of subsets, independent of any physical motivation.

Example 4. If the set S has 15 elements, then how many different subsets of 4 elements does S contain? Since subsets (sets) are different if and only if they have different elements, the number of different subsets is *not* the number of arrangements but the number of combinations. In the formula, n is 15 and r is 4. Therefore, $n - r$ is $15 - 4 = 11$, so the formula yields

$$_{15}C_4 = \frac{15!}{4!11!} = \frac{15 \cdot 14 \cdot 13 \cdot 12 \cdot 11 \cdot 10 \cdot 9 \cdot 8 \cdot 7 \cdot 6 \cdot 5 \cdot 4 \cdot 3 \cdot 2 \cdot 1}{4 \cdot 3 \cdot 2 \cdot 1 \cdot 11 \cdot 10 \cdot 9 \cdot 8 \cdot 7 \cdot 6 \cdot 5 \cdot 4 \cdot 3 \cdot 2 \cdot 1}$$

$$= \frac{15 \cdot 14 \cdot 13 \cdot 12}{4 \cdot 3 \cdot 2 \cdot 1} = 1{,}365$$

Example 5. As another example, suppose that a set S has 15 elements. How many different subsets of 11 elements does S contain? Again the formula for combinations applies, and $_{15}C_{11}$ is to be computed since $n = 15$ and $r = 11$. Thus,

$$_{15}C_{11} = \frac{15!}{11!4!} = \frac{15!}{4!11!} = {_{15}C_4} = 1{,}365$$

The example illustrates that for each subset of 4 elements there is a corresponding subset of 11 elements, and this is but a special case of the following equality between combinations:

For all nonzero counting numbers n and r $(0 < r < n)$, it is true that

$$_nC_r = {_nC_{n-r}}$$

Still another way to understand this relationship is to recall that each subset has a *complementary* set in a universal set. Naturally, for each subset of 4 elements, there is a complementary set with 11 elements.

To verify this claim by a computation technique, simply observe that

$$_nC_r = \frac{n!}{r!(n-r)!} = \frac{n!}{(n-r)!r!} = \frac{n!}{(n-r)![n-(n-r)]!} = {_nC_{n-r}}$$

It will aid in understanding the relationship between the two formulas to apply them to specific cases, and the first four exercises are directed to this subject. These remarks make possible the use of either the symbol $_nC_r$ or the symbol $_nC_{n-r}$ to represent the number of subsets with r elements in a set with n elements. There will be occasion in this book to use both notations, and you should become accustomed to both usages.

The formula for the number of combinations of r elements chosen from a set of n objects was motivated by some examples which made clear why the number of arrangements ${}_nP_r$ must be reduced by a *factor* of $r!$. There is another way of viewing such problems (there are usually several ways of viewing any problem) that may be helpful in understanding the final formula.

The purpose of the second explanation (which follows) of the formula for ${}_nC_r$ is to show still another way that the formula can be derived. The procedure will represent, in two ways, the number of permutations of r elements chosen from a set with n elements. The first, and trivial, way for such a representation is by the symbol ${}_nP_r$. The second way is to select a subset of r elements. The number of subsets that can be chosen has been represented by the symbol ${}_nC_r$. Now for each of these subsets of r elements, there are ${}_rP_r$ permutations possible. So the total number of permutations possible is the product of ${}_nC_r$ and ${}_rP_r$. This conclusion, together with the agreement about the symbol ${}_nP_r$, leads to the equality

$$ {}_nP_r = {}_nC_r\,{}_rP_r $$

Division of both numbers by ${}_rP_r$ leads to

$$ \frac{{}_nP_r}{{}_rP_r} = {}_nC_r $$

or

$$ {}_nC_r = \frac{{}_nP_r}{{}_rP_r} = \frac{n!}{r!(n-r)!} $$

which agrees with the formula stated earlier.

One of the reasons for including this last discussion, which, after all, only leads to a result already obtained, is that it illustrates an often used method of problem solving. Consideration of a problem situation from two (or more) viewpoints often results in two (or more) statements of the same relationship. These statements can then be equated to obtain a formula for the solution of the problem.

Exercises

1. Show ${}_nC_n = 1$. Interpret this result as the number of the subsets of n elements of a set which has n elements.
2. Show ${}_nC_1 = n$. Interpret this result as the number of the subsets which have a single element of a set which has n elements. Although ${}_nC_0$ has not been defined, can you give a set interpretation of what it means?
3. A personnel director has six applicants for two jobs. How many different pairs of employees can he hire? How many different sets of four persons can he hire?
4. From 100 items, a sample (subset) of 2 will be chosen. How many different samples are possible? How many samples of 98 each are possible?

5. A committee of three is to be composed of two members of a board of directors which has four members and a third member chosen from two vice-presidents. How many committees are possible?
6. Three of nine salesmen who have achieved fine sales records are to be selected for a special article in the business newsletter. In how many ways can the subset be selected?
7. A shipment of materials has 12 items, one of which is defective. In how many ways can a subset of two be selected such that
   a. The defective item is not selected?
   b. The defective item is included?
8. A salesman will carry, as a sample, two of the seven different-colored skirts, one of the five different-colored blouses, and two of the three different-colored jackets. How many different samples are possible? (HINT: Use the formula for combinations followed by the formula for a sequence of independent events.)
9. An airline has 200 people waiting to board two planes which can each hold 100 passengers. The ticket salesman is in a dilemma about how to assign people to the different planes. In how many ways can he assign people to the planes? (Assume that the plane seats are not numbered.)

Answers to problems

A. $${}_{10}C_2 = \frac{10 \cdot 9}{1 \cdot 2} = 45$$

B. $${}_{540}C_5 = \frac{540!}{535!5!} = \frac{540 \cdot 539 \cdot 538 \cdot 537 \cdot 536}{1 \cdot 2 \cdot 3 \cdot 4 \cdot 5}$$

2.4 The binomial formula

Problems

A. A set of 16 items will have 0, 1, 2, or 3 items selected for inspection. How many such distinct subsets can be selected for inspection?

B. Ten weeks before a space shot, the list of persons from which two astronauts will be selected has been narrowed to 10 names. These two men are to spend two days in space together in extremely confined quarters. Hence, compatibility is extremely important. Because of this, it is suggested that each possible pair of candidates should be confined together for two days to test their reaction to one another. Is this plan practical?

The binomial formula proves to be quite useful in certain computational problems, a few of which will be demonstrated in this section.

However, before stating the formula and also prior to its application to some new kinds of problems, we shall see how a familiar counting problem (already considered) is just a special case of the formula.

Consider again a counting problem which was first discussed in Sec. 1.2. Given a set S with n elements, how many subsets does S contain? An argument was given on page 16 to show that the number of subsets is 2^n. The mathematical tools now available are reason to review the situation, and as promised, another argument will be given to support the conclusion. To see how the argument proceeds, a few special cases are considered first. The crucial point in this counting process is the fact established at the end of Sec. 2.3:

If S is a set with n elements, then the number of subsets of S with r elements ($n > r$) is ${}_nC_r = n!/[r!(n-r)!]$.

Remember that it has been established that ${}_nC_r$ and ${}_nC_{n-r}$ are equal and that either notation can be used.

Example 1. If a set S has no elements (S is empty), then the number of subsets is one (the empty set). For convenience in notation, ${}_0C_0 = 1$.

Example 2. For the second case, if S has a single element, then the number of subsets is two; they are S and the empty set. Note that to count the subsets, first count the number of subsets with one element and then add the number of subsets with no elements. Again for convenience of notation, since

$$ {}_1C_1 = \frac{1!}{1!0!} = \frac{1}{1 \cdot 0!} $$

and there is one subset with one element, we define ${}_1C_1$ as 1, or equivalently we define $0!$ as 1. The number of subsets is

$$ {}_1C_1 + {}_1C_0 = 1 + 1 = 2^1 $$

Example 3. For the third case, if S has two elements (say a and b), then the subsets of S are $\{a,b\}$, $\{a\}$, $\{b\}$, $\{\ \}$. The number of subsets is four, and that result can be obtained from the sum

$$ {}_2C_2 + {}_2C_1 + {}_2C_0 = 1 + 2 + 1 = 4 = 2^2 $$

It should be clear that ${}_2C_2 = 1$; in fact,

${}_nC_n = 1$ *for all counting numbers n.*

Also ${}_2C_0 = 1$; in fact,

${}_nC_0 = 1$ *for all counting numbers n.*

Example 4. If a set S consists of three elements (say a, b, and c), then the subsets of the set S are

$$\begin{array}{llll} \{a,b,c\} & \{a,b\} & \{a\} & \{\ \} \\ & \{a,c\} & \{b\} & \\ & \{b,c\} & \{c\} & \end{array}$$

The manner in which the subsets are written is deliberate. First is the subset with three elements (and there is only one such subset); second, the subsets with two elements (there are three of these); third, the subsets with a single element (there are three of these); and finally, the subset with no elements (there is one). The number of subsets of a set with three elements can be written as a sum of the number of subsets with three, two, one, and zero elements. Hence, the number of subsets is

$${}_3C_3 + {}_3C_2 + {}_3C_1 + {}_3C_0 = 1 + 3 + 3 + 1 = 8 = 2^3$$

Example 5. As the final particular example, let S have four elements. Then S contains (1) one subset with four elements, (2) as many subsets with three elements as there are combinations of three objects chosen from a set of four objects, (3) as many subsets with two elements as ${}_4C_2$, (4) ${}_4C_1$ subsets with one element, and (5) the empty subset (${}_4C_0$). The number of subsets of a set with four elements can be written as a sum:

$${}_4C_4 + {}_4C_3 + {}_4C_2 + {}_4C_1 + {}_4C_0 = 1 + 4 + 6 + 4 + 1 = 16 = 2^4$$

The five examples are summarized in Table 1.

Table 1

Number of elements in S	*Number of subsets of S*	*Formula for determining the number of subsets of S*
0	$2^0 = 1$	${}_0C_0 = 1 = 2^0$
1	$2^1 = 2$	${}_1C_1 + {}_1C_0 = 1 + 1 = 2^1$
2	$2^2 = 4$	${}_2C_2 + {}_2C_1 + {}_2C_0 = 1 + 2 + 1 = 4 = 2^2$
3	$2^3 = 8$	${}_3C_3 + {}_3C_2 + {}_3C_1 + {}_3C_0 = 1 + 3 + 3 + 1 = 8 = 2^3$
4	$2^4 = 16$	${}_4C_4 + {}_4C_3 + {}_4C_2 + {}_4C_1 + {}_4C_0 = 1 + 4 + 6 + 4 + 1 = 16 = 2^4$

Examples 1 to 5 have occurred prior to this section, although perhaps not in this exact form, and are repeated here for a reason that should soon become apparent.

The sums in the expressions in the table are special cases of the formula known as the *binomial formula*. The *bi* means "two," and *nomen* is an algebraic expression; hence, *binomial formula* means a formula concerning two algebraic expressions. These two expressions may be simply represented as x and y, and so we have a formula involving x and y.

Formula: **(*Binomial formula*)** *If x and y are any real numbers and n is any counting number, then*

$$(x + y)^n = {}_nC_nx^n + {}_nC_{n-1}x^{n-1}y + {}_nC_{n-2}x^{n-2}y^2 + \cdots + {}_nC_rx^{n-r}y^r + \cdots + {}_nC_1xy^{n-1} + {}_nC_0y^n$$

The binomial formula concerns counting numbers and can be established by an argument based on *mathematical induction*, a topic not included in this book. Although no proof will be offered here for the binomial formula, perhaps the correctness of the expression ${}_nC_{n-r}x^{n-r}y^r$ will be seen intuitively if you reason as follows: The number of terms in the product $(x + y)^n$ which involve $(n - r)$ x's and r y's is the number of factors obtained by using $n - r$ of the x's and r of the y's, which is the number of subsets of n objects with $n - r$ of the x's. This is known to be ${}_nC_r$. Observe that the argument is symmetric since it is also known that ${}_nC_{n-r} = {}_nC_r$.

It will be helpful to reflect for a moment upon why some definitions were made for certain symbols. Recall that for any real number a, $a^0 = 1$, that ${}_0C_0 = 1$, and that ${}_nC_0 = 1$. These should be treated for what they are, definitions that make for symmetry in the notation. Since $a^0 = 1$ for any real number a, $(x + y)^0 = 1$ for any x and y. How this *convention* is helpful in displaying information will become apparent now.

This formula is, in fact, true for other values of n besides the counting numbers 0, 1, 2, 3, 4, ... But the elements of the set of counting numbers are sufficient as replacements for n in this book. The formula is also true when x and y are replaced by other than real numbers, but only real numbers will be used as replacements for x and y in this book.

Now let us determine how the binomial formula can be applied to count the number of subsets of a set with n elements. Since the formula applies for all real numbers x and y, the formula applies when $x = 1$ and $y = 1$. With these replacements, the formula becomes

$$(1 + 1)^n = {}_nC_n \cdot 1^n + {}_nC_{n-1} \cdot 1^{n-1} \cdot 1 + {}_nC_{n-2} \cdot 1^{n-2} \cdot 1^2 + \cdots + {}_nC_r \cdot 1^{n-r} \cdot 1^r + \cdots + {}_nC_1 1 \cdot 1^{n-1} + {}_nC_0 \cdot 1^n$$

Since $1^k = 1$ for any counting number k, the formula can be written more briefly as

$$(1 + 1)^n = {}_nC_n + {}_nC_{n-1} + {}_nC_{n-2} + \cdots + {}_nC_r + \cdots + {}_nC_1 + {}_nC_0$$

This is recognizable as the sum of the number of subsets with n elements, $n - 1$ elements, $n - 2$ elements, . . . , $n - r$ elements, . . . , 1 element, no elements. That is, the number of subsets of a set with n elements is $(1 + 1)^n$. But $(1 + 1)^n = 2^n$, and the argument is finished.

The numbers that occur in the front of each of the terms of the sum in the formula are called the *coefficients* of the binomial formula. Arrays of binomial coefficients have been computed for many values of n (and the display is called *Pascal's triangle*). A small portion of such a display is in Table 2.

Table 2

n																	
0									1								
1								1		1							
2							1		2		1						
3						1		3		3		1					
4					1		4		6		4		1				
5				1		5		10		10		5		1			
6			1		6		15		20		15		6		1		
7		1		7		21		35		35		21		7		1	
8	1		8		28		56		70		56		28		8		1

The triangle is not complete, and clearly can never be complete, since when the coefficients for $(x + y)^n$ are known, the coefficients for $(x + y)^{n+1}$ can be computed. The display even suggests how succeeding lines can be computed using only the results in the line above. To discover this rule of procedure, look at any entry in the display. Either it is a 1, or it is the sum of the two numbers just above and to the left and right. In terms of symbols,

$$_nC_r = {}_{n-1}C_{r-1} + {}_{n-1}C_r$$

with $n > r > 1$. A bit of computation verifies that this is the case, and Exercise 10 is directed at establishing this equality.

The remainder of the section will be devoted to special applications of the binomial formula. One special case of the binomial formula has already been considered:

If $x = y = 1$, *then* $(x + y)^n = 2^n$.

Example 6. Another special case of the binomial formula that will be important later is the case where $x + y = 1$. If $x + y = 1$, then $(x + y)^n$

$= 1^n = 1$. As an example of this special case, consider

$$\begin{aligned}(\tfrac{1}{3} + \tfrac{2}{3})^4 &= {}_4C_4(\tfrac{1}{3})^4 + {}_4C_3(\tfrac{1}{3})^3(\tfrac{2}{3})^1 + {}_4C_2(\tfrac{1}{3})^2(\tfrac{2}{3})^2 \\ &\qquad + {}_4C_1(\tfrac{1}{3})^1(\tfrac{2}{3})^3 + {}_4C_0(\tfrac{2}{3})^4 \\ &= 1 \cdot \tfrac{1}{81} + 4 \cdot \tfrac{1}{27} \cdot \tfrac{2}{3} + 6 \cdot \tfrac{1}{9} \cdot \tfrac{4}{9} + 4 \cdot \tfrac{1}{3} \cdot \tfrac{8}{27} + 1 \cdot \tfrac{16}{81} \\ &= \frac{1 + 8 + 24 + 32 + 16}{81} = \frac{81}{81} = 1\end{aligned}$$

It is repeated for emphasis:

If the sum $x + y$ is 1, then the sum obtained by expanding $(x + y)^n$ by the binomial formula is 1.

Example 7. Another special case of the binomial formula does not deal with special replacements for x and y but rather shows how partial information can be obtained from the formula without computing all the summands in the formula. To illustrate, consider the question: "What is the coefficient of x^6y^9 in the expansion of $(x + y)^{15}$?" One way to answer the question is to compute the first 10 terms in the expansion and observe the term involving x^6y^9. There is an easier way. By the binomial formula, the tenth term of $(x + y)^{15}$ is

$$_{15}C_9x^{15-9}y^9$$

But the formula for $_{15}C_9$ as given in Sec. 2.3 yields $_{15}C_9 = 5{,}005$, so the complete term is $5{,}005x^6y^9$.

Any particular term in the expansion of the binomial formula can be determined without computing the terms which precede it in the expansion.

Example 8. As another example of how to determine information by the use of only a portion of the binomial formula, consider the question: "If a set S has 20 elements, how many subsets of S have 14, 15, or 16 elements?" This is a question that can be answered by counting, and one way is to tabulate all the subsets and count those with 14, 15, or 16 elements. But there are over one million subsets of a set with 20 elements. Exactly how many subsets are there? (HINT: Compute 2^{20}.) So it becomes necessary to find an easier way to count the subsets. The binomial formula can be used, and the sum to be determined can be expressed as

$$\begin{aligned}{}_{20}C_{14} + {}_{20}C_{15} + {}_{20}C_{16} &= \frac{20!}{14!6!} + {}_{20}C_{15} + {}_{20}C_{16} \\ &= \frac{20 \cdot 19 \cdot \cdots \cdot 15}{1 \cdot 2 \cdot 3 \cdot \cdots \cdot 6} + {}_{20}C_{15} + {}_{20}C_{16} \\ &= 38{,}760 + 38{,}760 \cdot \tfrac{6}{15} + {}_{20}C_{16} \\ &= 38{,}760 + 15{,}504 + 15{,}504 \cdot \tfrac{5}{16} \\ &= 38{,}760 + 15{,}504 + 4{,}845 \\ &= 59{,}109\end{aligned}$$

If the computations are made, then it is easy to see why the factor of $6/15$ was used to compute ${}_{20}C_{15}$ once ${}_{20}C_{14}$ was known and why the factor of $5/16$ was used to compute ${}_{20}C_{16}$ once ${}_{20}C_{15}$ was known.

Example 9. We conclude the section with one more application of the binomial formula. Write as a decimal $(1.0075)^{80}$. There are two obvious and difficult ways to solve the problem. One method is to form a product with 80 equal factors of 1.0075. Such a computation is not very difficult with a desk calculator, but if the product is to be found by the usual process of multiplication, then only a few of the necessary steps in the computation will discourage the person attempting the process. Could you perform such a task in an hour? A day? A lifetime? Another method is to expand by the binomial formula and form the indicated sum. In this procedure, we write 1.0075 as a sum, $1 + 0.0075$, and apply the formula

$$\begin{aligned}(1.0075)^{80} &= (1 + 0.0075)^{80} \\ &= {}_{80}C_{80}1^{80} + {}_{80}C_{79}1^{79}(0.0075)^1 + {}_{80}C_{78}1^{78}(0.0075)^2 \\ &\qquad + {}_{80}C_{77}1^{77}(0.0075)^3 + \cdots\end{aligned}$$

where most of the terms of the sum have been indicated by $\cdots$. How many terms are omitted? Further computation yields

$$\begin{aligned}(1.0075)^{80} &= 1 \cdot 1^{80} + 80 \cdot 1^{79} \cdot 0.0075 + \frac{80 \cdot 79}{1 \cdot 2} \cdot 1^{78} \cdot 0.00005625 \\ &\qquad + \frac{80 \cdot 79 \cdot 78}{1 \cdot 2 \cdot 3} \cdot 1^{77} \cdot 0.000000421875 + \cdots \\ &= 1 + 0.006 + 0.17778 + 0.032 + \cdots \\ &= 1.21578 + \cdots\end{aligned}$$

Only 4 of the 81 terms have been considered, but only that many should be enough to make clear the tediousness of the task.

Exercises

1. How many subsets with 13 elements does a set with 18 elements have?
2. How many subsets with 8 elements does a set with 13 elements have?
3. Find the fifteenth term of $(1/2 + 1/2)^{20} (r = 14)$.
4. Find the eighteenth term of $(1/10 + 9/10)^{100} (r = 17)$.
5. Find the first, second, and third terms of $(1 + 0.005)^{100}$.
6. Find the sum of the first four terms of the expansion of $(1/10 + 9/10)^{40}$.
7. Use the binomial formula to expand $(2u - 3w)^4$. (HINT: Let $x = 2u$ and $y = -3w$.)

8. Construct a triangle in the same way that the triangle in Table 2 is constructed except instead of writing the entries write "even" or "odd." (The set of even counting numbers is 0, 2, 4, 6, 8, ...)
   a. What property do rows 1, 2, 4, and 8 have in common?
   b. What would you predict for rows 16, 32, 64, etc.?
9. In the expansion of $(2x - 3y)^7$, find the coefficient of the term containing as a factor:
   a. x^2y^5 *b.* xy^6 *c.* x^4y^3
10. The object of this exercise is to prove that for $n > r > 1$, the relationship ${}_nC_r = {}_{n-1}C_{r-1} + {}_{n-1}C_r$ is true. By definition,

$${}_nC_r = \frac{n!}{(n-r)!r!}$$

$${}_{n-1}C_{r-1} = \frac{(n-1)!}{[(n-1)-(r-1)]!(r-1)!} = \frac{(n-1)!}{(n-r)!(r-1)!}$$

$${}_{n-1}C_r = \frac{(n-1)!}{(n-1-r)!r!}$$

Observe that $r! = r[(r-1)!]$ and $(n-r)! = (n-r)[(n-r-1)!]$. Using these relationships, form the sum ${}_{n-1}C_{r-1} + {}_{n-1}C_r$ and, by using a common denominator, read the desired result, namely, ${}_nC_r$.

11. Rewrite Table 2 substituting the correct ${}_nC_r$ for each entry.

Answers to problems

A. ${}_{16}C_0 + {}_{16}C_1 + {}_{16}C_2 + {}_{16}C_3 + {}_{16}C_4$

$$= 1 + 16 + \frac{16 \cdot 15}{2} + \frac{16 \cdot 15 \cdot 14}{1 \cdot 2 \cdot 3} + \frac{16 \cdot 15 \cdot 14 \cdot 13}{1 \cdot 2 \cdot 3 \cdot 4}$$

$$= 1 + 16 + 120 + 560 + 1{,}820 = 2{,}517$$

B. The number of possible pairs is

$${}_{10}C_2 = 45$$

and the number of days needed is $45 \cdot 2 = 90$. There is not time to run the experiments. The plan is not practical.

2.5 The multinomial formula

Problem

A manufacturing plant produces 11 distinct products that are shipped to wholesalers. Because of other considerations, the management decides to ship each set of 11 items in three cartons—the first carton to contain 3 items, the second to contain 4, and the third to contain 4. They decide to experiment with all the various possibilities of crating. If they ship 10,000 of each item per month, can they try all the possibilities in a day? A week? A month? A year? A decade?

One way to think of the symbol ${}_nC_r$ (that has already proved fruitful) is as the number of different subsets, each with r elements, in a set with n elements. Since for each subset of r elements there is an associated subset, its complement, with $n - r$ elements, the symbol ${}_nC_r$ is also *the number of ways that a set with n elements can be partitioned into two subsets with r and n − r elements.* It may be helpful to reread the material on partitions of sets in Sec. 2.1 and the related materials in Secs. 2.3 and 2.4.

This section deals primarily with the problem of counting the number of ways that a set can be partitioned into three or more subsets. Again an example will aid in the motivation of the formula.

Example 1. The food-counter department of Cut-Rate Drug Store is to hire a new waitress. Twelve girls have applied for the job, and they are to be interviewed by some of the staff of the department. It is decided that Mr. Green, Mr. Harvey, and Mr. Irwin will each share in the job of interviewing the girls. After considerable discussion, it is decided that Mr. Green will interview two of the girls, Mr. Harvey will interview six, and Mr. Irwin will interview four. In how many different ways can the three men interview the 12 applicants?

A little reflection shows that the question can be answered if the number of distinct partitions of the set of 12 girls into three subsets—the first with two elements, the second with six elements, and the third with four elements—can be ascertained. But what constitutes a *distinct partition?* To aid in answering this preliminary question, some notation will be helpful.

As a concession to the methods of our time, let us refer to the girls not by name but by the use of symbols $a_1, a_2, \ldots, a_{12}$. One possible arrangement is

$$(1) \quad (a_1,a_2|a_3,a_4,a_5,a_6,a_7,a_8|a_9,a_{10},a_{11},a_{12})$$

where the vertical bars indicate the partition according to interviewer; that is, Mr. Green interviews a_1 first and a_2 second, and Mr. Harvey interviews a_3 first, a_4 second, a_5 third, etc. But this represents only one possible arrangement. Another possible arrangement is

$$(2) \quad (a_2,a_1|a_3,a_4,a_5,a_6,a_7,a_8|a_9,a_{10},a_{11},a_{12})$$

To interpret expression (2) correctly, note that the order in which Mr. Green conducts his interviews has been changed but that each of the three men still interviews the same applicants. Now each of the arrangements (1) and (2) represents a permutation of the 12 objects (girls); but since they result in the same subsets of girls being interviewed by the same men, we shall call them *equivalent permutations.* Another way to think of this is that the partitions indicated in each of the permutations are the same. This means that the question concerns combinations rather than permutations. Therefore, to answer the question, the number of permutations must be divided by the number of times each permutation occurs in equivalent form. As a first step toward determining the correct formula, it is necessary to divide the number of permutations by $2 = 2!$ since the permutations a_1, a_2 and a_2, a_1 result in equivalent permutations.

Refer again to (1) and (2), which are equivalent permutations. For each of the arrangements

$$(a_1, a_2, \ldots)$$

and

$$(a_2, a_1, \ldots)$$

how many equivalent ways can Mr. Harvey interview the members of the set $\{a_3, a_4, a_5, a_6, a_7, a_8\}$? The answer is that there are as many different equivalent arrangements as there are permutations of the symbols a_3, a_4, a_5, a_6, a_7, a_8, that is, ${}_6P_6 = 6!$.

As a second step toward determining the number of equivalent permutations in each distinct partition, the factor $6!$ must appear in the denominator. In summary, thus far there are $2! \cdot 6!$ equivalent permutations that arise from a change in the order of interviews of Mr. Green and Mr. Harvey.

By similar reasoning about the order of Mr. Irwin's interviews, it is seen that there are $2! \cdot 6! \cdot 4!$ different ways that the three men can interview the same subsets of applicants. In other words, $2! \cdot 6! \cdot 4!$ is the number of times that each permutation will occur in equivalent form in the total list of permutations. Since the total number of permutations is $12!$, the answer to the original question is

$$\frac{12!}{2! \cdot 6! \cdot 4!} = 13{,}860$$

The symbol ${}_{12}C_{2,6,4}$ is used to denote $12!/(2! \cdot 6! \cdot 4!)$. This symbol is a natural extension of the notation ${}_nC_r$, which has been used to count the distinct partitions of a set of n elements into subsets of r and $n - r$ elements. Thus ${}_{12}C_{2,6,4}$ represents the number of distinct partitions of a set of 12 elements into subsets of 2, 6, and 4 elements, respectively.

Only one example has been given, but it illustrates the formula that follows.

Formula: ***(Number of distinct partitions of a set*** S ***with*** n ***elements into subsets of*** $n_1, n_2, \ldots, n_r$ ***elements,*** $n_1 + n_2 + \cdots + n_r = n$.)

If S is a set with n elements and $n_1, n_2, \ldots, n_r$ are counting numbers such that $n_1 + n_2 + \cdots + n_r = n$ and if S is to be partitioned into subsets such that there are $n_1, n_2, \ldots, n_r$ elements in the subsets, respectively, then the number of distinct partitions is the number of permutations divided by the product of the number of permutations of each subset. Symbolically, we write ${}_nC_{n_1,n_2,\ldots,n_r}$ for the number of distinct partitions and

$$ {}_nC_{n_1,n_2,\ldots,n_r} = \frac{n!}{(n_1!)(n_2!)\cdots(n_r!)} $$

It must be observed that the partitions are *ordered*, in the sense that n_1 elements are in the first subset, n_2 elements in the second subset, etc.

The formula for the number of distinct partitions of a set into r subsets is a generalization of the formula for $r = 2$,

$$ {}_nC_r = \frac{n!}{n!(n-r)!} $$

To see this, note that if $r = 2$, then $n_1 + n_2 = n$, $n_2 = n - n_1$, and ${}_nC_{n_1,n_2} = n!/[n_1!(n - n_1)!]$, which is recognized as the formula for ${}_nC_{n_1}$ or, equivalently, for ${}_nC_{n_2}$.

An application in the previous section demonstrated the connection between the numbers ${}_nC_r$ and the binomial formula. It will be helpful to recall how the numbers ${}_nC_r$, which have one interpretation as the number of distinct partitions of a set into two subsets, have a second interpretation as the coefficients of the binomial formula.

If a set S with n elements is partitioned into two subsets, with $n_1, n_2, \ldots, n_r$ elements in each subset $(n_1 + n_2 + \cdots + n_r = n)$, then the number of distinct partitions is

$$ {}_nC_{n_1,n_2,\ldots,n_r} = \frac{n!}{(n_1!)(n_2!)\cdots(n_r!)} $$

But these numbers are also the coefficients of the terms in the *multinomial* expansion of $(x_1 + x_2 + \cdots + x_r)^n$.

Just as no proof was given for the binomial formula, no proof will be given for the multinomial formula. An intuitive argument was given to motivate the reason to expect the coefficients for the binomial formula to be the numbers ${}_nC_r$ (see page 117). Reread that argument to see how only a slight modification could be made so that the argument would apply when three or more real numbers are involved. A statement of the multinomial formula follows, and it should be compared with the binomial formula.

Formula: (***Multinomial formula***) *If $x_1, x_2, \ldots, x_r$ are real numbers, if n is a counting number, and if $n_1, n_2, \ldots, n_r$ are counting* *numbers such that* $n_1 + n_2 + \cdots + n_r = n$, *then the term in the expansion of* $(x_1 + x_2 + \cdots + x_r)^n$ *involving* $x_1^{n_1} \cdot x_2^{n_2} \cdot \cdots \cdot x_r^{n_r}$ *has the coefficient* ${}_nC_{n_1,n_2,\ldots,n_r} = n!/[(n_1!)(n_2!) \cdots (n_r!)]$.

Two examples are offered to illustrate how the multinomial formula can be used to obtain partial information without a complete expansion of the expression.

Example 2. Suppose it is desired to find the coefficient of the term $x_1^2x_2^6x_3^4$ in the product $(x_1 + x_2 + x_3)^{12}$. One way to compute this quantity is to determine the product with $x_1 + x_2 + x_3$ used as a factor 12 times. But computation of a few of these products will convince you that this becomes quite a task. Consider instead the product with this trinomial [$(x_1 + x_2 + x_3)$ is called a *trinomial* because it has three terms] as a factor 12 times as being written

$$(x_1 + x_2 + x_3)^{12} = (x_1 + x_2 + x_3)(x_1 + x_2 + x_3) \cdots (x_1 + x_2 + x_3)$$

with the dots $(\cdots)$ representing the missing nine factors. We are particularly interested in determining that summand in the product which involves $x_1^2x_2^6x_3^4$. But how many products that involve $x_1^2x_2^6x_3^4$ are possible? This is just another way to pose the question about interviewing secretaries which occurred in the first example of this section. The product $(x_1 + x_2 + x_3)^{12}$ has ${}_{12}C_{2,6,4} = 13{,}860$ products of the form $x_1^2x_2^6x_3^4$. The expression $13{,}860x_1^2x_2^6x_3^4$ is only one term of the expansion of $(x_1 + x_2 + x_3)^{12}$, but the argument reveals how to determine the coefficient of any summand in the expansion.

Example 3. As another example, the coefficient of the term that involves $x_1^3x_2^7x_3^2$ in $(x_1 + x_2 + x_3)^{12}$ is

$${}_{12}C_{3,7,2} = \frac{12!}{3!7!2!} = 7{,}920$$

See Exercise 2 for further questions of this nature.

The use of the multinomial formula in sampling problems will be explored in Sec. 3.8.

Exercises

1. Compute $(x_1 + x_2 + x_3)^{100}$ if:

   a. $x_1 + x_2 + x_3 = 1$ *b.* $x_1 + x_2 + x_3 = 0$

2. Consider $(x_1 + x_2 + x_3)^9$. Determine the coefficient of:

   a. $x_1^3x_2^3x_3^3$ *b.* $x_1^4x_2^4x_3$

   c. $x_1^7x_2x_3$ *d.* $x_2^6x_3^3$

3. There are five typists in the secretarial pool at a company. If one day one vice-president uses two typists, another uses one typist, and still another uses two typists, in how many ways can the girls be assigned to the vice-presidents?

4. In how many ways can six people be assigned to offices if:
   a. There are six offices and one person is assigned to each.
   b. There are three offices and two people are assigned to each.
   c. There are four offices, two accommodating two people and two accommodating one person each?

5. A construction company is to move 15 pieces of equipment from one job to another. In how many ways can this be achieved if:
   a. There are 15 trucks, and one piece of equipment is put on each.
   b. There are three trucks, and five pieces of equipment are put on each.
   c. There are three trucks, and six pieces of equipment are put on one of the trucks, four on another, and five on the third.
   d. There are three trucks, and seven, seven, and one pieces of equipment are put on each of the trucks, respectively.
   e. There are three trucks, and three, six, and six pieces of equipment are put on each of the trucks, respectively.

6. Compute $(x + y + z)^4$ by multiplication. Show that the coefficients are those given by the formulas of this section.

7. Consider the product $(x + y + z)(x + y + z)(x + y + z)(x + y + z)$. In how many different ways can you choose two x's, one y, and one z if you must always choose exactly one letter from each set of parentheses? Explain how this is similar to counting the number of different ways of partitioning a set of four objects into three subsets of two, one, and one elements, respectively.

Answer to problem

There are ${}_{11}C_{3,4,4} = 11!/(3!4!4!) = 11{,}500$ different ways. No; no; no; yes; yes.

3

INTRODUCTION TO PROBABILITY

3.0 Introduction

This chapter discusses several of the elementary concepts of probability. The content of each individual section will be treated in some detail a little later in this introductory section, but first there are some general comments about the subject of probability.

Many of the ideas that have been studied in the first two chapters of this book will be used in this chapter. This use of old ideas offers a dividend to those readers who have pursued with diligence the definitions and formulas that precede. It also probably means that the reader will need to refer back to refresh some of the ideas which, because of the time lapse in their usage, may have slipped away.

Probability is an old subject, having among its earlier contributors such very fine mathematicians as Pascal and Fermat, to mention but two. So it is relevant to refer to the history of probability and also to its current applications. Probability is certainly not a dead subject; for example, probability theory is still being applied to games to determine winning strategies. Much of the early work in probability theory came into being as a result of games of chance, and some of the applied work in probability theory is still motivated by such games of chance as blackjack, baccarat, and the stock market. Probability theory has also found application in such diverse fields as genetics (in biology), particle motion (in physics), and human behavior (in psychology).

The examples in this chapter are stated almost entirely in terms of business situations. This is in resistance to the (natural) tendency to use dice, cards, and coin tosses for the examples, as many books of this type do. The selection of dice, cards, etc., is natural since such events lend themselves easily to illustrations of the properties of probability to be established here. There is one notable exception to the restraint from using examples about gambling games: Section 3.9 is devoted to odds and mathematical expectation and uses a game for its motivation.

A distinction between probability and statistics (the subject of Chap. 4) should be drawn even at this early point in the development of the first of the two subjects. For *probability*, the population is known (or assumed to be known) and the probability of observing a particular sample is studied. For *statistical inference*, the roles are reversed. The population is assumed unknown, the sample is known, and inferences about the population are made. The reasoning is from the sample to the population for statistical inference. In this chapter, we shall assume the population known and shall compute numbers that tell the probability of drawing a particular type of sample.

As a final general remark, it may be surprising to find that probability—which seems to be an elementary natural concept—is rather difficult to define. Certainly probability plays an important role in the discussion of everyday affairs, and most people feel that they have some understanding about the "meaning" of probability. Nevertheless, there are obstacles in formulating the definition. Besides the problem of assigning meaning to the word, there are problems in assigning measurements to the elements of the population. As we shall see, there is a great deal of subjectivity in the assignment of probability to the population. Even after the assignment of measurement, it is necessary to exercise great care in computing the probability for complex events. We shall point out how to exercise this care in the computations in this chapter.

There are 10 sections in the chapter, and no real good can come of an enumeration of their titles here. Nor does it seem profitable to give a summary of each section now. However, as guidelines for study, it is remarked that the first four sections deal with elementary examples and use previously developed vocabulary and visual aids to establish an intuitive feeling for "how things ought to be." There are some formulas in these sections, but they are largely revisions or reinterpretations of old formulas. The summing up occurs in Sec. 3.5. A probability space is defined in Sec. 3.5, and properties of probability already exhibited for special examples are established for probability spaces. After the description of the mathematical system, probability space, the applications start and continue for the last half of the chapter. In every case, the applications chosen are as closely connected to business or social problems as seems feasible, except for those in Sec. 3.9, which deals largely with games.

As in previous chapters, formulas are developed. To acquire an understanding of these formulas and the ability to use them, the reader will need to work the exercises provided for just this purpose.

3.1 Equiprobability

Problems

A. Of 28 employees in the cost accounting department of the Stevens Furniture Store, one is to be transferred to the general accounting division. If each employee is equally likely to be chosen, what is the probability that Mr. Hafen will be selected?

B. Of the seven girls in the secretarial pool at the Acme Tire Company, five are excellent typists, four are excellent at taking shorthand, and three are excellent in both typing and shorthand. If Mr. Gates selects one of the secretaries at random, what is the probability that he will get an excellent typist who is not excellent in taking shorthand?

Everyday conversation includes many statements similar to the ones which follow:

He will probably take her to the dance.

The boss's nephew will undoubtedly get the job.

It is likely that we shall sell more mixers than blenders.

It may rain today.

The union will certainly negotiate a new contract.

These statements have an important property in common. Each statement refers to a situation in which the outcome is unknown; that is, there is an *uncertainty* with regard to the occurrence of the events to which the statements refer. This section, and indeed most of this chapter, will be concerned with such events. Associated with each event is the *probability* of the occurrence of the event, and it is the probability of the statement (event) that will be the main concern. But before defining a probability space, we shall consider some examples.

Example 1. Suppose that one of four vice-presidents of a company is to be added to the board of directors. Assume also that it is not known which one of the vice-presidents is to be chosen and, furthermore, that it is *equally likely* that any of the four vice-presidents might be the person chosen. Of course, what is meant by "equally likely" has not been defined, but undoubtedly each reader already has a natural understanding of what the phrase means. To be more specific, *equally likely* infers that no one of the four vice-presidents has an inside track, nor has the person who is to make the selection for the board of directors yet shown any favoritism toward any one of the four candidates. Any one of four events may occur: the first, second, third, or fourth vice-president may be added to the board of directors. An *open sentence* describes the event that is to occur.

x will be a new member of the board of directors.

In this statement, x represents any one of the four vice-presidents, and the statement is true or false depending upon which vice-president's name is inserted for x and, of course, upon the outcome of the appointment. Since it has been assumed that any one of the four events is equally likely to happen, each of the four persons is said to have a probability of $1/4$ of attaining the appointment. Note that the fraction $1/4$ has as its numerator the number of successful events (1) and as its denominator the number of possible events (4).

Example 2. Suppose that an assortment of 100 typewriters is in a display room and that 9 of the 100 typewriters have a defect in the material and 8 have a manufacturing defect. Suppose also that none of the typewriters has both types of defects. A diagram will be helpful, and Diagram 1 relates this information. The numbers in the diagram refer to the conditions stated about defects.

Buyers for office-supply stores go to the display room and look over the 100 typewriters. Normally, buyers inspect only one typewriter closely. If each typewriter has an equally likely chance of being selected for inspection by a buyer, then a probability can be assigned to the event, expressed by the following statement:

p: The buyer selects a typewriter with a material defect.

Since there are 100 typewriters, each of which has an equally likely chance of selection, there are 100 events which may happen. Of these events, there are nine which make statement p true. Consequently, the event described by statement p is assigned the probability of $9/100$. Again,

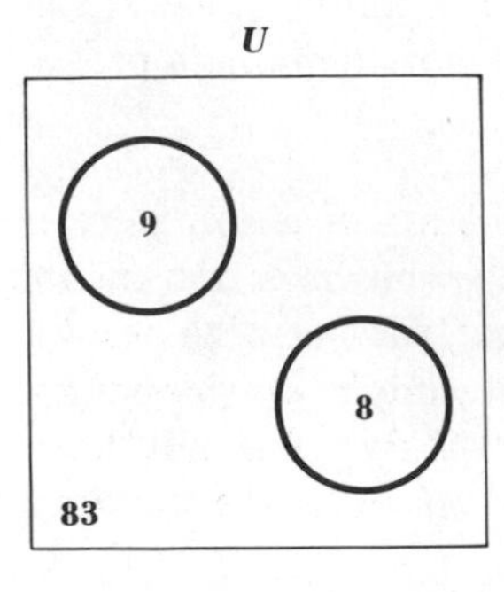

Diagram 1

the numerator is the number of successful events and the denominator is the number of possible events.

Example 3. Continuing with Example 2, if q is the statement

q: The buyer selects a typewriter with a manufacturing defect.

then the probability assigned to the event described by statement q is $8/100$, or $2/25$. Observe that there are 8 successful events and 100 possible events.

Note that it is *events* which have probability but *statements* which describe events. The probability of events or the probability of the statements that describe them will be used interchangeably when no confusion can arise from such an interchange. This subject will be discussed in more detail in Sec. 3.5 on probability spaces. In the meantime, we speak of both the probability of the event of a buyer selecting a typewriter with a manufacturing defect and the probability of the statement denoted by q.

Example 4. In this example, the circumstances of Example 2 will be altered slightly. Again assume that 100 items are on display. Nine of these have material defects, eight have manufacturing defects, and two have both defects. Diagram 2 illustrates these circumstances.

Again, we assume that the buyer will select one item and that each item has an equally likely chance of being chosen. The statements below describe events.

s: The buyer selects an item with a material defect.

t: The buyer selects an item with no defects.

v: The buyer selects an item with a manufacturing defect.

The probabilities assigned to the events described by the statements are $9/100$, $85/100$, and $8/100$. Is it still the case that the numerator of the probability is the number of successful events and the denominator is the number of possible events?

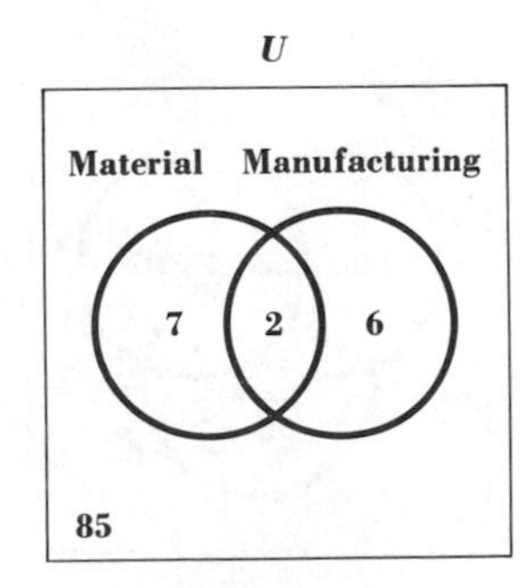

Diagram 2

Example 5. Let us assume that a sales promotion firm has made a survey of the purchasing habits of 100 people, with the following results:

5 people buy curtains, linens, and rugs each year.
11 people buy curtains and linens each year.
9 people buy curtains and rugs each year.
12 people buy linens and rugs each year.
37 people buy curtains each year.
42 people buy linens each year.
8 people buy none of the three.

It will be helpful in determining answers to questions to be asked with regard to this example to have a diagram. Diagram 3 consists of a universal set U and truth sets P, Q, and R, which correspond to the statements:

p: x buys curtains.

q: x buys linens.

r: x buys rugs.

Inside the eight disjoint subsets of U are inserted numbers which refer to the results of the survey. At this point, it may be helpful to reread Example 6 in Sec. 1.4 to see how the numbers in Diagram 3 are determined.

Now statements about the persons and their buying habits will be considered to see how a probability can be assigned to events that the statements describe.

The probability to be assigned to the event of statement p is $^{37}/_{100}$, and that was clear from one of the given facts. Similar remarks apply to statement q, which has the probability $^{42}/_{100}$. There is no necessity to draw the diagram to obtain this information. However, try to decide the probability to be assigned to statement r. The facts supplied in the example do not relate directly what probability should be assigned to statement r. But, of course, the probability is implied, and it can be read from Diagram 3. The probability of a person purchasing a rug is $^{40}/_{100}$.

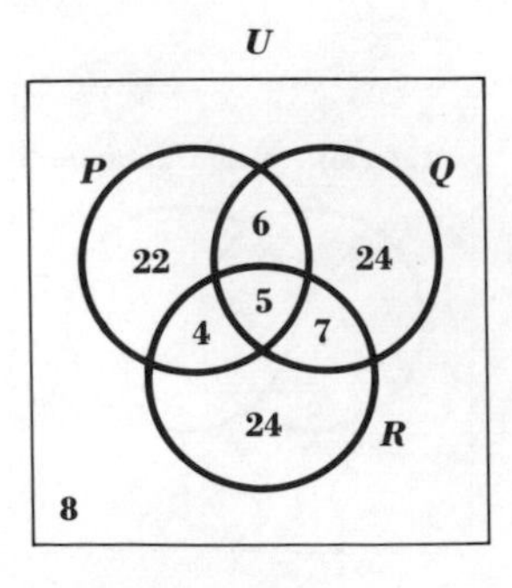

Diagram 3

It is customary in books of this type to take examples similar to the ones discussed here from card games or dice games. One reason for this is that such games afford excellent examples of circumstances of the type used in this section. Other reasons that such games are of importance will be discussed in later sections of this chapter. To provide variety, several exercises that deal with games are included.

Exercises

1. Assume that one card is to be chosen from a deck of playing cards. Assume that each card has an equally likely chance of being selected. Determine what probability should be assigned to each of the following events:
   a. The card selected is black.
   b. The card selected is a spade.
   c. The card selected is a four.
   d. The card selected is an ace.
   e. The card selected is a face card.
2. Assume that two dice are to be thrown simultaneously, and also assume *unbiased* dice. (Why is the second assumption necessary?) Determine the probability that should be assigned to each of the events below and save the results for later use (and for use in later exercises). What is the probability that the sum of the faces of the dice will be:

a. 2	*b.* 3	*c.* 4	*d.* 5
e. 6	*f.* 7	*g.* 8	*h.* 9
i. 10	*j.* 11	*k.* 12	

3. A survey is made of 100 people who are going to purchase a new automobile. The results show:
   3 are going to buy a two-door white car with air conditioning.
   9 are going to buy a two-door white car.
   5 are going to buy a two-door car with air conditioning.
   26 are going to buy a white car with air conditioning.
   49 are going to buy a white car.
   39 are going to buy a car with air conditioning.
   24 are going to buy a car with none of these.

   Assign probability to the event

   x is going to buy a two-door car.
4. What is wrong with the following conclusion? The probability that an American citizen chosen at random satisfies

   x was born in Alaska.

   is $1/50$ since there are 50 states.
5. Criticize: The outcomes of an experiment are equally likely if there is no physical reason which would cause one outcome to occur with a greater frequency than any of the others.

6. There are n outcomes for an experiment. What probability should be assigned to each outcome if it is desired to have an equiprobability?

7. Assign probability to each of the following:
   a. Event A or B will occur, and each is equally likely.
   b. If a counting number between 16 and 25 will be chosen, and each is equally likely to be chosen, then 18 will be chosen.
   c. Heads will show when a nonbiased coin is tossed.

8. A letter is chosen at random from the word *chosen*. What is the probability:
   a. That it is an h?
   b. That it is a consonant?
   c. That it is a vowel?

9. Of the seven girls in the secretarial pool at Acme Drug Company, six are typists, five take shorthand, and three can operate a calculator. Of the five who take shorthand all are typists, but only one can operate a calculator. If Mr. Speer selects one of the girls from the pool and it is equally likely that each girl will be chosen, what is the probability that he will get a calculator operator who can neither type nor take shorthand?

10. Of the seven girls in the secretarial pool at the Baker Drug Company, five are excellent at taking shorthand, four are efficient calculator operators, six are typists. Only two of the calculator operators take shorthand, and both of them are also typists. Only one of the typists does not take shorthand. If Mr. Greer selects one of the secretaries at random, what is the probability that he will get a secretary who can neither type nor take shorthand?

11. Product X comes in matched pairs. (The two items making up each pair are identical except for hidden defects.) In one pair, both are defective; in the second pair, both are good; and in the third pair, there is one good and one defective. If you selected one of the pairs at random and tested one of the elements and found it good, what would be the probability that the other element in that pair would be good?

12. Assume the following classification of employees of the Acme Company:

Where working	*Male*	*Female*	*Total*
Plant	1,540	160	1,700
Office	20	240	260
Sales	40	0	40
Totals	1,600	400	2,000

   a. What is the probability that an Acme Company employee selected at random is a male office worker?
   b. What is the probability that a woman employee selected at random is on the sales force?

c. What is the probability that a plant employee selected at random is a male?

d. What is the probability that a male employee selected at random is a plant worker?

Answers to problems

A. $\frac{1}{28}$

B. $\frac{2}{27}$. See diagram.

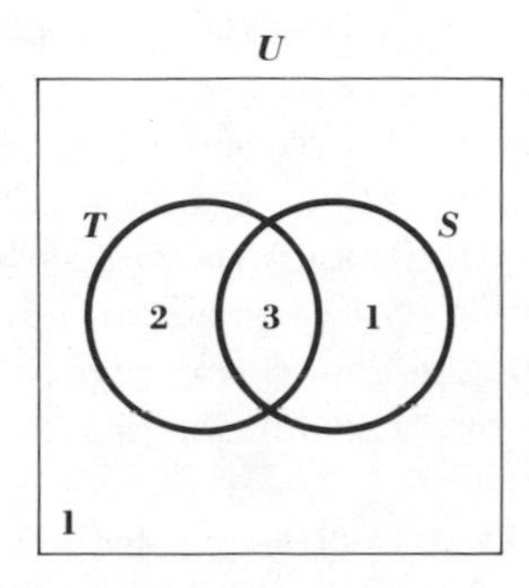

3.2 *Probability of the conjunction of statements*

Problems

A. A women's cosmetic firm sells 10 items in a grocery chain and 4 in a drugstore chain, with 2 items selling in both outlets. What is the probability that an item selected at random is sold in both the grocery stores and the drugstores?

B. Of the seven girls in the secretarial pool at the Acme Drug Company, five take shorthand, four are beautiful, and two are both beautiful and take shorthand. If Mr. Speer selects one of the secretaries and if it is equally likely that each girl will be chosen, what is the probability that he will get a beautiful girl who takes shorthand?

C. A firm has divided its sales territory into three sales districts, A, B, and C. It has 20 salesmen, including the sales manager. Three salesmen are assigned solely to each of the three districts. The sales manager works in all districts. There are 12 men who work at least partially in district A, 10 who work at least partially in district B, and 10 who work at least partially in district C. Five, including the sales manager, work in both A and B.

1. If two salesmen are selected from the regular salesmen (and each of the 19 has an equally likely chance of selection), what is the probability that both will work in both districts B and C?
2. Work part 1 for districts A and C.

Several terms in the previous section (*event, open sentence,* and *probability* are examples) have not been rigorously defined. These terms, along with others, will be more formally defined in Sec. 3.5. In the meantime, the procedure will be to work via examples toward the definitions of these terms and others in this and the next two sections.

This section deals with the conjunction of statements and how to assign probabilities to the compound statement.

Probabilities of simple statements, and hence of the events that they describe, have already been assigned in the previous section for some particular examples. The events were equally likely to occur, and the probability was informally called the *equiprobability.* In the absence of rigorous definitions, some common sense must prevail about what is meant with regard to the concepts of *equally likely* and *equiprobability.* Now we shall consider what probability should be assigned to simple statements that are compounded by the use of the connective *and.*

The examples of this section will draw from those of Sec. 3.1.

Example 1. Refer to Example 1 of the previous section; it deals with the probability of the event that one of a set of four vice-presidents of a company will become a member of the board of directors. If the first of these candidates is now designated by A and the second candidate by B, then the compound statement (conjunction by the connective *and*)

A becomes a member of the board of directors, and B becomes a member of the board of directors.

is, by reason of Sec. 1.3, a conjunction.

Since the truth set of this conjunction is empty (both A and B cannot simultaneously be chosen as a member of the board of directors) and since probabilities were assigned on the basis of truth sets in Sec. 3.1, the probability to be assigned to the conjunction is zero. Observe that the truth sets of the simple components of the compound statement have no elements in common. When two events cannot occur simultaneously, they are called *mutually exclusive events.*

Example 2. If statements p and q of Examples 2 and 3 of Sec. 3.1 are reconsidered, it becomes clear that the truth sets for these two statements have no elements in common. Hence, their conjunction has a void (empty) truth set. The probability assigned to the conjunction of such statements is zero. In other language, the events described by statements p and q of Examples 2 and 3 cannot occur simultaneously. See Diagram 1 and statements p and q for a visual picture of why this is true. These are also mutually exclusive events.

Example 3. Now an example of a conjunction will be given in which the intersection of the truth sets of the simple components is not empty.

Refer to Diagram 2 to see that the truth sets of statements s and v have elements in common. Consider the compound statement:

$s \wedge v$: The buyer selects an item with a material defect, and the buyer selects an item with a manufacturing defect.

Since the number of elements in the truth set of s and v, $S \cap V$, is 2 and the total number of possibilities is 100, the probability assigned to the statement $s \wedge v$ (equivalently, to the truth set $S \cap V$) is $2/100$. Still referring to the same example, observe that the statement $s \wedge v$ can also be expressed as

The buyer selects an item with both defects.

When the statement is written in this form, it is an example of a *compound statement.* This is a little different usage of the word *compound* since there is no connective in the sentence. But the example shows that the compound statement permits decomposition into two simple statements connected by *and.*

Example 4. The remaining example deals with the information given in Example 5 of the previous section, and the information is summarized in Diagram 3.

Consider the statement

$p \wedge q$: x buys curtains, and x buys linens.

Reference to Diagram 3 shows that the probability should be $11/100$. Consult the diagram to see that the probability of $p \wedge r$ is $9/100$ and of $q \wedge r$ is $12/100$. In Chap. 1, we discussed how to form compound statements like $p \wedge q \wedge r$, and the probability for the compound statement $p \wedge q \wedge r$ is $5/100$.

Examples have been given to show how reasonable assignments of probability can be made to the simultaneous occurrence of events. The word *simultaneous* (in its usual meaning) is important because in a later section the probability to be assigned to the occurrence of a *sequence* of events is discussed. The rules of procedure are different if the order of occurrence of events is a factor in the outcome. When sequences of events are discussed, you will be reminded of the results of this section and the differences between the situations.

In each example of this section, the events were assumed to be equally likely and the probability assigned was the equiprobability.

Exercises

1. Assume that one card is to be chosen from a deck of playing cards and that each card in the deck has an equally likely chance of being chosen. Use

the following statements:

p: The card selected is black.
q: The card selected is a spade.
r: The card selected is a four.
s: The card selected is an ace.
t: The card selected is a face card.

Determine what probability should be assigned to each of the following events:

a. $p \wedge q$ *b.* $p \wedge r$ *c.* $p \wedge s$ *d.* $p \wedge t$
e. $q \wedge r$ *f.* $q \wedge s$ *g.* $q \wedge t$ *h.* $r \wedge s$
i. $r \wedge t$ *j.* $s \wedge t$

2. Refer to Exercise 3 of Sec. 3.1 and assign probability to:
   a. x is going to buy a two-door car with air conditioning.
   b. x is going to buy a white two-door car.
   c. x is going to buy a white car with air conditioning.
   d. x is going to buy a white two-door car with air conditioning.

3. The employees of the Acme Company are distributed as shown in the following table.

Where working	***Male***	***Female***	***Total***
Plant (P)	1,540	160	1,700
Office (O)	20	240	260
Sales (S)	40	0	40
Totals	1,600	400	2,000

Suppose that one employee is to be selected and that each has an equally likely chance of selection. Use the following statements:

p: A plant employee is selected.
q: An office employee is selected.
f: A female employee is selected.
s: A sales employee is selected.
m: A male employee is selected.

Write the probabilities for:

a. p *b.* q *c.* s *d.* $p \wedge s$
e. $p \wedge f$ *f.* $p \wedge m$ *g.* $m \wedge f$ *h.* $s \wedge m$
i. $s \wedge f$ *j.* $q \wedge s$ *k.* $p \wedge q$ *l.* $f \wedge s$

4. Given the data in Exercise 3, what is the probability that an employee who is known to be a male:
   a. Works in the plant?
   b. Works in the office?
   c. Works in the sales force?

5. Given the following classification by school and sex of 1,100 students in a college:

Class	Engineering		Business		Humanities		Science	
	M	*F*	*M*	*F*	*M*	*F*	*M*	*F*
Freshman	40	10	60	15	35	70	90	5
Sophomore	30	5	70	10	20	70	80	5
Junior	30	0	70	15	15	50	75	10
Senior	25	0	60	20	0	40	70	5
Totals	125	15	260	60	70	230	315	25

Suppose that a student is to be selected and that each has an equally likely chance of selection. Let:

p: The student is a male.
q: The student is a female.
r: The student is an upperclassman.
s: The student is in Business.
t: The student is in Humanities.

Assign probabilities:

a. p *b.* q *c.* r *d.* s
e. $r \wedge s$ *f.* $r \wedge q$ *g.* $s \wedge t$ *h.* $p \wedge t$
i. $p \wedge q \wedge r \wedge s$

6. The Acme Company has 2,000 employees, who may be classified as follows: 1,700 plant employees, 260 office employees, and 40 salesmen. There are 400 women employees, of whom 240 work in the office and 160 work in the plant.
   a. What is the probability that an employee selected at random is a woman who works in the plant?
   b. What is the probability that a woman employee selected at random is an office worker?
   c. What is the probability that an office employee selected at random is a woman?

Answers to problems

A. $2/12 = 1/6$

B. $2/7$

C. 1. $2/19$
   2. $4/19$

3.3 Probability of the disjunction of statements

Problems

A. In a radio-assembly process, at one station the chassis is fitted with two turning knobs. The knobs come in two colors, red and green, to match the chassis and are intermingled in one large box. The foreman, without knowing it, has placed a color-blind operator at this station. What is the probability that a radio selected at random from the assembly line will not have color-coordinated turning knobs? Explain why this is not $\frac{1}{2}$ (if it is not), since the operator is essentially choosing each knob at random without regard to color.

B. A manufacturer of plywood found the following results in a survey of 1,000 sheets of his product:

1 defect	60
2 defects	300
3 defects	350
4 defects	200
5 defects	40
6 or more defects	20

1. What group has not been accounted for?
2. If this sample is typical of the total production, what percentage of the product has two or more defects?
3. What percentage of the product has fewer than two defects per sheet?

The previous section was concerned with the probability of the simultaneous occurrence of two (or more) events; this section deals with the occurrence of one event *or* another. Before reading this section, it will be helpful to reread Sec. 1.4, which explains the meaning of *or* with both truth tables and Venn diagrams. The familiar procedure of using examples to motivate the natural assignment of probability will be used.

Example 1. Suppose that Mr. Kole is going to call upon two potential customers on Monday and that it is (on the basis of past experience) equally likely that each customer will either buy or not buy. Consider the event described by the statement

p: Mr. Kole will make two sales on Monday.

Mr. Kole wishes to determine the probability that statement p is true. He reasons that there are three events that may occur: (1) neither of the customers will buy, (2) exactly one of the two customers will buy, (3) both of the customers will buy. Since there are three events, Mr. Kole decides to assign the probability of $\frac{1}{3}$ to each of these three events. This

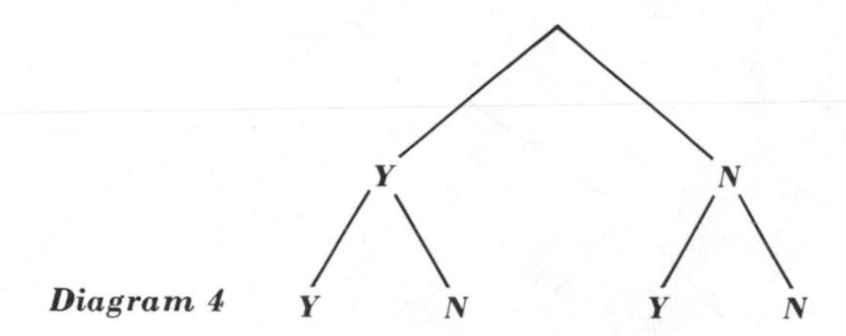

Diagram 4

is in accordance with the reasoning for previous examples, when the events were equally likely and equiprobability was assigned. In this situation, one of the three events is considered successful, so Mr. Kole reasons that the probability that he will make two sales is $\frac{1}{3}$. But Mr. Kole is wrong!

Perhaps a tree diagram will aid in determining his error. The tree in Diagram 4 shows there are four possible events, not three. The letter Y represents *yes*, hence a sale, and N represents *no sale*.

Since it is assumed that it is equally probable that each customer will buy or will not buy, the four events should have equal probability assigned, namely, $\frac{1}{4}$. Only one of these events is successful for p, so the correct probability is $\frac{1}{4}$.

The point of Example 1 is to draw attention to the fact that some care must be taken in deciding the number and nature of logical possibilities. But the main consideration in this section will be with the disjunction of simple statements, so the circumstances described in Example 1 will be used again but this time with study focused on a compound statement which can be written as a disjunction.

Example 2. Consider the event described by the statement

q: Mr. Kole will make exactly one sale on Monday.

This is a compound statement in disguise. An example has already been given (Sec. 3.2, Example 3) of a conjunction which permits an equivalent restatement by the use of *and*. This is a similar situation. Consider

q: Mr. Kole sells the first prospect and not the second prospect, or Mr. Kole does not sell the first prospect and sells the second prospect.

With the restatement, it becomes clear that the "correct" probability should be $\frac{2}{4} = \frac{1}{2}$ (see Diagram 4). If statement q is thought of in its second form, then the probabilities of its simple components are each $\frac{1}{4}$. The sum $\frac{1}{4} + \frac{1}{4} = \frac{2}{4} = \frac{1}{2}$ yields the "natural" result.

Example 3. If Mr. Kole calls on three customers and it is equally likely that each will buy, then the eight possible, equally likely events are illustrated by the tree in Diagram 5. Again the letters Y and N have been used to represent *yes* (a sale) and *no* (no sale), respectively.

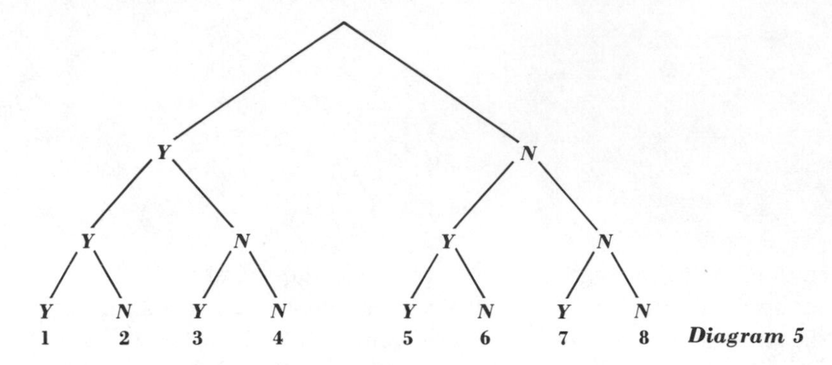

Diagram 5

The tree diagram shows that there are eight events that may occur. In only one of these events does Mr. Kole fail to make a sale (event 8), and in only one event does he make three sales (event 1). The number of events with exactly one sale is three (events 4, 6, and 7). The number of events with exactly two sales is also three (events 2, 3, and 5). If each of the eight equally likely events is assigned the probability of $\frac{1}{8}$, then some statements and their probabilities are

a: Mr. Kole makes no sales, $\frac{1}{8}$.
b: Mr. Kole makes one sale, $\frac{3}{8}$.
c: Mr. Kole makes two sales, $\frac{3}{8}$.
d: Mr. Kole makes three sales, $\frac{1}{8}$.

So that it will be more easily discernible why statement b is a disjunction and why the probability assigned to it can be thought of as a sum, the statement is repeated with the use of simple statements and the connective *or*.

b: Mr. Kole sells the first customer but not the second or third customer;
or Mr. Kole does not sell the first customer, sells the second customer, but does not sell the third customer;
or Mr. Kole does not sell the first customer, does not sell the second customer, but sells the third customer.

The simple statements that make up the compound statement b describe events 4, 6, and 7, respectively, each of which has probability $\frac{1}{8}$. The sum $\frac{1}{8} + \frac{1}{8} + \frac{1}{8} = \frac{3}{8}$ is the assigned probability.

Statement c is very similar to statement b, and Exercise 6 requires you to write statement c as the disjunction of the simple statements.

There is one aspect of the simple statements in the examples which has been specifically noted and which is of utmost importance. The events described are *mutually exclusive*. Refer to Sec. 3.2 to see that these are events that cannot occur simultaneously. The definition of this term will be delayed, but for Example 3 it means that no two of three successful events can occur at the same time. The occurrence of any one of the events excludes the occurrence of any one of the others.

So that the concept of mutually exclusive events can be emphasized, two other examples are offered.

Example 4. Suppose that Mr. Kole is again to call on his three customers and that the probabilities discussed in Example 3 have been assigned. Suppose now that the event to which probability is to be assigned is described by:

e: Mr. Kole will have either complete success or no success.

Since success (for Mr. Kole) is a sale, the statement can be rephrased in terms of the number of sales he makes. A restatement is:

e: Mr. Kole will make zero sales, or he will make three sales.

The probability assigned to selling all three customers is $1/8$ (event 1 of Diagram 5), and the probability assigned to selling no customers is also $1/8$ (event 8 of Diagram 5). Since these are mutually exclusive events (both events cannot occur simultaneously) and since both events are considered successful for statement *e*, the probability that is assigned the statement that one *or* the other of these events will occur is

$$1/8 + 1/8 = 1/4$$

Example 5. Let *f* be the statement

f: Mr. Kole sells two *or* fewer of the three customers.

The probability of selling no customers is $1/8$, of selling one customer is $3/8$, and of selling two customers is $3/8$; and these are mutually exclusive events (two of them cannot occur simultaneously). So the probability assigned to the event described in statement *f* is

$$1/8 + 3/8 + 3/8 = 7/8$$

The assignment of probability follows the pattern of determining the probability of each of the several events and adding the probabilities assigned to mutually exclusive events.

Now an example is given to emphasize the necessity of the *mutually exclusive* property in order to add probabilities to obtain the "correct" result.

Example 6. Let *g* be the statement

g: Mr. Kole does not sell all the customers or Mr. Kole sells to fewer than two of the customers.

The probability assigned to the event that not all the customers buy is $7/8$ (see Diagram 5). The probability assigned to the event that fewer than two of the customers buy is $\frac{4}{8}$ (see Diagram 5). The sum of these probabilities is

$$\tfrac{7}{8} + \tfrac{4}{8} = \tfrac{11}{8} = 1\tfrac{3}{8}$$

But the assignment of probabilities has been such that they never exceed the number 1, which is reserved for events that are *certain* to occur. In view of this, something is definitely wrong with the sum obtained. The connective *or* occurs in statements *e* and *f*, and the (correct) result was determined by adding probabilities. However, there is an important difference between those examples and Example 6. The events described by the simple statements in *e* are mutually exclusive, as are those described in statement *f*. However, the events described in statement *g* are *not* mutually exclusive.

As an aid in determining why the procedure used so successfully in Examples 1 to 5 did not work for Example 6, both a tree diagram and a set diagram will be used. First, the tree diagram shown in Diagram 5 is reproduced.

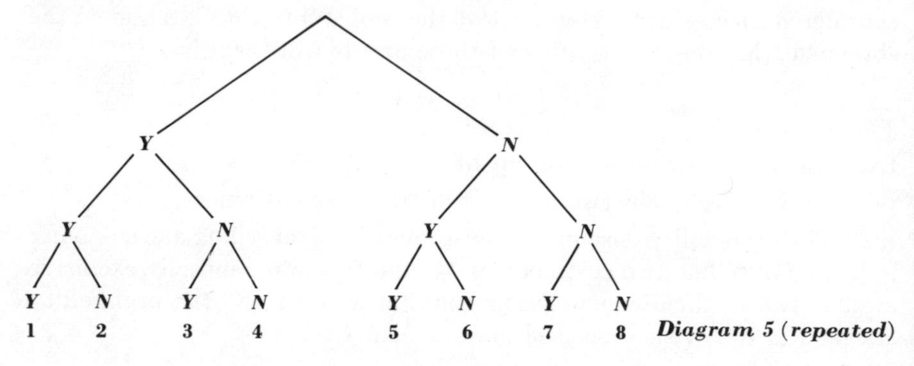

Diagram 5 (repeated)

Consider again statement *g* and observe that it consists of two other statements. The first of these is

Mr. Kole does not sell to all the customers.

and the second is

Mr. Kole sells to fewer than two customers.

Referring to the tree diagram, events 2 to 8 are successful for the first part and events 4 and 6 to 8 are successful for the second part. This means that events 4 and 6 to 8 have been counted twice in the sum $\frac{7}{8} + \frac{4}{8}$. The error committed in computing the probability assigned to statement *g* by adding the probabilities of the simple statements of *g* is shown in Diagram 6; the numbers refer to the events in Diagram 5.

The correct probability for the event described by statement *g* is

$$\frac{7 + 4 - 4}{8} = \tfrac{7}{8}$$

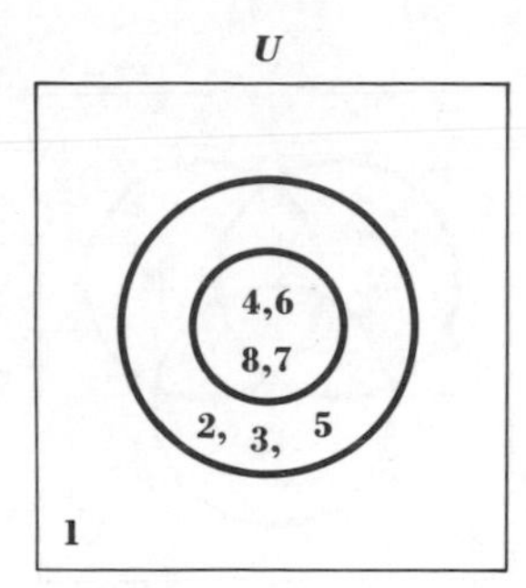

Diagram 6

The procedure used to arrive at the probability for statement g in Example 6 should appear familiar. Section 2.1 dealt with formulas for counting the number of elements in the union of two sets. If A and B are sets, then

$$N(A \cup B) = N(A) + N(B) - N(A \cap B)$$

and it is precisely this formula which is used to compute the "correct" probability in Example 6. But Sec. 2.1 also developed a formula for counting the number of elements in the union of three sets. Read the next example to see how it can be used in assigning probabilities.

Example 7. As the final example of this section, the data from Example 5 of Sec. 3.1 will be utilized. The pertinent facts of that example are relisted below for convenience.

A survey of the purchasing habits of 100 persons was made, with the following results:

5 people buy curtains, linens, and rugs each year.
11 people buy curtains and linens each year.
9 people buy curtains and rugs each year.
12 people buy linens and rugs each year.
37 people buy curtains each year.
42 people buy linens each year.
8 people buy none of the three.

It will be helpful to have a diagram in determining answers to the questions to be asked in regard to this example. Diagram 7 consists of a universal set U and truth sets P, Q, and R, which correspond to the statements p, q, and r.

p: x buys curtains.
q: x buys linens.
r: x buys rugs.

Inside the eight disjoint subsets of U are inserted numbers which refer to the results of the survey. It would be helpful to reread these results to see how the numbers arise from the statements about the survey.

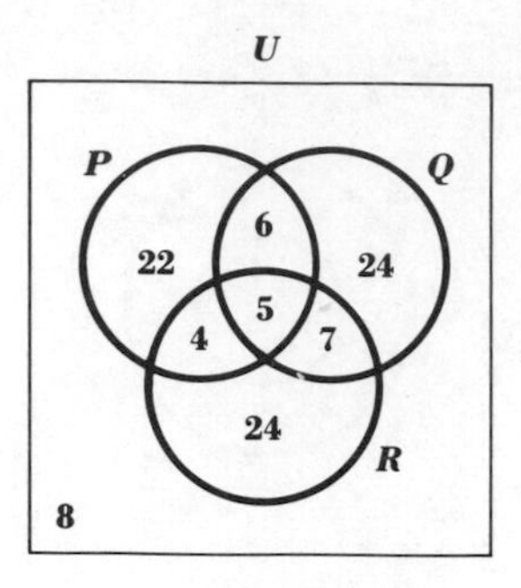

Diagram 7

What probability should be assigned to the event which is represented by the statement

h: x buys curtains *or* buys linens *or* buys rugs?

If the probabilities assigned to each of the sets P, Q, and R are added, the sum is

$$37/100 + 42/100 + 40/100 = 119/100 = 1\,19/100$$

But this sum exceeds 1, which is the probability that should be assigned to an event which is certain. The difficulty arises because the sets P, Q, and R are not disjoint (the events are not mutually exclusive). A formula for counting the number of elements in the union of three sets is available and will be used here to overcome the difficulty encountered.

Refer to Sec. 2.1 to see that the number of elements in the union of three sets is given by the formula

$$N(P \cup Q \cup R) = N(P) + N(Q) + N(R) - N(P \cap Q) - N(P \cap R) - N(Q \cap R) + N(P \cap Q \cap R)$$

If this formula is applied to statement h, then use of Diagram 7 yields

$$\begin{aligned} N(P \cup Q \cup R) &= (37 + 42 + 40) - (11 + 9 + 12) + 5 \\ &= 119 - 32 + 5 \\ &= 92 \end{aligned}$$

The probability assigned to the event described by the statement $h(p \wedge q \wedge r)$ is $92/100$.

As a result of the consideration of several examples, a pattern emerges as to the method used in assigning probability to disjunctive statements.

Sometimes statements that permit expression as the disjunction of simple statements are not so written. As an aid in assigning probability, they should first be rewritten (Examples 1 to 3). If the events described in the disjunction are mutually exclusive, then the probability is the sum of the probabilities of the components of the disjunction (Examples 4 and 5). If the events described are not mutually exclusive, then the formulas for counting the union of sets developed in Sec. 2.1 should be employed (Examples 6 and 7).

Exercises

1. Assume that one card is chosen from a deck of playing cards (consider an ace as 1 and face cards as 10). Assume that each card in the deck has an equally likely chance of being selected. Determine what probability should be assigned to each of the following events:
   a. The card selected is black.
   b. The card selected is a four.
   c. The card selected is a black four.
   d. The card selected is black or a four.
   e. The card selected is less than 10 or red.
   f. The card selected is odd and less than 10.
2. Refer to your results for Exercise 2 of Sec. 3.1 to decide the probability that the sum of the faces of the dice will be:
   a. Less than 4
   b. Greater than 7
   c. Not 7
   d. 2, 3, 4, 5, 9, or 10
   e. 7 or 11
   f. 2, 3, or 12
   g. 4, 5, 6, 8, 9, or 10
3. Suppose that Mr. Kole is going to call on four customers in a given day and that it is equally likely that each customer will either buy or not buy. Assign probabilities to the following statements:
   a. Mr. Kole makes three sales.
   b. Mr. Kole makes five sales.
   c. Mr. Kole makes two sales, or he makes no sale.
   d. Mr. Kole makes one sale, and he makes three sales.
   e. Mr. Kole makes less than four sales.
   f. Mr. Kole makes more than two sales.
4. Assume that it is known that one of four codes (DOG, FOX, CAS, NIK) is going to be used and that it is equally likely that CAS or DOG will be used. It is also known that DOG is twice as likely to be used as FOX and that NIK is three times as likely to be used as FOX.
   a. What is the probability that the code used will be CAS?
   b. That it will be NIK?
   c. That it will be FOX?
5. If the probability of event R is $1/4$, the probability of event S is $1/8$, and the probability of event R *and* event S is $1/8$, then what is the probability of event R or event S?
6. Write statement *c* of Example 3 as the disjunction of simple statements.
7. From every 20 customers, it is equally likely that 1 will be chosen from 12 who have purchased at least twice and from 5 who have annual incomes above $7,000. Of the five who have annual incomes above $7,000, there

are two who have purchased at least twice. What is the probability that:

a. The person chosen has either purchased at least twice or has an annual income above $7,000?

b. The person chosen neither has purchased at least twice nor has an annual income of $7,000?

c. The person chosen has purchased at least twice but does not have an income of $7,000?

8. The following data summarize the sales records for salesmen Alex and Baker during the past month:

Salesman	*Calls made*	*Calls successful*	*Result of call*	
			Battery sale	*Tire sale*
Alex	1,200	600	360	420
Baker	800	500	350	450

a. What percentage of Alex's calls resulted in sales?

b. How many of Alex's customers bought both tires and batteries?

c. If you selected one of Baker's calls at random, what is the probability that it is one which resulted in a sale of both tires and batteries?

d. Now, considering both salesmen, let:

p: The call resulted in a battery sale.

q: The call resulted in a tire sale.

What is the probability that a call resulted in both a tire and a battery sale?

e. Refer to part *d* and give the probability that a call resulted in either a tire sale or a battery sale.

Answers to problems

A. The knobs must match the chassis as well as each other. The probability that the set will be completely color-coordinated depends on the number of red and green sets produced, as well as on the number of red and green knobs. If the number of each color set is equal, as are the knobs, then the probability of complete color coordination is ¼ and the probability of some disharmony is ¾. The following chart shows the various possibilities.

RRR	Colors match
RRG RGR RGG GRR GRG GGR	Colors do not match
GGG	Colors match

B. 1. There are 30 sheets with no defects.
2. $(300 + 350 + 200 + 40 + 20)/1{,}000 = 91\%$
3. $90/1{,}000 = 9\%$, or $100\% - 91\% = 9\%$

3.4 *Probability of the negation of statements (and other compound statements)*

Problems

A. A machine produces pieces that have a probability of passing inspection of $^{98}/_{100}$. What probability should be assigned to the event that a piece will not pass inspection?

B. Of the nine senior officers of a company, five are men. There are four who are over 63 years old, and two of these are women. What is the probability that an officer chosen at random is *not* a man over 63?

C. Manufacturers who wish to avoid shutdowns on an assembly line often provide standby equipment which is used if machines in the line fail.

1. If the probability that a machine will not function properly is 0.10, how many standby machines would be required to reduce the probability of a shutdown due to failure of this machine to no more than 0.01, that is, not over 1 chance in 100?
2. If a more reliable machine can be purchased for twice the cost of the machine described above, how reliable would this machine have to be to provide protection equal to that of two of the other, less reliable machines?

This is another section about the assignment of probability to compound statements. Recall that conjunction and disjunction have already been discussed. Most of this section is about *negation* of statements, but since the previous sections and the first portion of this one will have made clear the general procedure, similar methods can be used to assign probability to the *conditional, biconditional,* and other compound statements. The easiest way to proceed would seem to be to use the examples already developed in Secs. 3.1 to 3.3, and we shall do so.

Example 1. In Example 1 of Sec. 3.1, a member of the board of directors is to be selected from four vice-presidents. If Mr. Adams is one of the vice-presidents and p is the statement

p: Mr. Adams will be selected as the member of the board of directors.

then the probability assigned the event of statement p is $\frac{1}{4}$. The negation of p is

$\sim p$: Mr. Adams will *not* be selected as the member of the board of directors.

and since there are three elements in its truth set, the probability assigned is $3/4$. Note that $1/4 + 3/4 = 1$.

Example 2. In Example 2 of Sec. 3.1, there are 100 typewriters on display; 9 have a material defect, and 8 others have a manufacturing defect. The probability assigned to the event of the statement

q: The buyer selects a typewriter with a material defect.

is, as agreed upon before, $9/100$. The negation is

$\sim q$: The buyer does *not* select a typewriter with a material defect.

and of the 100 possible events, there are 91 which result in the selection of a typewriter that does not have a material defect (all successful for $\sim q$). The assigned probability is $91/100$. Note that the sum of the probabilities of q and $\sim q$ is $9/100 + 91/100 = 1$.

By now it should be clear why the statement

r: The buyer selects a typewriter with a manufacturing defect.

has probability $8/100$, while the statement

$\sim r$: The buyer does *not* select a typewriter with a manufacturing defect.

has probability $92/100$. The sum of the probabilities is again 1.

Example 3. The modification of Example 2 to give Example 4 in Sec. 3.1 poses no new problems in the assignment of probability. See Diagram 2 for the new information, and see statements s, t, and v to see that their assigned probability is $9/100$, $85/100$, and $8/100$. The diagram should make clear why the probabilities assigned to the negations $\sim s$, $\sim t$, and $\sim v$ are $91/100$, $15/100$, and $92/100$, respectively.

Example 4. In this example, reference is made to Example 5 of Sec. 3.1. Since it is somewhat lengthy, it will not be reproduced here, but see Diagram 3 for a visual display. Reread statements p, q, and r and the accompanying discussion to see why they have probabilities of $37/100$, $42/100$, $40/100$. By now you have probably concluded that their negations have probabilities $63/100$, $58/100$, $60/100$, respectively. Consult Diagram 3 to see how these numbers are derived. Exercise 7 requires that some justification be given for these assignments.

The examples that precede deal with the negation of simple statements. Negation of compound statements was presented in Sec. 1.5. The next two examples discuss the negation of a disjunction and of a conjunction.

Example 5. Refer to Example 3 of Sec. 3.2, where the conjunction

$s \wedge v$: The buyer selects an item with a material defect, and the buyer selects an item with a manufacturing defect.

refers to the situation related visually in Diagram 2. The negation of $s \wedge v$ is $\sim(s \wedge v)$, which has a truth set with 98 ($7 + 6 + 85 = 98$) elements. The assigned probability is $98/100$. Note that the sum of $2/100$, the probability of $s \wedge v$, and $98/100$, the probability of $\sim(s \wedge v)$ is 1.

Example 6. Refer to Example 4 of Sec. 3.3 for the disjunction

e: Mr. Kole will make zero sales,
or he will make three sales.

The number of logical possibilities is 8, and they are displayed in the tree in Diagram 5. The probability for the first part of e is $1/8$, as it is also for the second part. The probabilities are added (for mutually exclusive events) to give the probability for statement e as $1/8 + 1/8 = 1/4$. Count the branches of the tree to see that the probability for the negation of e is $6/8 = 3/4$.

Example 7. We shall keep the same example to illustrate how to assign probability to a conditional statement. The following statement refers to the set of logical possibilities displayed in the tree in Diagram 5:

If Mr. Kole makes a sale on his first call, then he makes a sale on his third call.

For definiteness, let us denote this statement by $a \to b$. At this point, you may find it helpful to refer to Sec. 1.7, where the truth table for the conditional was first given. First, observe that events 5 to 8 are all true for $a \to b$. This is because the hypothesis is false (Mr. Kole does not make a sale on his first call), and the agreement (page 58) is that regardless of the truth of the conclusion, the conditional is true. What of the other events? Look at the branches of the tree to see that events 1 and 3 are also in the truth set of $a \to b$ (TT). But events 2 and 4 are not in the truth set (TF). Summarizing, events 1, 3, and 5 to 8 make $a \to b$ true; events 2 and 4 make it false. The assigned probability is therefore $6/8 = 3/4$.

Example 8. For a biconditional, the same example is used.

Mr. Kole makes a sale on his first call if and only if he makes a sale on his third call.

Notationally, we have $a \leftrightarrow c$. See Diagram 5 to see that events 1 and 3 are in the truth set of $a \leftrightarrow c$ (TT). But so are events 6 and 8 (FF). (You may find it helpful to consult the truth table on page 65.) Hence, the probability assigned is $4/8 = 1/2$.

The same procedure can be followed for compound statements like $\sim p \wedge q$, $\sim(p \vee q)$. Exercises are provided for the development of skills in the solution of these types of problems.

Exercises

1. Assume that one card is to be chosen from a deck of playing cards. Assume that each card in the deck has an equally likely chance of being selected. Assign probability to the following statements:
   a. The card selected is not black.
   b. The card selected is not a spade.
   c. The card selected is not a four.
   d. The card selected is not a heart.

2. Assume that two dice are to be thrown simultaneously, and also assume unbiased dice. (Why is the second assumption necessary?) Determine what probability should be assigned to the event that the sum of the faces of the dice is:
   a. Not 2   *b.* Not 3   *c.* Not 4   *d.* Not 5
   e. Not 6   *f.* Not 7   *g.* Not 8   *h.* Not 9
   i. Not 10   *j.* Not 11   *k.* Not 12

3. A survey is made of 100 people who are going to purchase a new automobile. The results show the following:
   3 are going to buy a two-door white car with air conditioning.
   9 are going to buy a two-door white car.
   5 are going to buy a two-door car with air conditioning.
   26 are going to buy a white car with air conditioning.
   49 are going to buy a white car.
   39 are going to buy a car with air conditioning.
   24 are going to buy a car with none of these.

   Assign probability to the following events:
   a. x is *not* going to buy a two-door car.
   b. x is *not* going to buy a two-door car with air conditioning.
   c. x is *not* going to buy a white two-door car.
   d. x is *not* going to buy a white car with air conditioning.
   e. x is *not* going to buy a white two-door car with air conditioning.

4. If p and q are statements such that the probability of $p \wedge q$ is $\frac{1}{3}$, the probability of $\sim p$ is $\frac{1}{2}$, and the probability of q is $\frac{2}{3}$, then what is the probability of:
   a. $p \vee q$ *b.* $\sim p \wedge \sim q$
   c. $p \rightarrow q$ *d.* $p \leftrightarrow q$

5. In a certain city, one-fourth of the people read paper A, three-eighths read paper B, and one-eighth read both A and B. If a person is randomly selected from this city, what is the probability that he:

a. Does not read paper A?
b. Does not read either A or B?
c. Reads either A or B, or both?

6. Refer to Exercise 5 in Sec. 3.2. Assign probabilities to each of the following:
   a. $\sim p$ *b.* $\sim t$ *c.* $\sim(s \wedge t)$
   d. $(q \vee r) \wedge \sim p$ *e.* $q \wedge \sim r$ *f.* $q \vee \sim r$
7. Justify the assignment of the probabilities in Example 4 of this section.

Answers to problems

A. $\frac{2}{100}$

B. $\frac{7}{9}$

C. 1. The probability of a malfunction is 0.10. The probability that two machines would both fail to function is $0.10 \cdot 0.10 = 0.01$. The probability of one or the other functioning is $1 - 0.01 = 0.99$. Two machines (one standby) would provide the required assurance.
2. Probability of malfunction would have to be no more than 0.01, or probability of operating properly would have to be 0.99.

3.5 Probability spaces

Before reading this section, take inventory of what has been discussed, so that what follows will appear as a summary. What precedes consists largely of quite a few examples, with some exercises of the same type to provide practice. The approach has been informal, nonrigorous, relatively easy, and deals with natural extensions of the ideas of Chap. 1. But in the discussion of a mathematical model, there comes a time to discard the specific and formulate some general rules of procedure. That is what will be done in this section.

The formalization of the ideas advanced in the previous sections will take on its final form as axioms for a probability space, followed by theorems that result from the axioms. However, do not become unduly alarmed. The promise made to you before that this is *not* a book about rigorous proof of mathematical theorems will be kept. But there is a dire need for some common understanding (definitions) about the meanings of technical words (remember the concern in Chap. 1 about precision in language?) as well as for some carefully stated rules of procedure about how to manipulate probabilities.

Behind considerations about probability are physical situations that require analysis and (remember Chap. 1?) decision making. Along with the circumstances, there are *data* which often arise as a result of an experiment or a series of experiments. These experiments and the accompanying data fall into two categories: (1) observation of uncontrolled events and (2) controlled experimentation in the laboratory (or elsewhere). So that you may have something specific to think about as an example of the first, consider the maximum daily temperatures in Philadelphia in 1968; for the second, consider the amount of pressure (determined by an experiment) necessary to crumble each of 100 bricks selected by a brickmaking firm.

Some more examples follow.

Example 1. Interviewing a voter about his preference on a school bond issue.

Example 2. The burning time (before burning out) of a light bulb.

Example 3. The height measurements of 100 freshmen at Drexel Institute of Technology.

Example 4. The force necessary to break each of 1,000 window panes.

Classify each of the above into one of the two categories.

Some sets of data are *quantitative*, as in the examples above. Others are *qualitative*. Almost all the examples in this book are of the quantitative type, and for the second kind to yield to the analysis described here, a quantitative measure must be applied.

Once the data are presented, there is a need for a vocabulary to permit intelligent discussion. A *population* is a set of measurements that has resulted from experiments—often a repeated experiment of the same kind. A *sample* of the population is a subset (see page 128) of the population chosen for special consideration. Regardless of the way in which the data are collected—whether by observation or by controlled experiment—the elements are called *events*, which are simply the outcomes of the observations or controlled experiments. Events are in turn classified as *simple* or *compound*, depending upon whether or not they permit decomposition. Section 3.1 dealt with simple events, and Secs. 3.2 to 3.4 used compound events as examples, with the connectives *and*, *or*, *conditional*, and *biconditional* used to compound simple events. There are exercises that you can review to make clear this distinction.

We want to reach a definition of probability. The study of examples has led to the practical definition that the probability of an event is the fraction of the population that results in the event. This is particularly true of equiprobable events—which are the kinds studied so far. In later sections and even at the end of this section, some nonequal probabilities will be discussed. But before that, the concept of probability space should be formalized in terms of axioms, that is, assumptions about the occurrence of the events. Some decision must be made about whether the elements of a probability space are to be treated as points (or subsets of points) or as statements that select events. Actually, a nebulous decision was already made in Sec. 3.1 when it was decided not to distinguish between the probability of an open sentence and the probability of its truth set.

The approach here will be to first talk in terms of the set vocabulary developed in Chap. 1, then relate this to statements, and eventually (as before) refer to the probability either of a statement or of its truth set. There is one other restraint to be imposed: Probability spaces are finite, countable infinite, or continuous. All the examples given have been finite, and for the purposes of this book, all probability spaces can be considered to be finite. So we agree to limit the discussion to the finite case.

Definition: **(Probability space)** *Let S be a finite set with points $E_1, E_2, \ldots, E_n$, which denote n simple events. Furthermore, let $m(E_1), m(E_2), \ldots, m(E_n)$ be nonnegative real numbers assigned to events $E_1, E_2, \ldots, E_n$, respectively, such that*

1. $m(E_1) + m(E_2) + \cdots + m(E_n) = 1.$

Then S is a probability space.

First, observe that this means that $m(S) = 1$. Next, note that it is not excluded that the probability of an event is zero, $m(E_i) = 0$. This convention is sometimes helpful. There are cases where the probabilities are not known in advance, and involved considerations are required to decide whether to completely disregard points of measure zero. These are beyond the scope of this book. We shall disregard elements that have a measure of zero.

The definition of probability space makes clear that to each element of a set someone assigns a measure. But what of the measure of subsets?

Definition: **(Probability of subset)** *If S is a probability space and T is a subset of S, then the probability of T is the sum of the probabilities of the elements of T.*

The next statement defines a property of probability spaces.

2. *If S is a probability space and T and R are nonempty subsets of S, then $m(T \cup R) = m(T) + m(R)$ if and only if $T \cap R = \Phi$.*

If $m(T \cup R) = m(T) + m(R)$, then T and R can have no elements in common (with the agreement that elements of measure zero are ignored). If T and R are disjoint sets, then the measure of their union is obtained by adding the measures of their respective elements. These are what have been called *mutually exclusive* events.

A different statement covers the case for two sets which are not disjoint.

3. *If S is a probability space and T and R are nonempty subsets of S, then $m(T \cup R) = m(T) + m(R) - m(T \cap R)$.*

Observe that this statement shows that the measure of common elements should be subtracted from the sum of the measures for sets that overlap.

Still another property of probability spaces is

4. *If S is a probability space and T is a subset of S, then $m(T)$ is such that* $0 \leq m(T) \leq 1$.

The assignment of nonnegative measures guarantees that $0 \leq m(T)$, while the property expressed in statement 1 guarantees that $m(T) \leq 1$.

Still another property is

5. *If S is a probability space and T is a subset of S, then $m(T) = 0$ if and only if $T = \Phi$.*

Verification of these two facts follows. If $T = \Phi$, we have

$$1 = m(S) = m(S \cup T) = m(S) + m(T) = 1 + m(T)$$

which implies that $m(T) = 0$. Why does $m(S) = m(S \cup T)$? [Be sure you understand why $m(S \cup T) = m(S) + m(T)$ for this case.]

If $m(T) = 0$, we have

$$1 = m(S) = m(S) + m(T) = m(S \cup T) + m(S \cap T)$$

or

$$1 = m(S) + m(S \cap T)$$

which shows that $S \cap T = \Phi$ since all sets of zero measure have, by agreement, been disregarded.

The final property to be listed is

6. *If S is a probability space and T is a subset of S, then*

$$m(T) = 1 - m(\tilde{T})$$

The above statement is true because $1 = m(S) = m(T \cup \tilde{T})$, and since $T \cap \tilde{T} = \{ \quad \}$, we have $1 = m(T) + m(\tilde{T})$, which leads to

$$m(T) = 1 - m(\tilde{T})$$

This concludes the listing of properties, and a summary is in order.

A physical situation gives rise to a finite set S of n events, or outcomes. A measure is assigned to the elements of the events so that the sum is 1.

$$m(E_1) + m(E_2) + \cdots + m(E_n) = 1$$

Then, for subsets T and R of S, it is true that:

1. $m(S) = 1$.
2. $m(T \cup R) = m(T) + m(R)$ if and only if $T \cap R = \{ \quad \}$.
3. $m(T \cup R) = m(T) + m(R) - m(T \cap R)$ for any T, R.
4. $0 \leq m(T) \leq 1$.
5. $m(T) = 0$ if and only if $T = \{ \quad \}$.
6. $m(T) = 1 - m(\tilde{T})$.

For all the examples given in the first four sections, the above rules were observed. You should verify this. Only events that are *certain* are assigned probability 1. If events are mutually exclusive, then their probabilities are added. If they overlap (can occur simultaneously), the probability of their disjunction must not contain the probability of common elements in the sum twice. Always the probability of an event is between 0 and 1. For impossible events, it is 0. For complementary events, the probabilities should add to 1 (certainty).

Now we turn to statements. A statement p is made about an uncertain circumstance. A finite set of logical possibilities is selected. These are the elements of the probability space S. To each element of the set, a nonnegative measure m is assigned such that $m(S) = 1$, which is arrived at by summing the measure of each element. But the statement has a truth set, P, which in turn has a measure. The measure of the truth set, $m(P)$, is the *probability* of the statement. For equiprobabilities, this is the number of successful events divided by the number of possible events. We denote the probability of p by $Pr(p)$, which is, by definition, $m(P)$. The properties that have been stated earlier for sets now become properties of probability of statements. They are:

1. $Pr(p) = 1$ if and only if p is a tautology.
2. $Pr(p \vee q) = Pr(p) + Pr(q)$ if and only if p and q are mutually exclusive.
3. $Pr(p \vee q) = Pr(p) + Pr(q) - Pr(p \wedge q)$.
4. $0 \leq Pr(p) \leq 1$.
5. $Pr(p) = 0$ if and only if p is self-contradictory.
6. $Pr(\sim p) = 1 - Pr(p)$.

Some difficulties arise because of the close connection between events and the statements that describe them, between statements and their truth sets, and between the measure of sets and the probability of statements. But these difficulties are largely problems of language and can be solved—although it must be admitted that some of the difficulties have been solved by avoidance.

In what follows, the probability of statements will be discussed. Often this is phrased as the probability of an event, which simply means the event that the statement describes.

In each of the examples in Secs. 3.1 to 3.4, the assumption was made that events were equally likely to occur. No attempt was made to explain, with any more precision, just what was meant by *equally likely*, but the reader was asked to supply the natural, intuitive meaning for such an expression. Certainly, there are circumstances in which *not* all the possible events are equally likely. An often used example is that if a coin is tossed, then three possible events are (1) it lands with heads showing, (2) it lands with tails showing, and (3) it stands on its edge. We now give an example that deals with events that are not equally likely to occur. We shall describe how probabilities can be assigned and discuss some pitfalls to be avoided in these assignments.

Example 5. Suppose that a salesman for a printing company knows that a department store is going to buy a shipment of letterheads, envelopes, or advertising fliers and that there is an equally likely chance of the store buying letterheads or envelopes. But he also feels that the chances for selling letterheads is two times as good as the chances for selling advertising fliers and that the chances for selling envelopes is two times as good as the chances for selling advertising fliers. He wants to assign a probability to the event described by the statement

u: The department store will make a purchase of advertising fliers.

As a first step toward the assignment of a probability to the statement u, an arbitrary fractional probability a/n is assigned to the event that the department store will buy advertising fliers. Then the probability $2a/n$ should be assigned to each of the statements

v: The department store will buy letterheads.

w: The department store will buy envelopes.

(Why?) Since the sum of these three probabilities should be 1 (the salesman feels that it is certain that the department store will place an order), it follows that

$$\frac{a}{n} + \frac{2a}{n} + \frac{2a}{n} = 1$$

which yields

$$\frac{5a}{n} = 1$$

If $a = 1$ and $n = 5$, then this condition is satisfied. Consequently, the probabilities $\frac{2}{5}$, $\frac{2}{5}$, and $\frac{1}{5}$ are assigned to the events v (purchase of letterheads), w (purchase of envelopes), and u (purchase of advertising fliers), respectively.

Note that this example might be viewed by the use of a tree diagram as follows:

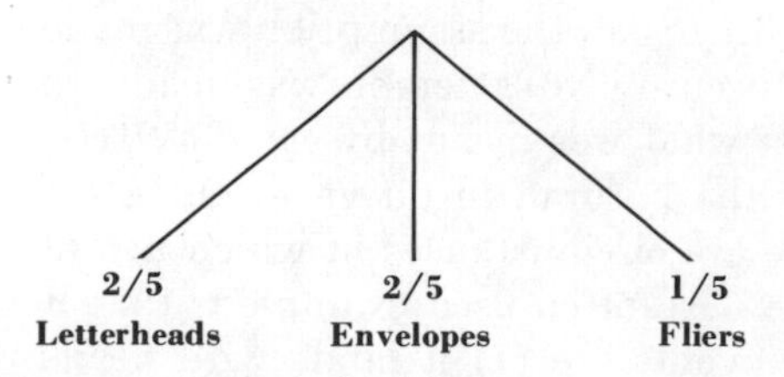

but if an equiprobability diagram is needed, then the following relays the same information:

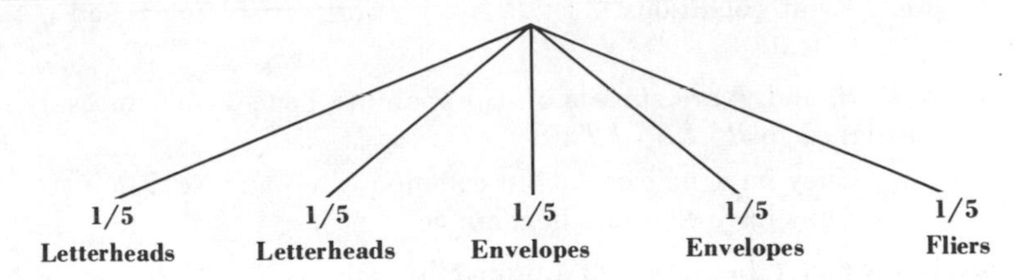

We shall close the discussion in this section with some general observations. First, notice that someone must decide upon the set of logical possibilities. The set of logical possibilities will depend in many cases upon data (or information) that are available about the problem. It has already been observed that the person who chooses the set of logical possibilities may have to decide from among several sets of logical possibilities. Consequently, the probability assigned will depend upon the set of logical possibilities chosen and hence will depend upon the person making the decisions. The set of logical possibilities may be chosen on the basis of some very definite facts (as it was in some of the examples), it may depend entirely upon the intuition of the person who is doing the deciding, or it may depend on a combination of fact and intuition (as some of the examples have indicated). In the examples, it was reasonably clear what probability to assign after the set of logical possibilities was chosen. It should be clear to you that there are situations where such decisions are not so clear-cut, and wise assignments depend on wise assigners.

Exercises

1. Give a set of logical possibilities for the following:
   a. A bond issue is to be voted upon.
   b. An election with three candidates is to be held.
   c. An election with one candidate is to be held.
   d. A living person is to be asked the year of his birth.
   e. A living senator is to be asked the year of his birth.
   f. A living wage earner not eligible for social security is to be asked the year of his birth.
2. A set of logical possibilities $U = \{a,b,c\}$ has been assigned a measure such that $m(a) = \frac{1}{2}$, $m(b) = \frac{3}{8}$, $m(c) = \frac{1}{8}$.
   a. Write the eight subsets of U.
   b. Write the probabilities assigned to each of the eight subsets of part *a*.

3. Let $Pr(p) = 2/3$ and $Pr(q) = 3/4$.
   a. Are p and q consistent?
   b. What conditions must $Pr(p) + Pr(q)$ satisfy for p and q to be consistent?
4. If R, S, and T are subsets of a probability space, what measure should be assigned to $R \cup S \cup T$ if:
   a. They have no elements in common (are pairwise disjoint)?
   b. They have elements in common?
5. Let R and T be subsets of a probability space.
   a. If R is a subset of T, what relation will hold for $m(R)$ and $m(T)$?
   b. If $m(R) \leq m(T)$, does it follow that R is a subset of T?
6. Use properties developed in Chap. 1 about sets to establish the following:
   a. $Pr(p \wedge q) + Pr(\sim p \wedge q) = Pr(q)$
   b. $Pr(p) = Pr(p \wedge q) + P(p \wedge \sim q)$
   c. $Pr(p \wedge q) = Pr(q \wedge p)$
7. Suppose that a coin is to be tossed until heads occurs. Write the events in the probability space.
8. Suppose a number is selected from the set $\{1,2,3,4,5\}$ and then another number is selected from the subset of the four remaining numbers.
   a. How many permutations are possible?
   b. Assign the equiprobability, and assign probability to the event that an odd number is selected the first time.
   c. Work part *b* for the probability of the selection of an odd number the second time.
   d. Work part *b* for the probability of the selection of an odd number both times.
9. A survey of 10,000 electrical gadgets showed that 200 had defective parts, 150 were improperly wired, and 12 had both defects. Let
   p: The gadget has a defective part.
   q: The gadget is improperly wired.
   Find the following:
   a. $Pr(p \wedge q)$ *b.* $Pr(p \wedge \sim q)$ *c.* $Pr(\sim p \wedge \sim q)$
   d. $Pr(p \vee q)$ *e.* $Pr(p)$ *f.* $Pr(q)$

3.6 An application of probability to sampling problems

Problems

A. Suppose that two of the spark plugs of an eight-cylinder car are worn out. If two are removed, with each pair having an equally likely chance of selection, what is the probability that the two defective plugs will be selected? That at least one of the defective plugs will be removed?

B. A company has bought 100 can openers. The management has decided to choose a sample of 10 and accept the shipment if none is defective. (Assume equiprobability of selection.) What is the probability of accepting the lot if none is defective? If one-tenth are defective? If one-half are defective?

C. Twenty aluminum castings were received in a shipment from the Acme Foundry. Past experience has shown that 30 percent of the castings received from this foundry will be defective. In a sample of five castings selected from this shipment, four were found to be defective.

1. How many different samples of five can be drawn from a shipment of 20 castings?
2. If the castings are really 70 percent good and 30 percent defective, how many good castings should be in the shipment?
3. In how many ways can you draw a sample which has four defective castings and one good casting if there are six defective and fourteen good castings in the shipment?
4. Considering the total number of samples of five which could be drawn and the number which would have four defective castings, would you agree that it is quite unlikely that there are really only six defective items in the shipment?

All considerations in this section about the occurrence of events will be limited to those events which have an equally likely chance of occurrence. The first example, which will be discussed at length, deals with a problem in sampling.

Example 1. A company has on display 10 of its blankets for inspection by the buyers for department stores. Assume that it is true that of the 10 blankets on display, there are 3 which have a minor defect, while the others are without defects. These conditions partition the universal set, which consists of 10 objects, into two subsets: a subset labeled G (good) and the complement of set G, which is labeled D (defective). Diagram 8

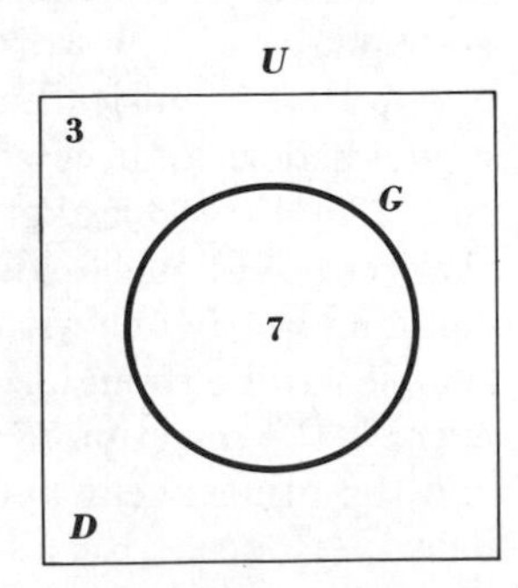

Diagram 8

represents this situation. The numbers in the diagram refer to the number of elements in the sets.

From experience, it is known that when buyers examine the blankets on display, they consider only 3 of the 10. There is some concern by the officials of the company that a choice of 3 out of 10 will cause the buyer to select as many as two blankets which have minor defects, and consequently, the buyer will be unfavorably impressed. However, experience also teaches the company that since the defects are minor, if the customer selects only one of the three which have defects, then he will be uninfluenced and will place an order as usual. Mr. Potter, who is in charge of marketing, is concerned with this problem, and he will consider the statement

p: The buyer examines a sample of three blankets which contains exactly one blanket with a minor defect.

(Although this consideration will not serve to answer all the questions that are relevant, read on.) The event that the statement describes is the selection of 3 objects from among 10. If it is assumed that each subset consisting of three elements has an equally likely chance of being selected, then the probability that is assigned to the selection of a *particular* subset is the fraction $1/n$, where n is the number of possible subsets with three objects of the set consisting of 10 elements. In Sec. 2.3, it was determined that the number of subsets containing exactly 3 elements in a set of 10 elements is given by the formula ${}_{10}C_3$. A computation, using the definition of the symbol ${}_{10}C_3$, yields

$$ {}_{10}C_3 = \frac{10!}{7!3!} = \frac{10 \cdot 9 \cdot 8}{1 \cdot 2 \cdot 3} = 120 $$

Therefore, to each of the events

y is a subset of three blankets selected.

is assigned the probability of $\frac{1}{120}$.

Thus far, the probability that is assigned to each of 120 equally likely events has been determined. Now a computation will be made of the number of events which may occur and be deemed successful for statement p. The word *successful* is sometimes used in such problems in a way that is misleading. But in this example, from the point of view of the seller, a successful event is one in which the customer selects a subset of three blankets with precisely one minor defect. The subsets that represent successful events are those subsets of three blankets which contain exactly one with a minor defect. But the selection of a successful sample can be thought of as consisting of two (*independent*) events. One event is the selection of two blankets from the set of seven good blankets, and the other event is the selection of one blanket from the set of three defective ones. Recall that in Sec. 2.2 a statement about independent events was advanced as a formula. That formula is specialized here to the case for only two events.

Formula: ***(The number of ways in which a sequence of two independent events may occur)*** *If event E_1 can occur in w_1 ways and event E_2 can occur in w_2 ways (and the events are independent), then the sequence of events E_1, E_2 can occur in w_1w_2 ways.*

So first it is essential to count the number of ways in which a subset of two elements can be chosen from a set of seven elements. This is just another application of the formula that was used for determining the number of subsets of 3 elements from a set with 10 elements. This time, an application of the formula for ${}_nC_r$ yields the number

$$ {}_7C_2 = \frac{7!}{2!5!} = \frac{7 \cdot 6 \cdot 5 \cdot 4 \cdot 3 \cdot 2 \cdot 1}{1 \cdot 2 \cdot 5 \cdot 4 \cdot 3 \cdot 2 \cdot 1} = \frac{7 \cdot 6}{1 \cdot 2} = \frac{42}{2} = 21 $$

The result of this computation shows that there are 21 different subsets of two elements in a set of seven elements. This is the w_1 referred to in the restatement of the formula for two independent events.

Second, the number of subsets of single elements in a set with three elements must be counted. Again an application of the formula ${}_nC_r$ yields

$$ {}_3C_1 = \frac{3!}{2!1!} = 3 $$

The formula for independent events asserts that the product of w_1 and w_2 should be computed, and it is

$$ 21 \cdot 3 = 63 $$

This is the number of different events which are successful in terms of statement p. The probability assigned to the event of the statement p is therefore

$$ Pr(p) = {}^{63}\!/_{120} = {}^{21}\!/_{40} $$

Example 2. Mr. Potter feels that the problem has been solved, but while this matter is being discussed at a board meeting, Mr. Ford observes that while the probability, $^{21}\!/_{40}$, of a buyer selecting a sample with one defective is helpful, this is not really all of the information needed to make the decision. If the buyer selects a subset consisting of three of the blankets which have either no defects or one defect, then the buyer's decision to buy will not be adversely influenced. Required, then, is the probability that the sample of 3 elements from the set of 10 will have none or one with minor defects. Mr. Potter sees that a different event must be considered, and he reasons that he must now consider the statement

q: The buyer selects a subset of three blankets which contains one blanket with a minor defect or a subset with no defects.

Statement q is compound; it is the disjunction of the statements

p: The buyer selects a sample of three blankets which contains exactly one blanket with a minor defect.

or r: The buyer selects a subset of three blankets which have no defects.

In other words, $q = p \vee r$; also p and r are mutually exclusive. In Sec. 3.5, the probability for disjunction for mutually exclusive events was determined to be the sum of the probabilities. The probability of p has already been established as $^{63}/_{120}$. Now $Pr(r)$ must be computed.

The selection of a subset containing three elements which are all in the subset labeled G is a successful event for r. The number of subsets of three elements that can be selected from a set containing seven objects is

$$_7C_3 = \frac{7!}{4!3!} = \frac{7 \cdot 6 \cdot 5}{1 \cdot 2 \cdot 3} = 35$$

The probability to be assigned to statement r is, therefore, $^{35}/_{120}$. We now have

$$\begin{aligned} Pr(q) &= Pr(p) + Pr(r) \\ &= {}^{63}/_{120} + {}^{35}/_{120} \\ &= {}^{98}/_{120} \\ &= {}^{49}/_{60} \end{aligned}$$

To state this only slightly less accurately, there are about 8 chances out of 10 that the customer will *not* be adversely influenced by a random choice of a sample of three blankets.

This example, like other examples, is just a special case. However, the problem and its solution can, and will, be used as a model for the derivation of a formula which can be applied to all such sampling problems.

Let a universal set U be given, and let U have n elements. The universal set U is partitioned into two (disjoint) subsets, which are denoted by S and $\tilde{S}$. Let the number of elements in the set S be r. The number of elements in set $\tilde{S}$, which is the number of elements in set U less the number of elements in set S, is $n - r$. Now suppose that the sample subset T of set U is to consist of m elements. Furthermore, suppose that the sample T is to contain s elements in S and that the remaining $m - s$ elements of S are in $\tilde{S}$.

To aid in understanding the number of elements in each of the subsets, diagrams will be given. The universal set is partitioned by the cross partition of $(S,\tilde{S})$ and $(T,\tilde{T})$, which is $(S \cap T, S \cap \tilde{T}, \tilde{S} \cap T, \tilde{S} \cap \tilde{T})$. Symbols representing the number of elements in each of the four subsets of the cross partition are also available, and they are included in Diagram 9.

Using the notation for the number of elements in a set adopted in

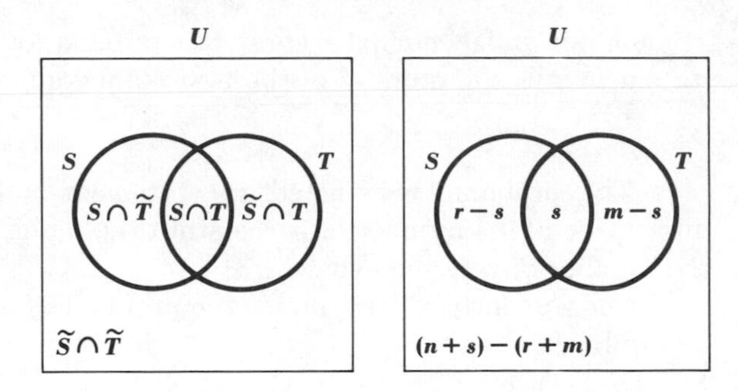

Diagram 9

Sec. 2.1, the number of elements in the members of the three partitions is given below for handy reference.

Partition	***Partition***
$U = S \cup \tilde{S}$	$U = T \cup \tilde{T}$
$N(U) = N(S) + N(\tilde{S})$	$N(U) = N(T) + N(\tilde{T})$
$n = r + (n - r)$	$n = m + (n - m)$

Cross Partition

$$U = (S \cap T) \cup (S \cap \tilde{T}) \cup (\tilde{S} \cap T) \cup (\tilde{S} \cap \tilde{T})$$
$$N(U) = N(S \cap T) + N(S \cap \tilde{T}) + N(\tilde{S} \cap T) + N(\tilde{S} \cap \tilde{T})$$
$$n = s + (r - s) + (m - s) + [(n + s) - (r + m)]$$

Now a probability will be assigned to the event described by the statement

p: The sample subset T of the universal set U contains exactly r elements of the set S.

First, we shall count the number of subsets of U containing m elements. In Sec. 2.3, a formula was derived to count these subsets, and when the formula is applied to this set of circumstances, it is seen that the number of subsets is ${}_nC_m$. Therefore, the probability

$$\frac{1}{{}_nC_m}$$

is assigned to the event

A particular subset of m elements is chosen from the set U.

Next, the number of subsets that make statement p true will be counted. Such subsets have been called the *successful events* of statement p. Each successful event for statement p consists of two independent events, which are:

1. The selection of s elements from the set S (which contains r elements)
2. The selection of $m - s$ elements from the set $\tilde{S}$ (which contains $n - r$ elements)

The number of subsets which make event 1 successful is ${}_rC_s$, and the number of subsets which make event 2 successful is ${}_{n-r}C_{m-s}$. By the formula for independent events, the number of successful events for statement p is the product of the two numbers (see page 163),

$$_{r}C_{s}\ _{n-r}C_{m-s}$$

The probability assigned to statement p is the fraction whose numerator is the number of successful events and whose denominator is the number of possible events.

Consequently, a formula for the probability of statement p has been determined.

Formula: **(Probability of a sample from a known set)** *If U is a set such that $N(U) = n$ and if $U = S \cup \tilde{S}$, $N(S) = r$, and $N(\tilde{S}) = n - r$, then the probability that a sample T such that $N(T) = m$ selected from U (such that each subset of m elements has an equally likely chance of selection) will have exactly s elements in S is*

$$\frac{_{r}C_{s}\ _{n-r}C_{m-s}}{_{n}C_{m}}$$

This set of probabilities is called the *hypergeometric probabilities.*

Although the formula was not available when the example was given, observe that with $n = 10$, $m = 3$, $r = 7$, and $s = 2$, the formula becomes

$$\frac{_{7}C_{2}\ _{3}C_{1}}{_{10}C_{3}} = \frac{21 \cdot 3}{120} = {}^{63}\!/_{120}$$

which agrees with the previous result. With $n = 10$, $m = 3$, $r = 7$, and $s = 3$, the formula becomes

$$\frac{_{7}C_{3}\ _{3}C_{0}}{_{10}C_{3}} = \frac{35 \cdot 1}{120}$$

which also agrees with the previous result.

Exercises

1. A set U of eight elements is such that $N(S) = 6$, $n(\tilde{S}) = 2$. A sample subset T of three elements is to be selected, and each such subset has an equally likely chance of selection. Assign probability to each of the events described by the statements below.
   a. $N(T \cap \tilde{S}) = 1$
   b. $N(T \cap \tilde{S}) = 2$
   c. $N(T \cap \tilde{S}) = 0$
   d. $N(T \cap \tilde{S}) = 3$
2. Work Exercise 1 for $N(U) = 10$, $N(S) = 8$, $N(\tilde{S}) = 2$.
3. Work Exercise 1 for $N(U) = 9$, $N(S) = 6$, $N(\tilde{S}) = 3$.
4. If a committee of two is selected from a set of four and each subset of two has an equally likely chance of selection:

a. What is the probability that Mr. Gordon is on the committee?
b. That Mr. Gordon or Mr. Hartler is on the committee?
c. That Mr. Gordon is not on the committee?
d. That neither Mr. Gordon nor Mr. Hartler is on the committee?

5. Let G be a set with six elements and assume that it is being partitioned into subsets of three elements each. What is the probability that any three specific members of G will end up as one of the smaller sets?

6. Let A be a set with eight elements and B be a set with four elements. How many different ways can you select a set of four elements which contains two from set A and two from set B?

7. Assume that a shipment of 20 electrical components contains 14 good and 6 defective items.
   a. How many different samples of five items can be selected from this population?
   b. The probability that a sample of five would all be good is $({}_{14}C_5\ {}_6C_0)/{}_{20}C_5$. Compute this number.
   c. The expression for the probability that a sample of five would contain four good items and one defective item is $({}_{14}C_4\ {}_6C_1)/{}_{20}C_5$. What is the probability?
   d. Write the expressions for the probability of:
   (1) Three good and two defective
   (2) Two good and three defective
   (3) One good and four defective
   (4) Zero good and five defective
   e. What is the sum of the probabilities in parts b to d?

8. Twenty persons are on a taste panel at a research laboratory. A test is being run to determine whether a spray used on growing beans to control disease causes a detectable difference in taste. The statistician in charge has stated that if there is no difference in taste, about 10 out of the 20 panel members would correctly identify the sprayed beans when asked to choose between two samples, one of which had been sprayed. (The probability of a correct identification, x, is $^{10}\!/_{20}$, or 0.50, and the probability of an incorrect identification, y, is equally likely.) The probability of exactly 17 correct identifications if there is no difference in taste is ${}_{20}C_{17}x^{17}y^3$.
   a. What is the probability of 18 or more correct identifications if there really is no difference?
   b. What would you conclude if 18 of the 20 made the correct identification?

Answers to problems

$$A.\quad \frac{{}_6C_0\ {}_2C_2}{{}_8C_2} = \frac{1 \cdot 1}{28} = \frac{1}{28}$$

$$\frac{{}_6C_1\ {}_2C_1}{{}_8C_2} + \frac{1}{28} = \frac{6 \cdot 2}{28} + \frac{1}{28} = \frac{13}{28}$$

B. 1

$$\frac{{}_{90}C_{10}\ {}_{10}C_0}{{}_{100}C_{10}}$$

$$\frac{{}_{50}C_{10}\ {}_{50}C_0}{{}_{100}C_{10}}$$

C. 1. $_{20}C_5 = \dfrac{20!}{5!15!} = 15{,}504$

2. $0.70 \cdot 20 = 14$ good (and 6 defective)

3. $_{14}C_1\ {}_6C_4 = \dfrac{14!}{1!13!}\ \dfrac{6!}{4!2!} = 14 \cdot 15 = 210$

4. The chances are 210/15,504, slightly more than 1 in 100 if there are really only six defective items in the shipment.

3.7 Binomial experiments

Problems

A. Miss Hardin types by selecting keys of the typewriter to strike. For each symbol to appear in the typing, she selects either the correct key or the incorrect key. Selection of the correct key to strike does not depend on previous selections, and Miss Hardin makes errors with probability $1/100$. If it is known that a particular piece of correspondence has 1,400 symbols, what is the probability that it contains exactly two errors?

B. It is known that a certain missile has a probability of $1/5$ of landing within a 1-mile radius of its target. If four of these missiles are aimed at the same target, what is the probability that exactly two of them will hit within a 1-mile radius of the target? That all four will? That none will?

C. If a machine is set properly, 90 percent of the product which it produces will be acceptable; if it is improperly set, only 40 percent of the product will be acceptable. If three out of a sample of five items tested are found unacceptable, can we form any reasonable conclusions about whether the machine is properly set?

Previously, we have dealt with the probability of the occurrence of a single event. Although many of the examples dealt with simple events, other examples dealt with compound events. But in either case, methods and formulas were developed which were adequate for assigning probabilities to the occurrence of the event.

Now, as a change of pace, a different type of event will be discussed —the type of event that occurs as a result of *repeated trials*. To see how repeated trials may be important, reflect on the fact that, in these days of automation and assembly-line production, both machines and humans are expected to repeat over and over (and for many hours) the same process. These repetitive processes are one reason that repeated-trial events are important in business. The repeated-trial events that will be the concern of this section are characterized by the fact that *previous trials do not influence the outcome of the next trial.*

Although no special attention was called to it at the time, one example of repeated trials, in which the probability of the next event in a sequence of events is not influenced by previous events, has already been studied. Recall Example 3 of Sec. 3.3, which dealt with Mr. Kole calling on three customers in one day. One of the assumptions that was made about this example (without emphasis) was that each time Mr. Kole called on a prospective customer, the probability of a sale was not influenced by events that had already occurred. But it seems likely that Mr. Kole would be discouraged (or encouraged) by what had happened in his prior attempts to sell to prospective customers. This would affect the zeal with which he approached the next prospect. However, it was assumed that the probability of his customer saying *yes* was exactly $\frac{1}{2}$, regardless of the events that had preceded.

Examples of repeated trials in which the probability of the occurrence of each event remains the same are easily obtained in certain games. For example, the outcome of throwing two dice is independent of what has happened in preceding throws of the dice. The streets of Las Vegas, Nevada, are full of empty-pocketed persons who do not believe this statement. To be specific, the probability that 2 sixes will occur in a single toss of a pair of dice is $\frac{1}{36}$ regardless of what has happened in preceding tosses of the dice. For another example that deals with games, if successive draws of a single card from a deck of cards are made (with replacement), then the probability of obtaining a red card on a single draw is $\frac{1}{2}$ regardless of whether or not a red card has been selected on the previous draw. In the discussions in Sec. 3.10 it will be shown how events which have occurred previously do affect not only the outcome of some games of chance but also the outcome of business situations. But such considerations are not of immediate importance since it is the major concern of this section to assign probability to events which occur as a result of repeated trials and in which the probability is unchanged by previous events.

Example 1. As a first example of repeated trials with the probability unchanged by previous events (this example is popular in texts), consider the tossing of a nonbiased coin. By *nonbiased coin* is meant a coin which has equally likely chances of falling heads or tails. If a fair toss is given (a toss high enough and with enough turns so that the toss does not influence the fall of the coin), then the probability $\frac{1}{2}$ is assigned to the event "heads shows" and the probability ½ is assigned to the event "tails shows." The probability on any trial is not influenced by the outcome of previous trials.

But enough about particular examples, for the moment. Consider now a more general situation.

Example 2. Suppose that some sort of repeated trial is made in which $Pr(p)$ for some event E is fixed. Understand that p is a statement that describes event E. The event E is a success, and "not E" is a failure. (So that some concrete example may be thought of in this connection, think of "heads shows" as a success and "tails shows" as a failure.) By agreement (see the axioms about probability spaces on page 157),

$$Pr(p) + Pr(\sim p) = 1$$

For reasons that will be clear presently, the notation

$$x = Pr(p) \qquad y = Pr(\sim p)$$

will be adopted. Tree Diagram 10 illustrates repeated trials with the probabilities x, y.

The probabilities shown at the ends of the branches (and at the intermediate points) are permissible because the events occur independently. Recall that it has already been asserted in the formula for independent events that the probability of a sequence of independent events is obtained by multiplying. This formula has already been stated twice, in one form or another, but it is stated again in a slightly different form that fits this situation.

Formula: ***(Probability of a sequence of two independent events)***
If E_1 and E_2 are independent events described by statements p_1 and p_2 with probabilities $Pr(p_1)$ and $Pr(p_2)$, then the probability of the sequence of events E_1, E_2 is the product of the probabilities. Symbolically, if

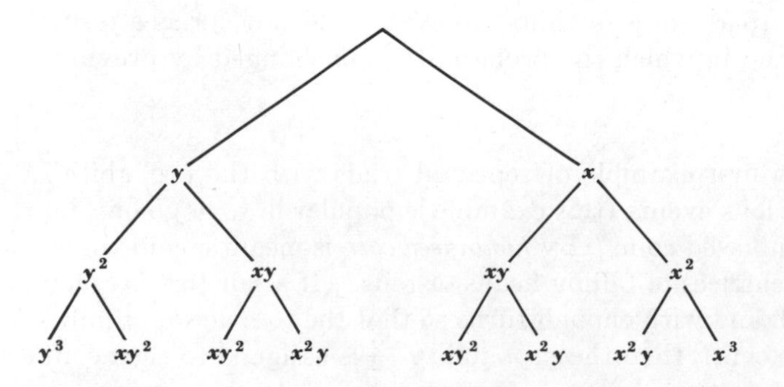

Diagram 10

$Pr(p_1,p_2)$ represents the probability of the sequence of events E_1 followed by E_2, then

$$Pr(p_1,p_2) = Pr(p_1) \cdot Pr(p_2)$$

The tree diagram drawn previously (see page 142) to illustrate Mr. Kole's visit to his customers is a special case of this formula, with $x = y = \frac{1}{2}$.

The symbols at the ends of the branches of the tree in Diagram 10 represent probabilities but should be familiar from previous considerations in this book. They are expressions that were encountered in the study of the binomial formula in Sec. 2.4. Observe that one branch terminates in x^3 and another in y^3 and that three branches terminate in x^2y and three in xy^2. The sum of these probabilities is given by

$$x^3 + 3x^2y + 3xy^2 + y^3$$

which is the binomial formula for $(x + y)^3$.

Diagram 10 relates the probabilities after three repeated trials to the binomial formula $(x + y)^n$ for $n = 3$. If Diagram 10 is extended by representing the probabilities after four repeated trials, the branches will then correspond to the terms in the expansion of $(x + y)^4$. More generally, after n repeated trials, the probabilities correspond to the terms of $(x + y)^n$. For the relation between a particular term of $(x + y)^n$ and the probability of a number of outcomes of a repeated-trial process, see the formula below.

***Formula:* (*The probability of r occurrences of event E in n independent repeated trials*)** *If, in a repeated-trial process, an event occurs with probability y [and $Pr(\sim E) = x = 1 - y$] independent of previous trials, then the probability of exactly r occurrences of E in n trials is*

$$_nC_r x^{n-r} y^r$$

So that there are some rules for determining when the formula should be applied, the characteristics of a binomial experiment are enumerated.

A binomial experiment is one such that:

1. The experiment consists of n identical trials.
2. Each trial results in a successful outcome or in a failure.
3. The probability of success on any trial is y (and remains the same). The probability of failure is $x = 1 - y$.
4. The trials are independent.
5. The number $_nC_r x^{n-r} y^r$ is the probability of exactly r successful occurrences.

Some discussion of repeated (independent) trials as they might occur in practice have been given, and a formula for computing the probability has been derived without reference to specific examples. This seemed to be a proper way to proceed since all the mathematical information necessary to make the formula "natural" was already at hand. But now that the repeated-trial process has been linked to the binomial formula, by way of a formula for probability, these tools can be put to work to compute the answers to questions about repeated trials. Read the next two examples with a view toward understanding how this latest formula can be used.

Example 3. Salesman Parsons is much more energetic than Mr. Kole, but not so good a salesman. While Mr. Kole is calling on three persons, Mr. Parsons calls upon nine. However, Mr. Parsons calls upon his customers with probability of only $\frac{1}{3}$ of receiving an order. *Determine the probability of Mr. Parsons' selling precisely three out of nine persons upon whom he calls.*

To arrive at the answer, use is made of the formula for exactly r (3) occurrences of event E (a sale) in n (9) repeated trials. A computation yields

$$ {}_9C_3(\tfrac{2}{3})^{9-3}(\tfrac{1}{3})^3 = 84 \cdot \tfrac{64}{729} \cdot \tfrac{1}{27} = \frac{5,376}{19,683} $$
$$ = \frac{1,792}{6,561} \cong \tfrac{27}{100} $$

The symbol $\cong$ means that $\frac{27}{100}$ is approximately the same fraction as 1,792/6,561, and the symbol $\cong$ is used because $\frac{27}{100}$ is more easily understood than 1,792/6,561. It can be said that Mr. Parsons has about 27 chances out of 100 of selling exactly three customers out of nine calls.

Example 4. The facts of the previous example will be used to answer a different, but related, question. *What is the probability that Mr. Parsons will sell at least six of the nine customers he calls upon?* It will be instructive to hazard a guess about this probability before the computation is made. A restatement of the question is: Determine the probability that Mr. Parsons will make six sales, seven sales, eight sales, or nine sales. Since these events are mutually exclusive (for example, he cannot sell six *and* seven of his customers), the probability for each event is computed and the probabilities of the four events are added to determine the final result. Since this type of problem frequently occurs in considerations about probability and since computations of this type are routine (and tedious), tables for such probabilities have been computed. The exact result is

$$
\begin{aligned}
&{}_9C_6x^3y^6 + {}_9C_7x^2y^7 + {}_9C_8xy^8 + {}_9C_9x^0y^9 \\
&\quad = \frac{9 \cdot 8 \cdot 7}{1 \cdot 2 \cdot 3}(2/3)^3(1/3)^6 + \frac{9 \cdot 8}{1 \cdot 2}(2/3)^2(1/3)^7 + 9(2/3)^1(1/3)^8 + (1/3)^9 \\
&\quad = \frac{224}{6{,}561} + \frac{16}{2{,}187} + \frac{2}{2{,}187} + \frac{1}{19{,}683} \\
&\quad = \frac{672 + 144 + 18 + 1}{19{,}683} = \frac{835}{19{,}683} \cong 0.04
\end{aligned}
$$

If once again the answer is approximated, it is seen that Mr. Parsons has about 1 chance out of 25 of selling as many as six out of nine of his customers.

Exercises

1. Compute ${}_nC_rx^{n-r}y^r$ for each of the following values:

   a. $n = 3, r = 2, x = 1/5$
   b. $n = 4, r = 3, x = 1/4$
   c. $n = 5, r = 3, x = 1/4$
   d. $n = 6, r = 3, x = 1/10$
   e. $n = 7, r = 2, x = 1/2$
   f. $n = 8, r = 6, x = 1/2$
   g. $n = 9, r = 4, x = 7/20$

2. Apollo sheets are packaged with one dozen sheets to a package. If defects caused by poor manufacturing processes occur independent of previous events (a machine could go bad and stay bad, you know) and if defective sheets occur with probability 0.03:

   a. What is the probability that a package will have one defective sheet?
   b. Two defective sheets?
   c. More than half of the sheets defective?

3. Work Exercise 2 again with probability 0.1.

4. See Example 3 in this section and find the probability that Mr. Parsons will sell:

   a. Exactly four customers
   b. Fewer than three customers
   c. Two or seven customers

5. If it is known that either event A or event B will occur and that event A is three times as likely to occur as event B, then what is the probability that in three trials, event A will occur twice?

6. If event A has a probability of $1/10$ of occurring in a single trial, is it possible for event A to occur two times in two trials? If so, assign a probability to this occurrence.

7. Many firms use taste panels to ensure a uniform taste for successive batches of their product. Assume that such a panel is asked to attempt to identify the new batch after being given a reference taste of the old batch and two unlabeled servings, one new and one old. If 15 or more of the panel of 20 make correct identifications, the company reworks the new batch to try to improve its flavor.
   a. What is the probability of 15 correct identifications if the new batch tastes exactly like the old (which makes the probability of a correct identification equal to 0.50)?
   b. Write the expression which would give the probability of exactly 16 correct identifications.
   c. How would you compute the probability of 15 or more correct identifications out of 20?
   d. The probability that all 20 of the panel members would make correct identifications if there is no difference is 0.000 (correct to three decimals). Does this mean that such an outcome is impossible?

8. In a hypothetical automobile assembly line, 3 of every 100 master brake cylinders have some defect that must be remedied.
   a. What is the probability that out of a sample of 10 cars, exactly 1 car will have a defective master cylinder? (There is only one on a car.)
   b. What is the probability that 2 cars out of 10 will have defective master cylinders?
   c. What is the probability that the first and third cars of the 10 will have defective master cylinders? Why is this not the same as your answer to part *b*?

9. In a box of 100 rounds of ammunition, it is known that there are exactly 10 defective rounds.
   a. If an M-1 clip is loaded from this box (eight rounds), what is the probability that over one-half of the rounds are defective?
   b. What is the probability that all the rounds are functional (contain no defects)?

10. A production process is known to produce 10 percent defective parts in its operation. If a random sample of six is taken from the set of finished parts, what is the most probable number of defects that will be found in the sample? (HINT: Compare the probabilities of 0, 1, 2, . . . , 6 defects.)

11. A device used in a space vehicle functions properly 99 percent of the time. Because of the difficulty of producing a more reliable device, two devices of the same type are hooked in so that if one does not work, the other will be activated. What is the probability that the function of this device will be successfully completed?

Answers to problems

A. ${}_{1,400}C_2(1/100)^2(99/100)^{1,398}$

B. ${}_4C_2(1/5)^2(4/5)^2 = 16/625$

${}_4C_4(1/5)^4(4/5)^0 = 1/625$

${}_4C_0(1/5)^0(4/5)^4 = 256/625$

C. If the machine is properly set, the probability of three unacceptable items in a sample of five is

$$_5C_3(0.9)^2(0.1)^3 = 10(0.81)(0.001) = 0.0081$$

If the machine is improperly set, the probability of the specified outcome is

$$_5C_3(0.4)^2(0.6)^3 = 10(0.16)(0.216) = 0.3456$$

Since such an outcome is much more likely if the machine is improperly set, the reasonable conclusion is that it needs adjustment.

3.8 *Multinomial experiments*

Problems

A. The articles produced by an assembly line are perfect, have minor defects, or are unsalable, with probabilities $6/10$, $3/10$, and $1/10$, respectively. If the condition of an article is independent of the condition of the ones produced previously, what is the probability that for 10 consecutive articles, 8 will be perfect, 1 will have a minor defect, and 1 will be unsalable?

B. The probability that an assembly-line product will be free of defects is $3/4$. The probability that the inspector will check any particular product is $1/3$. For five consecutive products, what is the probability that two defective items will be checked, one defect-free item will not be checked, zero defect-free products will be checked, and two defective products will not be checked?

Repeated trials with two outcomes (independent of previous events) have an obvious generalization—repeated trials with three or more outcomes. (In the coin-tossing example, such possible outcomes as the coin standing on edge, falling in a crack, or remaining in the air were ignored because the probabilities of these events are so small that they generally need not enter the considerations.) As we shall see in this section, repeated trials with more than two outcomes are treated in a fashion quite similar to that of the previous section. Since the generalization and formula derivation so closely follow the pattern established in the previous section, examples will be postponed until after the formula is derived. This discussion depends on the ideas advanced in Sec. 2.5.

First, the tree diagram illustrating binomial experiments can be generalized to three or more outcomes at each branch. For reasons of simplicity, the diagram for trinomial (three) experiments is the one given (Diagram 11). The probabilities assigned to the three events are x_1, x_2, and x_3 to agree with the notation of Sec. 2.5. In fact, the formula for trinomials should be reviewed before Diagram 11 is studied. An observation, repeated here, is that symbols such as x_1^3, $x_1^2x_2$, and $x_1^2x_3$ correctly

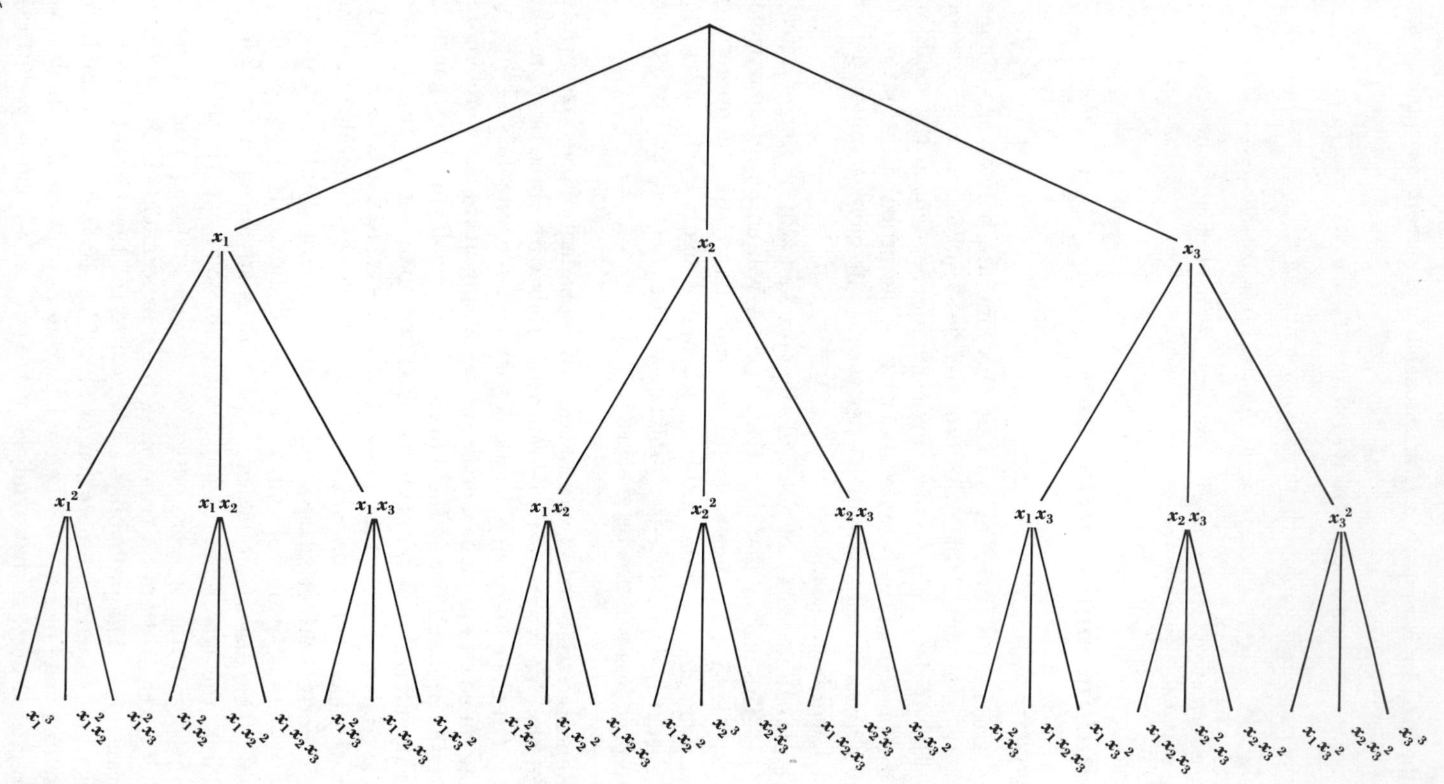

x_1
x_2
x_3
x_1^2
x_1x_2
x_1x_3
x_1x_2
x_2^2
x_2x_3
x_1x_3
x_2x_3
x_3^2
x_1^3
$x_1^2x_2$
$x_1^2x_3$
$x_1^2x_2$
$x_1x_2^2$
$x_1x_2x_3$
$x_1^2x_3$
$x_1x_2x_3$
$x_1x_3^2$
$x_1^2x_2$
$x_1x_2^2$
$x_1x_2x_3$
$x_1x_2^2$
x_2^3
$x_2^2x_3$
$x_1x_2x_3$
$x_2^2x_3$
$x_2x_3^2$
$x_1^2x_3$
$x_1x_2x_3$
$x_1x_3^2$
$x_1x_2x_3$
$x_2^2x_3$
$x_2x_3^2$
$x_1x_3^2$
$x_2x_3^2$
x_3^3

Diagram 11

appear at the ends of branches since it has been assumed that the repeated-trial process involves independent events and, by the formula for sequences of independent events, probabilities are multiplied.

A count of the number of times that the terms x_1^3, x_2^3, x_3^3, x_1^2, x_2, x_3, $x_1x_2x_3$, etc., occur will serve to show that (at least for three outcomes and three repeated trials) the number of times $x_1^{n_1}x_2^{n_2}x_3^{n_3}$ occurs is

$$_nC_{n_1,n_2,n_3}$$

At any rate, this is the case, and a formula which generalizes the repeated-trial formula stated on page 171 follows. Although no proof is given, it is hoped that it is apparent that this formula is an obvious generalization of the formula for two outcomes.

Formula: [***The probability of*** n_i ***occurrences of*** E_i $(i = 1, 2, \ldots, r)$ ***in*** n ***independent trials***] *In a repeated-trial process, independent events* E_1, E_2, . . . , E_r *occur with probabilities* x_1, x_2, . . . , x_r. *If statement* P_i *is that event* E_i *will occur, then*

$$Pr(p_1) + Pr(p_2) + \cdots + Pr(p_r) = 1$$

The probability of n_1 *occurrences of* E_1, n_2 *occurrences of* E_2, *etc., in* n *repeated trials* $(n_1 + n_2 + \cdots + n_r = n)$ *is*

$$_nC_{n_1,n_2,\ldots,n_r}x_1^{n_1}x_2^{n_2} \cdots x_r^{n_r}$$

If r is 2, then this formula reduces to the formula for binomial experiments of the previous section (and the multinomial formula becomes the binomial formula). One of the main applications of the multinomial formula (as well as of the binomial formula) is to sampling with replacement, where the elements in the set are classified into two (or more than two) categories.

A set of characteristics of a binomial experiment was formulated (see page 171), and it is possible to do this for a multinomial experiment. The characteristics are:

1. The experiment consists of n independent trials.
2. Each trial results in one of r events, E_1, E_2, . . . , E_r.
3. The probability of success on any trial for E_1, E_2, . . . , E_r is x_1, x_2, . . . , x_r, respectively (and remains that). The probabilities add to 1.
4. The trials are independent.
5. The number $_nC_{n_1,n_2,\ldots,n_r}x_1^{n_1}x_2^{n_2} \cdots x_r^{n_r}$ is the probability of exactly n_1, n_2, . . . , n_r successful occurrences of E_1, E_2, . . . , E_r, respectively $(n_1 + n_2 + \cdots + n_r = n)$.

Just as the formula for multinomial experiments is a generalization of the formula for binomial experiments, so examples for multinomial experiments are easily obtained as generalizations of examples for binomial experiments. Continue reading to see the meaning of this statement.

Example 1. Mr. Parsons knows that his clients buy sheets, blankets, and comforters with probabilities $\frac{1}{2}$, $\frac{1}{4}$, and $\frac{1}{4}$, respectively:

$$\tfrac{1}{2} + \tfrac{1}{4} + \tfrac{1}{4} = 1$$

If he calls on 12 customers, each of whom will buy exactly one of the three products, what is the probability that 3 will buy sheets, 8 will buy blankets, and 1 will buy comforters?

The formula for multinomial experiments can be used, with $x_1 = \frac{1}{2}$, $x_2 = \frac{1}{4}$, $x_3 = \frac{1}{4}$; $n_1 = 3$, $n_2 = 8$, $n_3 = 1$, $n = 12$. The formula yields

$$\begin{aligned} {}_{12}C_{3,8,1}(\tfrac{1}{2})^3(\tfrac{1}{4})^8(\tfrac{1}{4})^1 \\ &= \frac{1 \cdot 2 \cdot 3 \cdot 4 \cdot 5 \cdot 6 \cdot 7 \cdot 8 \cdot 9 \cdot 10 \cdot 11 \cdot 12}{1 \cdot 2 \cdot 3 \cdot 4 \cdot 5 \cdot 6 \cdot 7 \cdot 8 \cdot 1 \cdot 2 \cdot 3 \cdot 1} (\tfrac{1}{2})^3(\tfrac{1}{4})^8(\tfrac{1}{4})^1 \\ &= \frac{495}{532{,}288} \cong \frac{1}{1{,}000} \end{aligned}$$

The exercises provide more questions of the same nature.

We conclude the section with one example of an application of the multinomial formula that is of a different nature. Suppose that there are two sequences of binomial trials occurring at the same time and that they are independent. Suppose, furthermore, that the probabilities of success and failure of the second binomial experiment are x_2 and y_2. Tree Diagram 12 illustrates the probabilities for the first two trials of each sequence of events.

If these two sequences of trials are independent, they can be considered as one compound experiment with four possible outcomes in each trial. There may be success in each of the two trials (S,S), there may be success in the first and failure in the second (S,F), there may be failure in the first and success in the second (F,S), or there may be failure in each trial (F,F). The notation adopted uses S for success, F for failure, the first of the ordered pairs for the first experiment, and the second of the ordered pairs for the second experiment. If the two sequences of trials

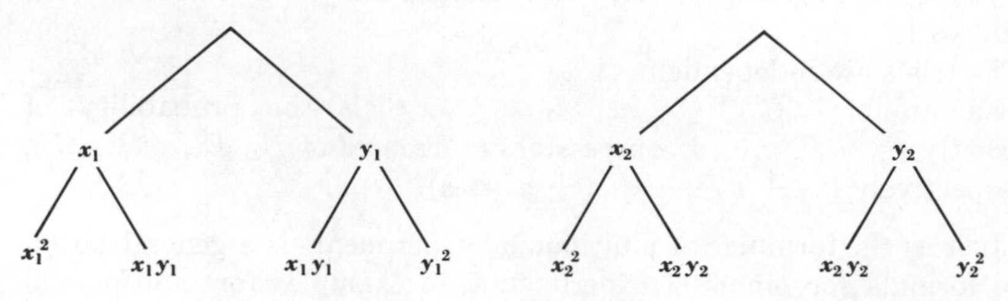

Diagram 12

are independent, then the principle of multiplication of probabilities can be used to assign probabilities to the four cases, and they are x_1x_2, x_1y_2, y_1x_2, and y_1y_2. Now consider a sequence of n trials, and let n_1, n_2, n_3, n_4 be such that $n_1 + n_2 + n_3 + n_4 = n$. The probability that in n trials (S,S) will happen n_1 times, (S,F) n_2 times, (F,S) n_3 times, and (F,F) n_4 times is

$${}_nC_{n_1,n_2,n_3,n_4}(x_1x_2)^{n_1}(x_1y_2)^{n_2}(y_1x_2)^{n_3}(y_1y_2)^{n_4}$$

Written in another fashion, we have

$${}_nC_{n_1,n_2,n_3,n_4}x_1{}^{n_1+n_2}y_1{}^{n_3+n_4}x_2{}^{n_1+n_3}y_2{}^{n_2+n_4}$$

Although the discussion of the previous paragraphs may seem somewhat involved, an example of sampling which uses the facts that have been stated above will illustrate the relative simplicity and value of this special application of the multinomial formula.

Suppose that an assembly-line production produces n items. The item as it comes off the production line is good or defective with probabilities x_1 and y_1. As an item comes off the assembly line, it may be inspected or not inspected with probabilities x_2 and y_2. Since products are inspected without knowledge of their quality, the trials are independent and the formula can be applied. To illustrate the kind of questions that can be answered, a numerical example is given.

Example 2. At a certain plant, the probability that a product coming off the assembly line will have no defects is $^{88}\!/_{100}$ (and the probability that it will have defects is $^{12}\!/_{100}$). The probability that a particular product will be inspected is $10/1{,}000 = {}^{1}\!/_{100}$ (and the probability of its not being inspected is $^{99}\!/_{100}$). If five products come off the assembly line, what is the probability that three good products will *not* be inspected, one good product will be inspected, no defective products will not be inspected, and one defective product will be inspected?

In this example, $x_1 = {}^{88}\!/_{100}$, $y_1 = {}^{12}\!/_{100}$, $x_2 = {}^{1}\!/_{100}$, $y_2 = {}^{99}\!/_{100}$, $n_1 = 1$, $n_2 = 3$, $n_3 = 1$, and $n_4 = 0$. Using the formula above, the probability is

$${}_5C_{1,3,1,0}({}^{88}\!/_{100})^{1+3}({}^{12}\!/_{100})^{1+0}({}^{1}\!/_{100})^{1+1}({}^{99}\!/_{100})^{3+0}$$

If we perform the computation indicated in the formula above, we obtain

$$\frac{5!}{1!3!1!0!}\left(\frac{88}{100}\right)^4\left(\frac{12}{100}\right)^1\left(\frac{1}{100}\right)^2\left(\frac{99}{100}\right)^3$$

$$= \frac{5 \cdot 4 \cdot 88 \cdot 88 \cdot 88 \cdot 88 \cdot 12 \cdot 99 \cdot 99 \cdot 99}{100^{10}}$$

$$\cong \frac{240 \cdot 5{,}761{,}536}{100^6}$$

$$\cong \frac{0.13}{100} \cong 0.0013$$

The question answered is typical of a large number of similar questions that can be answered by the same technique. Other examples of these questions are provided in the Exercises.

Exercises

1. Products coming off the assembly line of a certain company are marked good, poor, or bad with probabilities $\frac{92}{100}$, $\frac{6}{100}$, $\frac{2}{100}$, respectively. If 10 products come off the line, what is the probability that there will be 8 good, 1 poor, and 1 bad ($n = 10$, $n_1 = 8$, $n_2 = 1$, $n_3 = 1$)?
2. Solve Exercise 1 for:
   a. $n = 12$, $n_1 = 11$, $n_2 = 0$, $n_3 = 1$
   b. $n = 5$, $n_1 = 3$, $n_2 = 1$, $n_3 = 1$
   c. $n = 5$, $n_1 = 5$, $n_2 = 0$, $n_3 = 0$
3. How long must a sequence of random digits be in order that the probability of the digit 5 appearing is at least $\frac{9}{10}$?
4. What is the probability that the birthdays of five persons will fall in two months?
5. Irons are produced by assembly-line methods at an electric company with a $\frac{9}{10}$ probability of no defects. One iron out of two hundred is inspected. What is the probability of five consecutive irons satisfying:
   a. Three good, not inspected; one defective, not inspected; one defective, inspected?
   b. Four good, not inspected; one bad, inspected?
6. The probability of a tire made by the Retread Manufacturing Company wearing out after less than 5,000 miles is $\frac{2}{10}$, after 5,000 to 10,000 miles is $\frac{6}{10}$, after 10,000 to 15,000 miles is $\frac{1}{10}$, and after more than 15,000 miles is $\frac{1}{10}$. Suppose that you put four new Retread tires on your car.
   a. What is the probability that two will wear out after 5,000 to 10,000 miles, one will last for more than 15,000 miles, and one will last less than 5,000 miles?
   b. What is the probability that you will be so extremely lucky that all four tires will last for more than 15,000 miles?
7. The probability of tires produced by Retread Manufacturing Company having a defect is $\frac{3}{10}$, and the probability that any particular tire will be inspected is $\frac{1}{5}$. Taking a sample of 80 tires, what is the probability that 10 defective ones will be inspected, 30 good ones will not be inspected, 25 good ones will be inspected, and 15 defective ones will not be inspected?
8. A tile manufacturer makes ceramic wall tile which comes in three grades (standard, commercial, and economy), depending upon the number and kind of flaws in each piece. Experience has shown that in a sample of 100 tiles, we should expect 50 standard grade, 30 commercial grade, and 20 economy grade. (The probabilities are $S = 0.5$, $C = 0.3$, and $E = 0.2$.)

a. What is the number of ways in which you could get one standard, zero commercial, and one economy if the sample size is two?
b. What is the probability of getting zero standard, one commercial, and one economy?
c. In a sample of 20, what is the probability of 12 standard, 5 commercial, and 3 economy?

Answers to problems

A. ${}_{10}C_{8,1,1}(6/10)^8(3/10)^1(1/10)^1 = 226{,}848{,}160/10^{10} \cong 0.022$

B. ${}_{5}C_{0,1,2,2}(3/4)^{0+1}(1/4)^{2+2}(1/3)^{0+2}(2/3)^{1+2} \cong 30(3/4)^1(1/4)^4(1/3)^2(2/3)^3 = 5/1{,}982$

3.9 Odds and mathematical expectation

Problems

A. Your friend says to you, "I have three coins, and I will throw them into the air and let them fall. Since there are only four possible ways they can land (three heads, three tails, two heads and one tail, or two tails and one head), I will pay you $4 each time the result is three heads if you will pay me a dollar each time it is anything else." Is your friend a friend indeed or a friend in need? (That is, if he is a friend in need, will this arrangement help alleviate his need?)

B. Mr. Archer has a business that is having financial problems. He wants to move to a new location and buy some new equipment for his business. He has interested Mr. Ball in financing his new venture. They have both studied the new business and agree that its probability of success is $7/10$ (and of failure is $3/10$). Mr. Archer wants Mr. Ball to put $50,000 in the proposed business. Mr. Ball will lose his money if the business fails. If the business succeeds, how much should Mr. Archer pay (in addition to the $50,000) in order for the proposition to be fair to both parties?

C. There are 10 TV picture tubes in a lot. It is known that four of them are defective and worthless. The others are worth $30 each. How much could you afford to pay for the right to take three of the tubes selected at random?

D. Assume that you are considering a purchase of a piece of land which costs $40,000. If a highway being planned passes through the land, the land will be worth $100,000; but if it does not, the land will be worth only $20,000. The probability that the highway will pass through the land is estimated to be 0.30. Evaluate the investment in light of the probabilities and values given.

Throughout this chapter, the impulse to discuss probability in terms of games of chance and gambling has been largely resisted. The examples, exercises, and applications have been slanted toward the world of business, as is proper in a book addressed primarily to people who plan business as a career. However, as many business executives will confirm, business itself is a gamble, and it would be unfair to the reader if the chapter on probability did not include at least some discussion of odds. To see how odds and mathematical expectation might enter into conversations about business, consider statements such as the following:

1. The *odds* are against me on that arrangement.
2. There is little *expectation* that the merger will work.
3. The stock was purchased because it is *expected* to rise.
4. He is an *odds-on* favorite to get the bid.

Neglect of the topics of odds and expectation would also be unfair from the historical point of view since much of the early theory of probability grew out of problems arising from gambling, and these terms normally carry a gambling connotation. Although the subject is still in its infancy, relatively speaking, much mathematical research is being done with regard to those special games of chance: the stock market and economic theory. In this section, we shall deal with two topics which have to do with gambling but which are also mathematical in nature, odds and mathematical expectations. The topics will be treated in that order.

Example 1. One of the games of chance that is played in many casinos where gambling is legal (and perhaps in other places) is played with two dice. Since the reader may not be familiar with this game, detail will be given. The dice have the numbers 1 to 6 on the six faces of each die. The two dice are thrown simultaneously. The sum of the numbers that appear on the two faces which are turned up varies from 2 to 12. This game will be discussed completely a little later in the section, but right now consider a particular problem. Probability will be assigned to the statement

p: The sum of the faces of the dice will total 9 before they total 7.

As can be seen, the game has already been limited since any toss of the dice which results in the sum of the faces totaling anything except 9 and 7 is being disregarded. Of course, it is conceivable that the dice could be thrown forever without either the total 9 or the total 7 occurring. However, it will be assumed that this is a *finite* game and that after a finite amount of time, either a 9 or a 7 will occur. If the numbers on the faces of the dice that are up are represented by ordered pairs, then the events which result in a total of 9 on the faces of the dice are

$$(3,6),\ (4,5),\ (5,4),\ (6,3)$$

Again using ordered pairs (with the first component representing the face of the first die and the second component representing the face of the second die), the events which result in a total of 7 on the face of the dice are

$$(1,6),\ (2,5),\ (3,4),\ (4,3),\ (5,2),\ (6,1)$$

The universal set of logical possibilities that are relevant to statement p has a total of 10 elements, 4 of which are successful for statement p. These totals are obtained from counting the events just mentioned. Consequently, the probability that is assigned to statement p is $4/10 = 2/5$, and it is denoted $Pr(p) = 2/5$.

Example 2. The procedure of the argument just given should be commonplace by now since similar procedures have been used repeatedly. However, the concern here is with the odds that should be given if gambling about the truth of p is to occur (and this is one of the statements upon which bets are made in casinos). The example just given will be generalized, and a formula derived for the odds.

So that a formula can be discovered, assume that the amount w will be won if the total of 9 on the face of the dice occurs before the total of 7. Additionally, assume that the amount l will be lost if the total of 7 occurs before 9. If the bet is to be fair (the odds are even), then the condition

$$Pr(p) \cdot w = Pr(\sim p) \cdot l$$

must be satisfied. When $2/5$ replaces $Pr(p)$ and $3/5$ replaces $Pr(\sim p)$ $[Pr(p) + Pr(\sim p) = 1]$, then the condition becomes

$$2/5 w = 3/5 l$$

So that the formula may take on a form more easily assessed, it is proper to multiply both sides of this condition by 5 so that

$$2w = 3l$$

Finally, division by $3w$ yields

$$2/3 = \frac{l}{w}$$

The ratio $2/3$ is called the *odds* of this particular bet. To have a fair bet that a 9 will occur before a 7, a wager of $2 against $3 should be made. To reverse this situation, a fair bet that 9 will not occur before a 7 is a wager of $3 against $2.

Example 3. But the odds computed deal, quite unnecessarily, with a proposition about a game. Odds are of interest under circumstances that have nothing to do with dice games, so the procedure just followed will be used as a model to compute a formula for the odds in a more general set of conditions.

To generalize, let t be a statement, with

$$Pr(t) = x \qquad \text{and} \qquad Pr(\sim t) = 1 - x$$

Fair bets result if the condition

$$xw = (1 - x)l$$

holds, where w and l represent the amount won and lost, respectively. Division by the number $(1 - x)w$ yields the condition

$$\frac{x}{1 - x} = \frac{l}{w}$$

The ratio

$$\frac{x}{1 - x}$$

is the odds of the proposition given by statement t. This result is of sufficient importance to be stated in a formula.

Formula: (**The odds for a fair bet**) *If t is a statement with probability x, if $\sim t$ has probability $1 - x$, if w represents the amount won if the event described by t occurs, and if l represents the amount lost if the event described by $\sim t$ occurs, then a fair bet results if the condition*

$$xw = (1 - x)l$$

is satisfied. The odds of the fair bet is the ratio

$$\frac{x}{1 - x}$$

The formula $x/(1 - x)$ will be applied to two other situations.

Example 4. In a dice game, one of the events that may occur upon the single toss of two dice is that each face of the two dice shows the number 2. This event is called "making 4 the hard way." The expression refers to the fact that a total of 4 can be achieved by any one of the three ways (stated in the usual notation) (3,1), (1,3), (2,2). To determine what odds should be given for a fair bet on this event, observe first that the number of events that may occur on a single toss of two dice is 36. Only one of these 36 events is considered successful for the proposition that (2,2) will occur. The probability assigned to the event (2,2) on the basis of these considerations is $1/36$. Use of the formula for odds on fair propositions yields

$$\frac{x}{1 - x} = \frac{1/36}{1 - 1/36} = \frac{1/36}{35/36} = 1/35$$

The interpretation of these odds is that the person wagering on the event (2,2) should wager \$1 and should receive \$35 if the event occurs (that is, if this is to be a fair bet). On all the dice tables that we have seen, the "payoff" is 30 for 1. There is even a gimmick to this payoff. The odds are not 30 *to* 1 because in the payoff, the dollar that was bet originally is kept, so that the actual payoff is 29 to 1.

Example 5. For the second application of the formula for odds, consider a new game. A roulette wheel has 37 small holes around the outside of a wheel, and the holes are numbered from 0 to 36. The holes numbered from 1 to 36 are colored alternately red or black, and the hole numbered 0 is not colored. A ball rolls around the outside of the wheel and drops into one of the holes. A standard bet is on the statement "The ball will drop into a hole colored red." The number of events is 37, and the number of successful events is 18. The probability of the ball dropping into a hole colored red is $^{18}/_{37}$, and the odds computed by the formula for a fair bet is

$$\frac{^{18}/_{37}}{1 - {^{18}/_{37}}} = \frac{^{18}/_{37}}{^{19}/_{37}} = {^{18}/_{19}}$$

For each \$18 bet on the proposition stated, the bettor should receive back \$19 if the event occurs. The payoff at most casinos is \$18. In order to give themselves a little more the better of it, some casinos include both a 0 and a 00 on the wheel (neither of which is colored).

The discussions above have centered around computing the odds for a single event to occur in a game of chance. However, in many games, the wagers are usually on compound events, and computing odds becomes a more complicated affair. In such instances, an amount called the *mathematical expectation* of the game is determined.

Formula: **(*Mathematical expectation*)** *If* $E_1, E_2, \ldots, E_r$ *are events with assigned probabilities of* $Pr(E_1), Pr(E_2), \ldots, Pr(E_r)$, *respectively, and if there are associated amounts* $a_1, a_2, \ldots, a_r$ *then the expectation is the sum of the products of the amounts and the probabilities. Symbolically, we write*

$$E = a_1Pr(E_1) + a_2Pr(E_2) + \cdots + a_rPr(E_r)$$

This formula will be illustrated with a complete description of the dice game as it is ordinarily played. While no definition of the *amount of an event* will be attempted, the example that follows will make clear what is meant by the expression.

Example 6. The rules (and procedure) of the game are given first. One person has a set of two dice; he is called the *shooter*. The shooter tosses the two dice, each of which has six faces with the numbers 1 to 6 on the faces. The sum of the numbers on the faces which are up will be one of the numbers 2 to 12. If, on the first toss of the dice, a sum of 2, 3, or 12 appears, then the shooter loses. For convenience, assume that wagers are for one unit. So the numbers a_2, a_3, and a_{12} in the formula for mathematical expectation are each -1. We shall compute $Pr(E_2)$, $Pr(E_3)$, and $Pr(E_{12})$ a little later. If on the first toss of the dice, a 7 or 11 occurs, then the shooter wins $+1$; and in the formula for expectation, a_7 and a_{11} are $+1$. The computation of $Pr(E_7)$ and $Pr(E_{11})$ is delayed, but it is made later. If the total on the face of the dice is 4, 5, 6, 8, 9, or 10, then the dice are returned to the shooter and the game goes on. This number is called the shooter's *point*. All the numbers except the shooter's point and 7 are now ignored. There are now only two events that cause a payoff. The shooter continues to toss the dice and receives one unit if his point occurs before 7 occurs. He loses one unit if 7 occurs before his point. The probabilities of each of the 11 numbers possible on the first toss of the dice are indicated in Diagram 13.

The denominator of each fraction is 36 since this is the total number of possible events. The numerator in each case is the number of ways that the total can be obtained. One illustration should suffice. The probability assigned to the sum 8 is $5/36$ because there are five events that result in the sum of the faces being 8, namely, (6,2), (5,3), (4,4), (3,5), and (2,6). If the total 2, 3, or 12 occurs, then the shooter loses; so in the formula, $E_2 = 2$, $E_3 = 3$, $E_{12} = 12$, $Pr(E_2) = 1/36$, $Pr(E_3) = 2/36$, $Pr(E_{12}) = 1/36$, $a_2 = a_3 = a_{12} = -1$. Similarly, $E_7 = 7$, $E_{11} = 11$, $Pr(E_7) = 6/36$, $Pr(E_{11}) = 2/36$, $a_7 = +1$, $a_{11} = +1$. These numbers will be substituted into the formula later.

Suppose the total 4 shows on the first toss of the dice. Now two events can occur: $E_{4,S}$ (4 occurs again before 7) or $E_{4,F}$ (7 occurs before 4 occurs again). The probabilities are $Pr(E_{4,S}) = 3/36 \cdot 3/9$, $Pr(E_{4,F}) = 3/36 \cdot 6/9$, and $a_{4,S} = +1$, $a_{4,F} = -1$. The probabilities are multiplied since these are a sequence of two independent events. The probability $3/9$ is used in the computation of $Pr(E_{4,S})$ because the first three of the nine events (1,3), (2,2), (3,1), (1,6), (2,5), (3,4), (4,3), (5,2), (6,1) are successful. This also explains why the probability $6/9$ is used in the computation of $Pr(E_{4,F})$. In a similar fashion, the probabilities for other

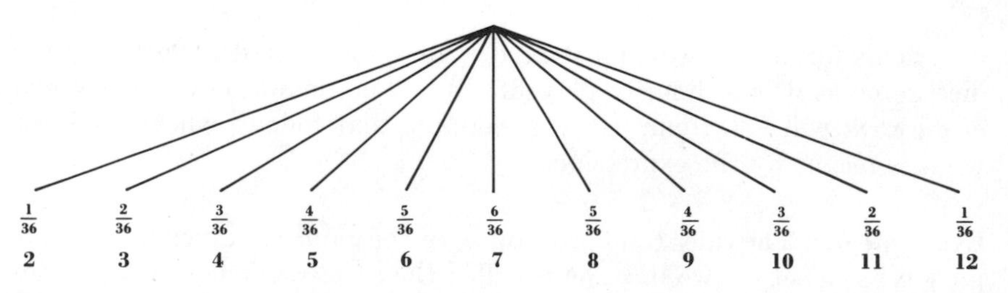

Diagram 13

events are determined, and the mathematical expectation is

$$
\begin{aligned}
E = {} & a_2 Pr(E_2) + a_3 Pr(E_3) + a_{4,S} Pr(E_{4,S}) + a_{4,F} Pr(E_{4,F}) + a_{5,S} Pr(E_{5,S}) \\
& + a_{5,F} Pr(E_{5,F}) + a_{6,S} Pr(E_{6,S}) + a_{6,F} Pr(E_{6,F}) + a_7 Pr(E_7) \\
& + a_{8,S} Pr(E_{8,S}) \\
& + a_{8,F} Pr(E_{8,F}) + a_{9,S} Pr(E_{9,S}) + a_{9,F} Pr(E_{9,F}) + a_{10,S} Pr(E_{10,S}) \\
& + a_{10,F} Pr(E_{10,F}) \\
& + a_{11} Pr(E_{11}) + a_{12} Pr(E_{12}) \\
= {} & (-1) \cdot 1/36 + (-1) \cdot 2/36 + (+1) \cdot 3/36 \cdot 3/9 \\
& + (-1) \cdot 3/36 \cdot 6/9 + (+1) \cdot 4/36 \cdot 4/10 + (-1) \cdot 4/36 \cdot 6/10 \\
& + (+1) \cdot 5/36 \cdot 5/11 + (-1) \cdot 5/36 \cdot 6/11 + (+1) \cdot 6/36 \\
& + (+1) \cdot 5/36 \cdot 5/11 + (-1) \cdot 5/36 \cdot 6/11 \\
& + (+1) \cdot 4/36 \cdot 4/10 + (-1) \cdot 4/36 \cdot 6/10 + (+1) \cdot 3/36 \cdot 3/9 \\
& + (-1) \cdot 3/36 \cdot 6/9 + (+1) \cdot 2/36 + (-1) \cdot 1/36 = -7/495
\end{aligned}
$$

Since the expectation is negative, the game is a losing one for the shooter. At a casino, a bet against the shooter offers a positive expectation. But if you bet against the shooter, the rules are changed so that a total of 12 is no longer a losing event for the shooter (this event is just ignored in the payoff). The techniques above can be used to compute the mathematical expectation of such a wager. It is a better wager (you lose less) to bet on the shooter, but this bet also has negative expectation for the shooter (of course, these are positive expectations for the casino).

It cannot be denied that the examples in this section were all about games. If some apology for such emphasis (while discussing odds) is in order, the apology is attempted in the Exercises, where there are some questions whose answers are normally found by use of the odds formula and yet which do not concern games.

Exercises

1. Refer to Example 5 and compute the fair odds that should be given on a wager on the statement "The ball will fall into a hole colored red if the roulette wheel has 38 holes, 2 of which are painted neither red nor black."
2. Refer to Example 6 and compute the fair odds that should be given on a wager on the statement "The event (4,4) will occur before a total of 7."
3. One wager on the dice game is called a *field bet*. The payoff is +1 if any one of the numbers 2, 3, 5, 9, or 10 occurs; the payoff is −1 if one of these numbers does not occur. Is this a fair bet? Explain your answer.
4. Since you would win if you bet against the shooter in a dice game under the usual rules, the casino changes the rules for such bets. Under the revised rules, the toss totaling 12 is ignored (instead of being a loss for the shooter). With these revised rules, what is the mathematical expectation of betting against the shooter?

5. Some years ago in a casino in a Midwestern state (where gambling was illegal), the rules for betting against the dice shooter (see Exercise 4) were changed so that an initial toss resulting in a total of 12 was a win for the shooter. This increased the casino odds. What is the mathematical expectation under these rules?
6. A ship with a valuable cargo is sunk at sea, and salvage operations are about to begin. The probability that the cargo can be salvaged is $\frac{1}{4}$. You are offered a chance to invest \$24,000 in the salvage operation in return for \$100,000 if the cargo is salvaged. What is your expected gain or loss under these conditions? If the odds were fair, what should your return be if the salvage operation was successful?
7. A farmer is going to use a certain fertilizer. He knows from past experience that one-fourth of the time it increases his yield by 20 percent, one-half of the time it increases his yield by 10 percent, and one-fourth of the time it increases his yield by 8 percent. If the normal profit from the field that he is going to use the fertilizer on is \$1,000 and if the amount of fertilizer needed will cost \$100, what is his expected gain (or loss) from this action?

Answers to problems

A. Your friend is a friend in need. $x/(1 - x) = \frac{1}{8}/\frac{7}{8} = \frac{1}{7}$. The payoff should be 7 to 1 rather than 4 to 1.

B.
$$\begin{aligned} \tfrac{7}{10}w &= \tfrac{3}{10} \cdot 50{,}000 \\ w &= \tfrac{3}{7} \cdot 50{,}000 \\ w &= 21{,}428.57 \end{aligned}$$

C.
$$\begin{aligned} E &= \frac{{}_6C_0\,{}_4C_3}{{}_{10}C_3}\,0 + \frac{{}_6C_1\,{}_4C_2}{{}_{10}C_3}\,30 + \frac{{}_6C_2\,{}_4C_1}{{}_{10}C_3}\,60 + \frac{{}_6C_3\,{}_4C_0}{{}_{10}C_3}\,90 \\ &= \tfrac{4}{120} \cdot 0 + \tfrac{36}{120} \cdot 30 + \tfrac{60}{120} \cdot 60 + \tfrac{20}{120} \cdot 90 \\ &= 9 + 30 + 15 = 54 \end{aligned}$$

D. $E = 0.3 \cdot 60{,}000 - 0.7 \cdot 20{,}000 = 4{,}000$. The investment is good.

3.10 Conditional probability

Problems

A. Mr. Fain has four children. If the probability is $\frac{1}{2}$ that each child born is a boy, then what is the probability that Mr. Fain has exactly two sons? If it is known that Mr. Fain has at least one son, what is the probability that he has exactly two sons?

B. A large chain of retail stores makes a survey with the following results: Out of 100 people interviewed, 70 did not buy from them at all, 15 made purchases from them by mail, 20 made purchases from them at a store, and 5 made both mail-order purchases and store purchases. If it is known that Mr. X makes store purchases, then what is the probability that he also makes mail-order purchases?

C. A production process turns out a product which is classified into two grades: choice and standard. The process produces an equal number in each grade. Two items are selected at random from the production line.

1. What is the probability that at least one of them is choice grade?
2. The first item was checked and found to be of standard grade. What is the probability that the second is standard grade also?
3. If the first item is checked and found to be standard grade, what is the probability that at least one of the two is choice grade?

The natural impulse to use games to illustrate problems in probability was resisted until the previous section. In this section, such examples will again be used for illustrative purposes. But to compensate, a large number of business problems which employ the same principles will be included in the Exercises. The similarity in the applications of the formulas to games and to business illustrates that business decisions are also based on probability.

The dice game that was discussed at some length in the previous section has the feature that the probabilities for any of the possible events on any one throw are the same regardless of what the preceding throw or throws may have been. This is despite the fact that there are those bettors who believe, for example, that since a 6 has not been thrown on any of the previous 10 throws of the dice, the odds must be better that a 6 will occur on the next throw. There is nothing in mathematical theory or empirical observation to bear out this idea.

Example 1. On the other hand, there are games in which events that have occurred previously affect the outcome of the game. One such game is blackjack, or twenty-one. This game is played with an ordinary deck of cards, which consists of 52 cards with four suits: spades, clubs, hearts, and diamonds. Each suit consists of 13 cards: 2 to 10, jack, queen, king, and ace. In this game, there are a dealer and a player (or players). The dealer and the players are each given two cards to start the game. The player can see his two cards and one of the two cards that the dealer possesses. One possible deal of the initial sets of cards will result in the situation in Diagram 14, with D representing dealer and P representing player.

The object of this game is to acquire a total of points on the cards as close to 21 as possible. The player wins if his total is closer to 21 than

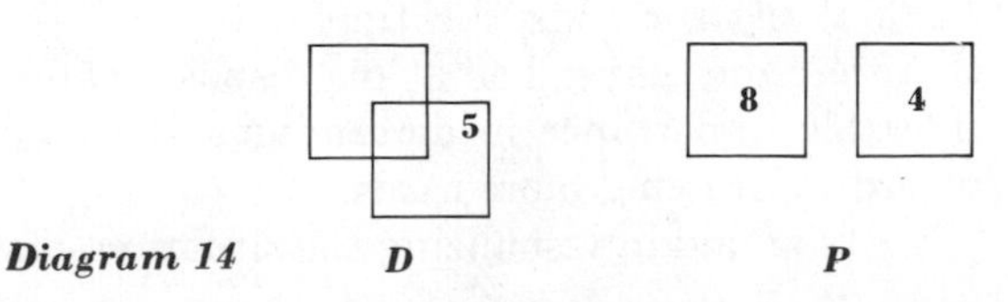

Diagram 14

the dealer's. A player may decide to receive more cards in an attempt to get a total closer to 21. However, if in the drawing of more cards, a total of more than 21 occurs, the player loses. Note the parallel between this situation and previous ones in which decisions must be made. On the other hand, the dealer, who is the representative of the casino, must draw more cards until his total reaches or exceeds 17. With a count of 17 or more, he cannot draw more cards. The jack, queen, and king each count as 10, and the ace counts either 1 or 11; the person possessing an ace can make the decision as to which of the two numbers it is to represent. (There are alternate strategies, such as doubling down or splitting, which affect the outcome of the game, but these will not be discussed here.)

Consider now the plight of the player who has a total of 12 and must decide whether to receive more cards. If this is the beginning of the game, the player knows only three cards (his two and one of the dealer's); the remaining 49 cards occupy unknown positions in the deck. Although the dealer possesses 1 of these 49 cards, the player does not know what the card is, and it must be treated just as any other member of the deck. In the situation that is under consideration here, 16 of the 49 cards have value 10. They are 10, J, Q, K of each suit. The probability that a draw of one card will cause the player's total to exceed 21 is $16/49$ (which is approximately $1/3$). Using the formula for odds obtained in the previous section, the odds are 1 to 2 that the player will exceed the permissible total on the draw of one card. Verify this statement by letting $x = 16/49$ in the ratio $x/(1 - x)$.

Suppose, however, that the game has been going on for some time and that many of the cards have been "burnt." This expression means that the used cards have been placed beneath the deck and are no longer in the game. Suppose, furthermore, that as the cards have been used, the player has counted the cards used (and we know from personal experience that the ability to count the cards that are used can be developed in a two-month period with only occasional play). Now if the player's count is such that he knows that 5 non-tens and 12 tens have been used, then he knows that there are 32 cards remaining in the deck and that only 4 of them are tens. ($5 + 12 = 17$ used cards, $49 - 17 = 32$ remaining cards, $16 - 12 = 4$ remaining tens, $33 - 5 = 28$ remaining non-tens.) The probability that the player's total will exceed 21 on a single draw of the cards now becomes $4/32 = 1/8$, which is approximately 0.13. The odds (using the formula of the preceding section) that a single draw of a card will cause the player's total to exceed the permissible total is 1 to 7. Again verify the odds, this time letting $x = 1/8$. Obviously, the information that the player has at his disposal about previous events is of considerable importance in determining the strategy that he employs with regard to drawing more cards.

As an aid in visualizing the situation, diagrams will be given so that something definite is at hand. Let

p: x is a card whose value is 10.

be a statement, with a universal set consisting of 49 possible events, 16 of which are in the truth set P of statement p. Diagram 15 represents this situation. The numbers in the diagram refer to the number of elements in the sets when no information is available about prior events.

Now, to illustrate the additional information whose existence has been assumed, another statement is needed. It is

q: x is a card that has not been used in previous play.

The assumption is that the truth set Q of statement q has 32 elements and, furthermore, that 28 of the 32 are in $\tilde{P}$, with the remaining 4 in P. There are two partitions $(P,\tilde{P})$ and $(Q,\tilde{Q})$ of the universal set, and these two partitions induce a third partition of the universal set, which can be denoted by $(P \cap Q,\ P \cap \tilde{Q},\ \tilde{P} \cap Q,\ \tilde{P} \cap \tilde{Q})$. It will probably be helpful to reread page 95, which deals with cross partitions. The cross partition is such that $N(P \cap Q) = 4$, $N(P \cap \tilde{Q}) = 12$, $N(\tilde{P} \cap Q) = 28$, $N(\tilde{P} \cap \tilde{Q}) + 5$. Diagram 16 illustrates the cross partition, with the numbers in the diagram representing the number of elements in the subsets.

If the player knows the number of elements in the truth set Q, then the sets $P \cap Q$, $\tilde{P} \cap Q$ (the number of cards that have not been used) are used to compute the probability of the next card being one that will cause the total to exceed 21. This is an example of a probability computed with information given by a known statement; such examples are called *conditional probabilities*.

It is clearly helpful in card games to know what cards have been used in previous play. Conditional probability has been applied to card games for many years, but an analysis of the game of blackjack was first attempted (to our knowledge) in 1946, and the authors of the paper decided that there was no winning strategy for the players. However, the game was reconsidered in 1961, and a winning strategy was devised.

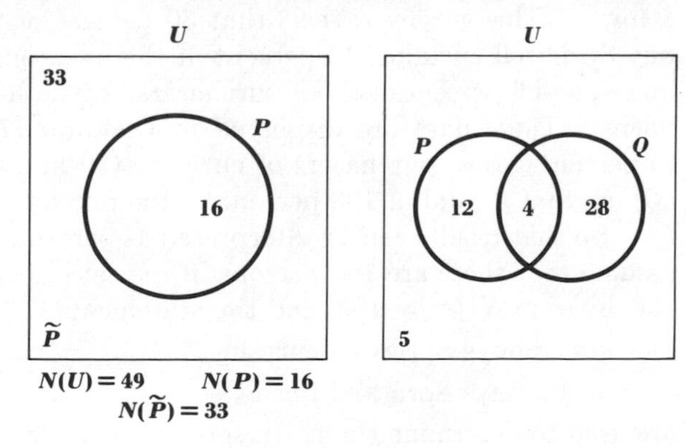

Diagram 15 *Diagram 16*

A summary of the conditions under which the principle of conditional probability applies will now be given, and a formula for its use in such cases will be stated. First, as is usual, statement p is given, a universal set of logical possibilities U is determined, and a subset of U, denoted by P, is established as the truth set of statement p. Additional information in the form of "statement q is true" is received. Statement q and the conditions under which it is true determine a truth set Q of U. This, in turn, determines the conditional probability, which is denoted by the symbol $Pr(p|q)$. So that no misunderstanding can occur and so that a usable formula is available, some of these facts are repeated in the formula that follows.

Formula: ***(The conditional probability of statement** p**, given statement** q**)*** *Let p be a statement with universal set of logical possibilities U, truth set P, and probability $Pr(p)$. Let the statement q be given, with truth set Q and probability $Pr(q)$. Then the conditional probability of p, given q, is the probability of p and q divided by the probability of q. Symbolically, we write the conditional probability of p given q as $Pr(p|q)$, and*

$$Pr(p|q) = \frac{Pr(p \text{ and } q)}{Pr(q)}$$

To see how this formula applies to Example 1, observe that $Pr(p \text{ and } q)$ is $4/49$ (see Diagram 16) and that $Pr(q)$ is $32/49$. Hence $Pr(p|q)$ is

$$\frac{4/49}{32/49} = 4/49 \cdot 49/32 = 4/32 = 1/8$$

This section is concluded with an application of the formula for conditional probability to a business problem.

Example 2. Mr. Evans has concluded a study of purchasing habits in a town. This survey reveals that 30 percent of the curtain purchasers buy Sprigwell curtains, 15 percent of the linen purchasers buy Sprigwell linens, and 3 percent of these purchasers buy both Sprigwell curtains and linens. These data are displayed in Diagram 17. The numbers refer to percentages of purchasers of curtains (P) and of purchasers of linens (Q), so that a total of 100 percent of the purchasers is illustrated.

So that results can be interpreted as percentages, it is convenient to assume that there are 100 persons in the set of persons interviewed. In the diagram $N(U) = 100$, and the statements

p: x buys Sprigwell curtains.

q: x buys Sprigwell linens.

are used to determine the truth sets P and Q, which have 30 and 15 elements, respectively. Mr. Archer asks: "If X buys Sprigwell linens, what

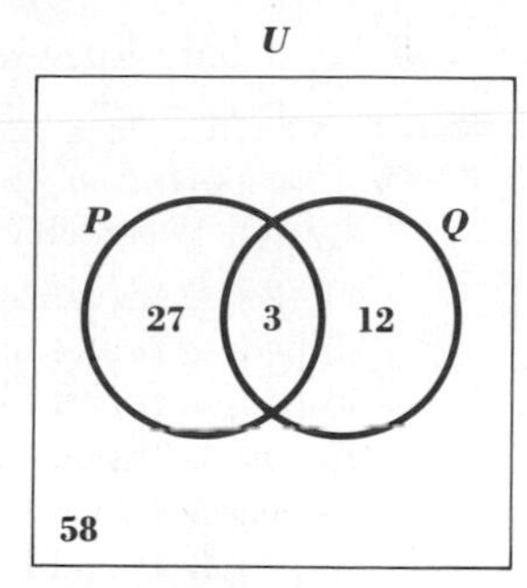

Diagram 17

is the probability that X buys Sprigwell curtains?" Mr. Evans recognizes this as an application of conditional probability, and by use of the formula for conditional probability, he is able to compute that

$$Pr(p|q) = \frac{Pr(p \wedge q)}{Pr(q)} = \frac{3/100}{15/100} = 3/15 = 1/5$$

Only 3 percent of all purchasers buy both products, but 20 percent of Sprigwell linen purchasers buy both products.

Exercises

1. Of 100 purchasers of Sprigwell rugs, 65 bought brown and 48 bought blue.
   a. If a purchaser bought a brown rug, what is the probability that she bought a blue rug?
   b. If a purchaser bought a blue rug, what is the probability that she bought a brown rug?

2. Recall Exercise 4 of Sec. 2.1. A survey of 580 housewives reveals that:
   70 purchase pink, blue, and striped towels.
   110 purchase pink and blue towels.
   100 purchase pink and striped towels.
   120 purchase blue and striped towels.
   200 purchase pink towels.
   260 purchase blue towels.
   150 purchase none of the three.
   How many purchase striped towels?
   a. What is the probability that x bought both pink and blue towels?
   b. What is the probability that x bought both pink and blue towels if x bought a pink towel?
   c. What is the probability that x bought both pink and blue towels if x bought a blue towel?
   d. What is the probability that x bought a pink, a blue, and a striped towel?
   e. What is the probability that x bought a pink, a blue, and a striped towel if x bought a pink towel?

f. What is the probability that x bought a pink, a blue, and a striped towel if x bought a pink and a blue towel?

3. A transportation commission makes a study of 100 people who commute daily between city A and city B. The results of the study are as follows:
   30 people travel by bus.
   10 people travel by train.
   50 people travel by auto.
   27 people travel exclusively by forms other than bus, train, or auto.
   5 people travel by both train and bus.
   5 people travel by both train and auto.
   10 people travel by both bus and auto.
   3 people travel by train, bus, and auto.

   Draw a diagram showing the above relationships, and answer the following questions:

   a. If Mr. Fox is a commuter who travels by train, then what is the probability that he also travels by bus?
   b. If Mr. Ball is a commuter who travels by auto, then what is the probability that he also travels by bus?
   c. If Mr. James is a commuter who travels by bus, then what is the probability that he travels by train? (Compare this with your result for part *a* above.)
   d. If Miss Allen is a commuter who travels by auto, then what is the probability that she travels also by bus and train?
   e. If Mr. Jacobs is a commuter who travels by bus but not by train, then what is the probability that he also travels by auto?
   f. If Miss Scott is a commuter who does not travel by both bus and train, then what is the probability that she travels by bus? That she travels by train? That she travels by auto?

4. In a study to determine the cause of an exceptionally large number of defective parts, the Acme Company prepared the following table, showing the number of good and bad items classified by source of raw materials and by machine on which the items were produced:

Raw-material source	***Machine 1***		***Machine 2***		***Machine 3***		***Totals by supplier***
	Good	***Bad***	***Good***	***Bad***	***Good***	***Bad***	
Supplier A	600	60	580	20	700	40	2,000
Supplier B	800	70	700	20	400	10	2,000
Totals	1,400	130	1,280	40	1,100	50	4,000
Totals by machine	1,530		1,320		1,150		4,000

   a. What is the probability that an item selected at random from the Acme Company production line will be defective?

b. Given that the item selected was made from materials from supplier A, what is the probability that it will prove defective?

c. What is the probability that an item selected at random will be made of materials from supplier B and produced on machine 1?

d. Given that an item is bad, what is the probability that it was made from materials from supplier A?

e. Given that an item is bad, what is the probability that it was made on machine 1?

f. If the probability that an item is defective and was made on machine 1 is 130/4,000 and the probability that it was made on machine 1 is 1,530/4,000, what is the probability that it is bad, given that it was made on machine 1? (Solve without referring to the table.)

5. If a machine on the production line of the Acme Company is properly adjusted, it will turn out a product which will average 80 percent acceptable. If the machine is not properly adjusted, only 15 percent will be acceptable. The machine has just been adjusted by Mr. A, who has a record of adjusting it properly 90 percent of the time. Five items were selected at random during the first hour the machine was in operation, and only one of them was acceptable.

   a. What is the probability that the machine is not properly adjusted?

   b. Why would you probably increase the sample size before having it readjusted?

6. *a.* In Prob. A, it was assumed that Mr. Fain had four children. If he had *five* children and the probability that each child born is a boy is ½, what is the probability that Mr. Fain has exactly two sons?

   b. If it is known that his oldest child is a son, what is the probability that he has exactly two sons?

   c. If it is known that at least one of his children is a boy, what is the probability that he has exactly two sons?

7. Two teams are playing in the World Series. Although the teams are considered equal, one team has already won the first two games and needs to win only two more to win the Series. If the teams are really equal, what is the probability that the team which has already won two games will:

   a. Win the third game?

   b. Win the third and the fourth games?

   c. Win the Series?

Answers to problems

A. $Pr(p) = 6/16; Pr(p|q) = \dfrac{6/16}{15/16} = 6/15 = 2/5$

B. $\dfrac{Pr(p \text{ and } q)}{Pr(q)} = \dfrac{5/100}{20/100} = 1/4$

C. 1. ${}_2C_1(1/2)^2 + {}_2C_2(1/2)^2 = 2/4 + 1/4 = 3/4$

2. ${}_1C_1(1/2) = 1/2$

3. ½, since the probability is ½ that the remaining one is choice grade

4

DESCRIPTIVE STATISTICS AND STATISTICAL INFERENCE

4.0 Introduction

The introductory sections of the chapters of this book have been written with one goal, that of giving a preview of what is to come, and that is one of the reasons for this section. However, because the title of the chapter describes such a large area of mathematical knowledge, it seems appropriate to offer an explanation for the briefness of the chapter and to make an attempt to put in proper perspective those topics which have been included.

Much of what is to follow in this chapter has to do with the description of data. It is probably not necessary to convince the reader of the existence of huge masses of data and, hence, of the necessity for thoughtful handling and analysis of data. The first few sections of this chapter discuss methods for organization of data and illustrate how quantitative information can be used for descriptive purposes.

But analysis without inference is often inadequate, and an attempt will be made to show how data collected from samples can be used to make *inferences* about total populations. The area of mathematics known as *statistical inference* is much too large to be summarized in this chapter—or in a book, for that matter. Perhaps Secs. 4.9 to 4.13 will assist the reader to see how statistics can be used for making inferences about a population.

Before data can be analyzed and conclusions drawn from the analysis, the data must be collected in a reasonable manner. One application which illustrates the need for care in the selection of data is *acceptance sampling*, and a discussion of this topic is included in Sec. 4.10. Again, because of space limitations, the discussion is brief, but the topics are chosen to typify the kinds of problems encountered in this area.

Since sampling plays a large role in statistical inference and hence in such specific areas as design of experiments, hypothesis testing, and confidence levels, to mention a few, the difference in the samples to be studied in this chapter and those studied in Chap. 3 (on probability) should be carefully noted. The difference is largely one of attitude. In the study of probability, facts are known, or assumed to be known, about the total population (set of elements to be sampled) and predictions are made about samples to be selected. With this statement in mind, it may be helpful to pause and reflect on the sampling problems of Chap. 3. For statistical inference, the roles are reversed; information is known about the sample, and predictions are made about the total population. This is a distinguishing feature between probability and statistics. But even so, the necessity of applying the properties of probability to make a successful study of statistics is obvious and should become even more apparent as we proceed.

4.1 Organization and presentation of data

Problems

A. A set of raw data is obtained in which the lowest value is 65 and the highest value is 325. What is the range of this particular set of data?

B. The data referred to in Prob. *A* are grouped into intervals, and a graph of these data is given below. From the graph, answer the following questions:

1. Which interval has the smallest number of elements?
2. Which interval has the largest number of elements?
3. Which intervals contain 20 elements?
4. How many elements are there in the set of data?

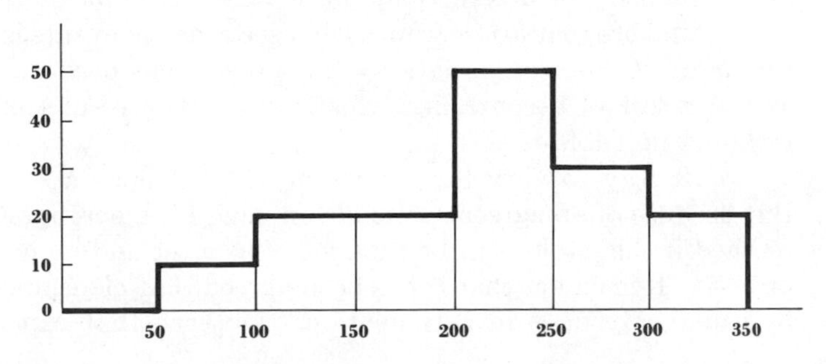

C. A company which is trying to select a source (or supplier) of metal parts is considering two possible companies. Company *A*'s product has an average strength of 320 lb, while company *B*'s product has an average strength of 350 lb. A minimum strength of 300 lb is required. What additional information would be required before a sound choice could be made?

D. The life in miles of a sample of 50 tires of company *A* had a range of 21,000 miles, while an equal sample of tires from company *B* had a range of 22,000 miles.

1. What other information would be required to select the better of the two?
2. What effect would average life expectancies of 20,000 and 40,000 miles for *A* and *B*, respectively, have on your evaluation of the range data?

The reader may not have found it necessary to organize a set of data for presentation, but he most certainly has seen such presentations since they abound in communications media such as newspapers, magazines, and television. The results of all sorts of polls, ranging from such diverse topics as the popularity of political leaders to favorite brands of soap, are periodically reported. Charts, graphs, and histograms (a special kind of graph) are frequently used in newspapers, and examples of several kinds of displays are included in this section.

Diagrams of the type that will appear in this chapter are used to present data; so a set of data is needed for use in the examples. The data given in the first example will be used throughout the chapter for illustrative purposes.

Example 1. As the first example, suppose that a company is concerned about the strength and durability of the items it manufactures, as any manufacturer is likely to be. The company assigns to one of the staff the task of testing the strength of one of the items manufactured. A sample consisting of 33 of the items is obtained. A machine is designed which applies an increasing pressure to each item until it is destroyed; at the moment of destruction, the machine records the pressure. The set of numbers that arises from such a sequence of events is an example of *raw data*. Collection of data by this process has been referred to earlier as the result of a controlled experiment. The results of the tests are recorded in Table 1.

One way to view the entries in Table 1 is as a set (of numbers). But because of an agreement made on page 13, a new agreement must be reached if this table is to be thought of as a set and no information is to be lost. Remember that it has been agreed that elements of a set would be listed only once in a tabulation. Observe that experiments 1, 14,

Table 1

1. 161	12. 213	23. 217
2. 70	13. 113	24. 141
3. 296	14. 161	25. 195
4. 172	15. 85	26. 122
5. 111	16. 190	27. 153
6. 43	17. 193	28. 184
7. 180	18. 230	29. 192
8. 177	19. 220	30. 141
9. 136	20. 55	31. 150
10. 210	21. 102	32. 214
11. 85	22. 161	33. 141

and 22 each resulted in the same number, 161. Clearly, it will distort the results if experiments are to be disregarded simply because they result in the same number. Hence, for the purposes of this discussion, we make the agreement that *a number may occur more than once in a set.*

There is another point about the data that have been displayed that should be commented upon now, although a longer discussion about the matter will be given in a later section. There must of necessity be a *selection process* in securing the manufactured items to be tested. In practice, this might vary from selection as the items come off the assembly line to selection from the warehouse, but the important point (mathematically) is that there should be an aspect of randomness about the selection procedure. The topic of *random processes* is a complicated one, but it has been discussed previously in terms of equally likely events.

The term *raw data* (to describe such a table) is used advisedly since no processing of the set of numbers has been made; the numbers are simply listed as they occurred as the result of the experiments. In at least one sense, the employee who was assigned the task has completed the job; a sample was selected and tested, and the results were recorded. However, if the 33 numbers that have been accumulated are submitted with no further comments by the employee to his employer, then he can expect to be reprimanded since the list of numbers, as such, has little or no meaning. (For another example, which illustrates the same point at a different level of magnitude, suppose that all the information gathered by the United States Census Bureau in a census is simply put on sheets of paper and that is the end of the matter. The numbers themselves—and there would be millions of them—would have little or no meaning in such a presentation.) The data must be *organized* and *presented* in a better way so that the table (or chart) easily lends itself to interpretation.

Consider now some important aspects of the presentation of a set of numbers and how the data can be presented to show these aspects.

Certainly, one of the facts that will be of interest is the *least amount* of pressure recorded. A scan of the entries in the table shows that the sixth entry, 43 units, is the smallest number in the set. Also of interest is the *largest amount* of pressure, and a look at the table reveals that the third entry, 296 units, is the largest number. The difference of these two numbers is $296 - 43 = 253$, and this is called the *range* of the set of numbers. For a more general statement of the range of a set of numbers, see the following definition.

Definition: (***Range of a set of data***) *Let S be a set of real numbers and let a be the least element of S and b the greatest element of S; then the range of S is $b - a$. Symbolically, if $S = \{x_1,x_2,x_3, \ldots ,x_n\}$ and if $a \in S$, $a \leq x_i$ for all $x_i \in S$, and if $b \in S$, $b \geq x_i$ for all $x_i \in S$, then $b - a$ is the range of S.*

Since there are some symbols in the definition which may not be familiar, a short explanation of their meaning will be given. The symbol $\leq$ is a combination of two familiar symbols, $=$ and $<$. It is used as the shorthand way of combining two statements into one statement. For example, $a \leq b$ means that either a is less than b ($a < b$) or a is equal to b ($a = b$). A little reflection will convince you that *or* is being used in the exclusive sense (see page 30) since $a < b$ and $a = b$ cannot *both* be true. Some examples of the symbols $\leq$ and $\geq$ are

$$\begin{array}{ccc} 3 \leq 4 & 41 \geq 17 & 11 \geq 1 \\ 16 \leq 23 & 32 \geq 32 & 12 \geq 6 \\ 5 \leq 5 & 64 \geq 0 & 7 \geq 7 \end{array}$$

The second symbol in the definition that may be unfamiliar is the number at the bottom of x to distinguish the elements of the set. This number is called a *subscript*. The symbol x_2 is read "x subscript 2"; x_{11} is read "x subscript 11"; x_n is read "x subscript n." The third symbol, x_i, is a slight variant of the second notation and is read "x subscript i." This symbol was used, and will continue to be used, to represent an arbitrary member of the set S.

A second way that the set of numbers can be displayed is by the use of a *number line*. If some of the points of a line are assumed to be in one-to-one correspondence with the counting numbers, then points which correspond to the entries in the table can be recorded in a second type of display called a *dot frequency chart*. The dot frequency chart for the 33 outcomes of the tests in this example are recorded in Diagram 1.

At least two advantages of such a chart are immediately apparent: (1) the chart gives the information in a visual manner not possible in a table, and (2) the distribution of the results over the range is more easily read off the chart. Any set of numbers will distribute over its range in

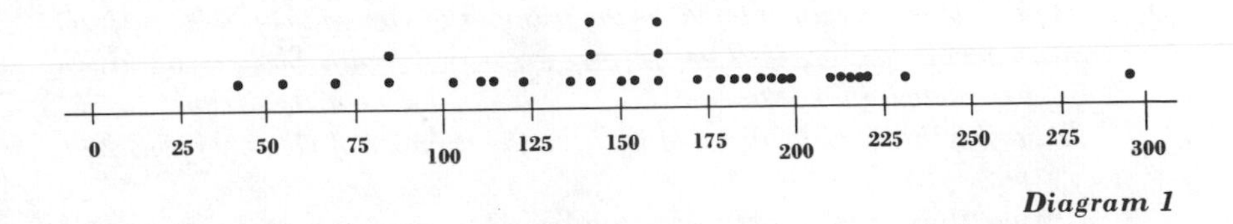

Diagram 1

some fashion; hence sets of numbers are often called *distributions*. This topic will continue to demand our attention as we proceed.

Balanced against the advantages of the dot frequency chart are its disadvantages, one of which is also immediately apparent: The dot frequency chart requires time and effort to prepare. The time used in recording the results in this manner is well spent since one outcome of such a chart is that the numbers have now been arranged in *increasing* (*ascending*) *order*. The new arrangement according to size is

43, 55, 70, 85, 85, 102, 111, 113, 122, 136, 141,
141, 141, 150, 153, 161, 161, 161, 172, 177, 180, 184,
190, 192, 193, 195, 210, 213, 214, 217, 220, 230, 296

The dot frequency chart, by its very nature, tends to partition the set of numbers into subsets. Frequently, the set of numbers has so many elements that such a partition is necessary to make an analysis of the data. You are familiar with such examples as the partition of a university enrollment by grade points, city populations by earnings, national populations by heights. The elements in sets with a large number of elements are grouped, and the grouped data are then considered. The number of elements in the set of data under consideration is not large; in fact, for convenience, it is small, but it will be used to illustrate how grouping procedures work.

First, the range is divided into *intervals*. It is not necessary to have the intervals equal in size, but for reasons not given here, it is almost always the case that they are chosen to be equal. Of course, the counting numbers from 0 to 300 can be partitioned in many ways, but one convenient choice (for this example) is 0 to 50, 51 to 100, 101 to 150, 151 to 200, 201 to 250, 251 to 300. The *frequency* of the set of numbers in each interval can now be counted. This simply means that the number of elements in the set that occur in each interval is counted. The reader should use Diagram 1 to verify the information contained in Table 2.

So that this process can be referred to in what follows in this section, as well as in later sections, some notation and a summary are included in a definition.

Definition: (**Frequency**) *Let S be a set of real numbers and let some set containing S be partitioned into intervals of equal length. Then the*

frequency of S in an interval is the number of elements of S in an interval. Symbolically, if $S = \{x_1,x_2, \ldots ,x_n\}$, *with the set containing S partitioned into intervals* $I_1, I_2, \ldots, I_r$, *then the frequency of S in the* ith *interval* I_i, *denoted by* f_i, *is the number of elements of S in* I_i.

Notice that in the definition, subscripts have again been used to distinguish the elements of S. The set has a first element, denoted x_1, a second element, denoted x_2, and so forth, until the last element, x_n, is reached. The number of intervals is r, with I_1 representing the first, I_2 the second, etc. Suggestively, f is chosen to represent *frequency*, and f_i represents the frequency in the ith interval. The symbols can (and will) be interpreted for the example.

$$S = \{43,55,70,85,85,102, \ldots ,230,296\}$$

$$x_1 = 43,\ x_2 = 55,\ x_3 = 70,\ x_4 = 85,\ x_5 = 85,\ x_6 = 102,\ \ldots,\ x_{32} = 230,\ x_{33} = 296$$

The range is partitioned into six intervals, with frequencies

$$f_1 = 1 \qquad f_2 = 4 \qquad f_3 = 9 \qquad f_4 = 12 \qquad f_5 = 6 \qquad f_6 = 1$$

The results of grouping the data are easily displayed in a type of chart called a *histogram*. A histogram is a bar chart; the length of the base of each bar represents an interval, and the height of the bar represents the frequency for that interval. A histogram for the data currently under discussion is in Diagram 2. Since histograms are quite frequently used to display data in newspapers, familiarity with such charts can probably be assumed.

To illustrate that the choice of the size of the interval will change the appearance of the histogram, Diagram 3 gives the same data with the length of the interval chosen as 25 instead of 50. Consult the arrangement of the set of numbers on page 201 to verify the frequencies listed.

One set of numbers which can be viewed as a set of grouped data is the set of coefficients in the binomial formula (page 117). For a particular case, consider $(x + y)^5$. The terms in this product are grouped by

Table 2

	Interval 1	*Interval 2*	*Interval 3*	*Interval 4*	*Interval 5*	*Interval 6*
Interval	I_1	I_2	I_3	I_4	I_5	I_6
Range of interval	0–50	51–100	101–150	151–200	201–250	251–300
Frequency	1	4	9	12	6	1

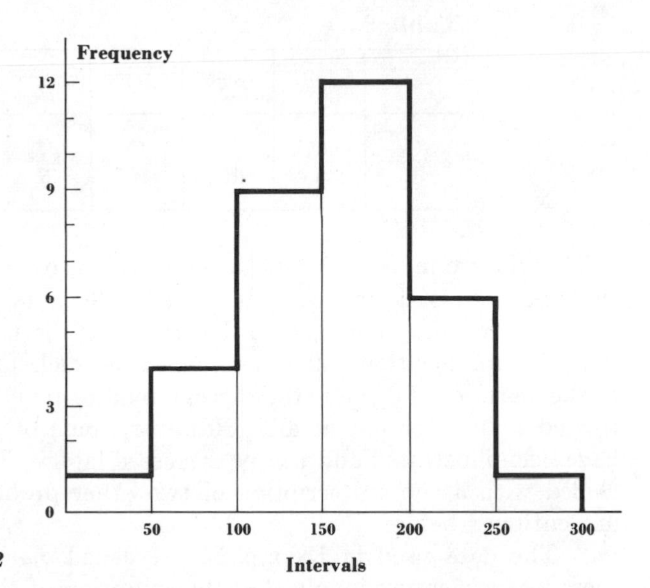

Diagram 2

those which have x^5 as a factor, x^4y as a factor, x^3y^2 as a factor, etc. The numbers which occur as coefficients of these products are the *frequencies* of the terms in the total expansion of the formula. It may be helpful to reread page 118 to see why the frequencies in Table 3 are as listed.

The histogram for this set of grouped data is given in Diagram 4.

A comparison of Diagrams 2 and 4 shows a marked resemblance between them, and while we readily admit that it was planned this way, it is known that many sets of data do have histograms that are close to the histogram for a set of binomial coefficients. There will be more on this subject in the following sections, but in the meantime, the Exercises

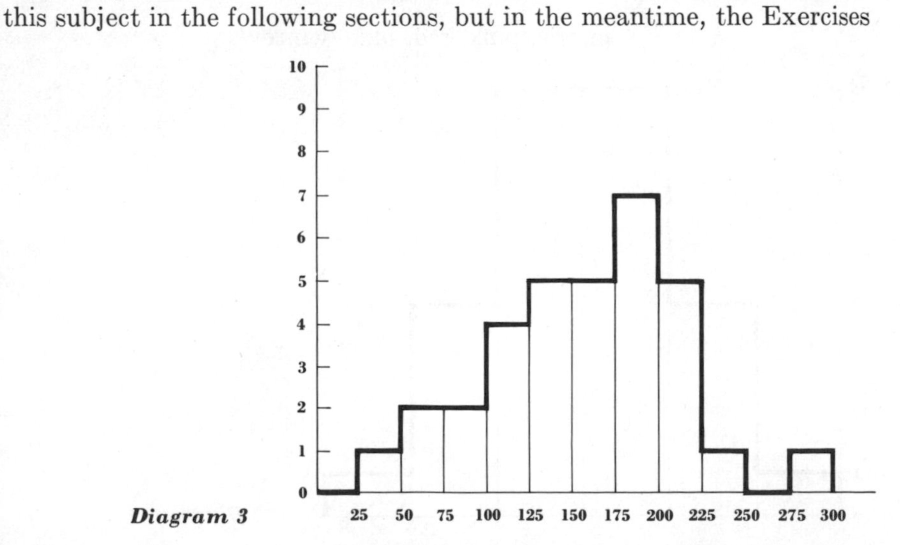

Diagram 3

Table 3

x^5	x^4y	x^3y^2	x^2y^3	xy^4	y^5
${}_5C_0$ 1	${}_5C_1$ 5	${}_5C_2$ 10	${}_5C_3$ 10	${}_5C_4$ 5	${}_5C_5$ 1

will lay the groundwork for such considerations by requiring the construction of some histograms using binomial coefficients.

It is certainly not claimed that this brief discussion of the presentation of data mentions all the problems of such presentations. Many of the parts of the presentation that require much thought have been treated lightly, or not at all. However, some of the basic techniques have been illustrated and will be discussed later. The discussion is concluded with a short description of two other problems that have gone unmentioned before.

The data used in Example 1 are called *quantitative data* because there was a measure involved in the collection of the data. By this we mean that the strength of each of the items was measured and numbers were used to indicate the strength. Sometimes it is desirable to accumulate data that do not lend themselves to quantitative description; such data are called *qualitative data.* As an example, a *preference list* may give valuable information yet not involve a set of numbers.

Example 2. Suppose a company has conducted a survey and has discovered that the preference for various colored towels is well established. As a result of the survey, the preference list is

purple, pink, red, blue, white

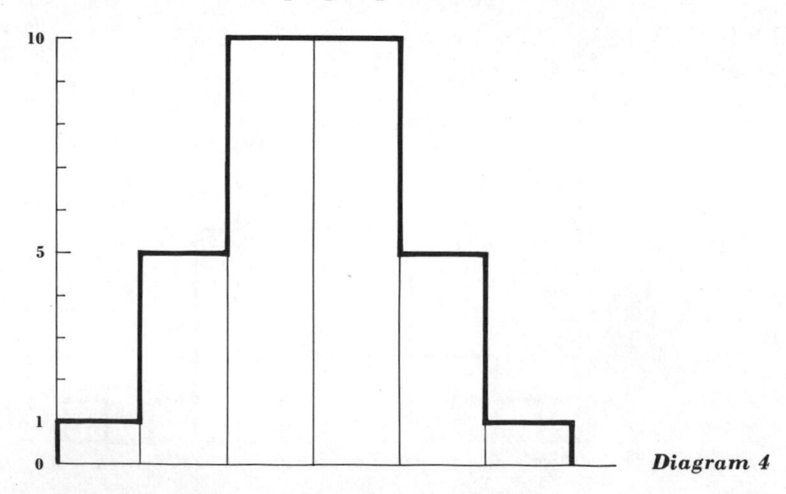

Diagram 4

This arrangement means that purple is least liked, pink is next least liked, and so forth, with white the most preferred.

For a final comment, think again about the set of numbers in Example 1. The quantitative data were gathered by a member of the company for use by members of the company. When the data are gathered by the company itself, within itself, this is referred to as *internal data* and is opposed to *external data,* which come from some other organization. Not only must the members of any company be aware of the quality of their own products, but they must keep up with such things as the general economic condition of the country. Data about conditions outside the company must obviously come from outside organizations (i.e., from the Department of Commerce, Bureau of Labor Statistics, and other such organizations). Therefore, companies have occasion to use both internal and external data. The study of problems connected with obtaining and using the services of such outside organizations is properly conducted in courses in business administration and does not belong in a book like this one. If the data to be considered are internal, that is, if they are gathered *by* the company *for* the company, then it is clear that this involves some bookkeeping within the company. Many problems can be studied by the use of data taken directly from the records of the company.

Example 3. As an example, suppose that an analysis of the company payroll is to be made. Then the payroll bookkeeping procedure would come under scrutiny, with a view toward organizing it in such a way that the data can be collected in the easiest possible way.

The discussion of the previous paragraph shows that the bookkeeping methods of the company may influence its ability to gather internal data. This brings to our attention the fact that there are many new ways of storing data today. Modern machines make it possible to store data which previously had to be destroyed or which were not available owing to the lack of such machines. These subjects are properly included in business administration courses and will not be discussed here.

Exercises

1. $A = \{50, 100, 220, 200, 145, 2, 4, 32, 47, 82, 12, 160\}$

   $B = \{10, 1{,}172, 84, 12, 30, 2, 112, 23, 360, 1{,}235, 1{,}663, 1{,}800, 1{,}625, 742, 13\}$

   Listed above are two examples of sets of data. For each of the examples, perform the following:

   a. Determine the range.

   b. Make a dot frequency chart.

   c. Make a frequency histogram.

2. Construct a histogram for the coefficients of $(x + y)^n$ patterned after Diagram 4.

   a. $(x + y)^2$ *b.* $(x + y)^3$ *c.* $(x + y)^4$
   d. $(x + y)^6$ *e.* $(x + y)^7$ *f.* $(x + y)^8$

3. What size intervals do you feel would be appropriate for a histogram if the number of elements n and the range $b - a$ are as follows?

	n	$b - a$		n	$b - a$
a.	20	20	*b.*	20	100
c.	40	20	*d.*	5	500
e.	200	5			

4. Give three examples of qualitative data other than those in this book.
5. Give three examples of internal data in which a company might be interested.
6. Give three examples of sources of external data for each of the following:
   a. A bakery
   b. A bank
   c. A farm
   d. A hardware store
   e. An automobile manufacturer
7. Given the dot frequency charts shown, perform the following:
   a. Determine the range.
   b. Make a frequency histogram.

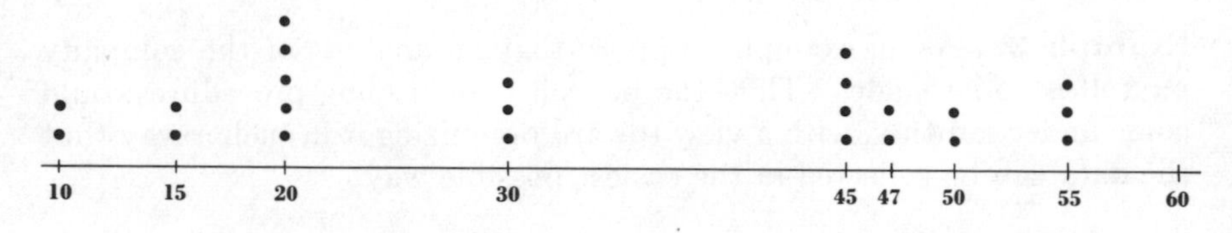

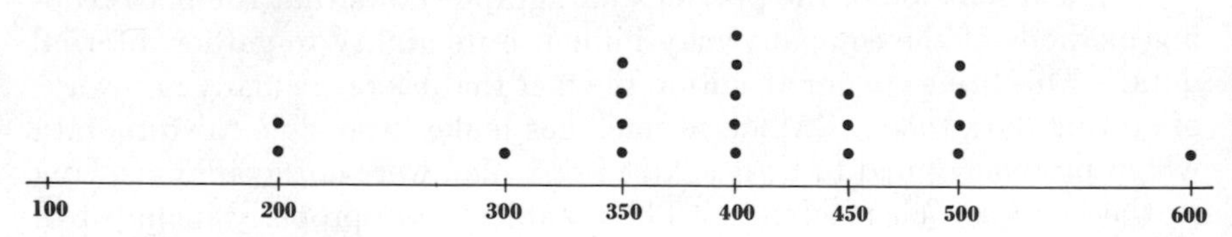

Answers to problems

A. The range is $325 - 65 = 260$.

B. 1. The interval 0 to 50
   2. The interval 201 to 250
   3. The intervals 101 to 150, 151 to 200, and 301 to 350
   4. There are $10 + 20 + 20 + 50 + 30 + 20 = 150$ elements in the set of data.

C. Information regarding the range and distribution of the strengths. It would be possible for the product with the lower average strength to have a large proportion above the minimum requirement if the product is more uniform.

D. 1. Their average life and a clearer picture of the dispersion. The range is a highly unstable measure.
2. Company *B*'s tires are *relatively* less variable and would be the better buy if the prices are comparable.

4.2 *Measures of central tendency*

Problems

A. The three dye vats in a chemical plant have the temperature gauges set at 86, 92, and 77°C. The relation between centigrade and Fahrenheit temperatures is

$$F = (9/5)C + 32$$

Find the average Fahrenheit temperature of the gauge settings without changing the numbers to the Fahrenheit scale.

B. What is the average of the set of numbers {12,16,18,20,24}? What is the average of the set of numbers

$$\{3 \cdot 12, 3 \cdot 16, 3 \cdot 18, 3 \cdot 20, 3 \cdot 24\} = \{36,48,54,60,72\}?$$

C. An examination of the current ratios of 16 firms operating under similar conditions and in the same industry resulted in the following observations:

1.9	2.3	2.6	4.6
1.9	2.3	2.6	5.8
2.0	2.3	2.8	6.2
2.1	2.4	3.2	17.4

The median ratio was 2.5, while the mean ratio was 3.9. Why do accountants almost invariably use the median rather than the mean to describe the current ratio?

D. In testing two makes of tires, it was found that a sample of brand *A* had a mean of 32,000 miles and a median of 34,400, while brand *B* had a mean of 33,200 and a median of 33,800. Assuming that the two brands are of equal uniformity, which brand would you prefer? Why?

A key word in describing data is *distribution*, and this word has already been used with reference to the dot frequency chart of the previous section. It is of concern here because of the tendency of some sets of numbers that arise experimentally to distribute in such a way that there are clusters of numbers toward the center of the range. Words are being used in this discussion which have not been defined, but as the discussion proceeds, it should become progressively clearer what is meant by *cluster*, *center*, and *distribution*.

Several different kinds of measures have been developed to describe the tendency of data toward a center, and three of these measures will be discussed in this section. They are the *mean*, the *median*, and the *mode*. The most important of these is the *mean*, an expression which will be defined, and supporting evidence will be offered to show why this particular measure will serve purposes not served by the other measures. To inaugurate the discussion, consider the following definition.

Definition: (***Mean***) *If S is a set of real numbers, then the mean of S is the average of the numbers, which is determined by adding the elements of S and dividing by the number of elements in S. Symbolically, if* $S = \{x_1, x_2, \ldots, x_n\}$, *then the mean of S, denoted by* $\bar{x}$, *is the sum of the elements of S, denoted by* $\sum_{i=1}^{n} x_i$, *divided by n:* $\bar{x} = \left(\sum_{i=1}^{n} x_i\right)/n$.

Some remarks about notation are also relevant.

Since $\left(\sum_{i=1}^{n} x_i\right)/n$ is equal to $\sum_{i=1}^{n} \frac{x_i}{n}$, the two symbols will be used interchangeably.

It is convenient to have symbols to represent ideas, and the symbol $\bar{x}$ has been chosen (temporarily) to represent the mean. Recall that in Sec. 3.5, the terms *population* and *sample* were adopted for the universal set and subsets, respectively. Now both the population and the sample have means. In recent years, it has become somewhat standard to refer to the mean of the population as μ (Greek mu) and the mean of a subset of the population as $\bar{x}$. It will be imperative to adopt this notation when both are considered simultaneously (as in Sec. 4.10). In the meantime, since most of the considerations are directed toward sample sets with at least an implied universal set, the notation $\bar{x}$ is adopted. Become accustomed to reading "average" or "mean" when the symbol $\bar{x}$ occurs, since it will always refer to these ideas. A bit more involved is the symbol Σ, or more specifically, $\sum_{i=1}^{n}$. This is a capital sigma, the Greek alphabet symbol equivalent to S. So read Σ to mean "sum," read the $i = 1$ at the bottom of the symbol to mean "start the sum with the first element," and read the n above Σ to mean "conclude the summation process with the nth element." Summations need not begin with the first element of a set, nor must they conclude with the last element. Since this may be the reader's first encounter with the summation symbol, listed below are six different elements in a set of data and some of the summations which occur using the symbol that is given in the definition of mean.

Example 1. $x_1 = 5, x_2 = 3, x_3 = 4, x_4 = 2, x_5 = 7, x_6 = 4$

$$\sum_{i=1}^{3} x_i = x_1 + x_2 + x_3 = 5 + 3 + 4 = 12$$

$$\sum_{i=2}^{5} x_i = x_2 + x_3 + x_4 + x_5 = 3 + 4 + 2 + 7 = 16$$

$$\sum_{i=1}^{6} x_i = x_1 + x_2 + x_3 + x_4 + x_5 + x_6$$

$$= 5 + 3 + 4 + 2 + 7 + 4 = 25$$

$$\sum_{i=3}^{6} x_i = x_3 + x_4 + x_5 + x_6 = 4 + 2 + 7 + 4 = 17$$

Example 2. The mean of a given set of numbers is easily achieved, although if n (the number of elements of S) is large or if the elements themselves are large, some help from a desk calculator may be useful. Applying the definition to the 33 results of the hypothetical experiments of the previous section, as they are arranged on page 201, yields

$$\sum_{i=1}^{33} x_i = 43 + 55 + \cdots + 230 + 296 = 5{,}214$$

and

$$\frac{\sum_{i=1}^{33} x_i}{n} = \frac{5{,}214}{33} = 158$$

The number 158 is the mean of the set of numbers.

There are at least three very good reasons why the mean is an important measure of a set of numbers. Each of these reasons will be given and will be illustrated by an example. Then a formula will be given for computing means of data under a set of circumstances that apply to the examples.

Example 3. To illustrate the first property, assume that not only have the 33 experiments of Example 1 of the previous section been conducted (and the results observed and recorded) but also two other sets of data have been collected in a similar fashion. If a second set of numbers has been collected as a result of 27 experiments such that the average (mean) of the set of numbers is 162.9 and a third set of numbers has been accumulated as a result of 38 experiments such that the mean is 154.1, then what is the mean of the set of data that arises from consideration of all the collected data? Certainly, the question can be answered if the individual elements of the second set and third set are known. One simply adds the elements (there are $33 + 27 + 38 = 98$ of them) and divides by $n = 98$. But suppose that the required information (the tabulation of the second and third sets) in these processes is not accessible. Is it still possible to determine the *overall mean* of the total set? The answer to this question is *yes*, and it is obtained in an obvious and simple manner. Some notation will be helpful in showing how to accomplish this result. For that purpose, subscripts on the symbols which represent the means of the three sets will be used so that the mean of the set of 33 elements discussed at length in the previous section is now written as

$$\bar{x}_1 = \sum_{i=1}^{33} \frac{x_i}{33} = \frac{x_1 + x_2 + \cdots + x_{33}}{33} = 158$$

In similar notation, the other two given means are written as

$$\bar{x}_2 = \sum_{i=1}^{27} \frac{y_i}{27} = \frac{y_1 + y_2 + \cdots + y_{27}}{27} = 162.9$$

and $$\bar{x}_3 = \sum_{i=1}^{38} \frac{z_i}{38} = \frac{z_1 + z_2 + \cdots + z_{38}}{38} = 154.1$$

It is possible to write each of the three sums using x_i to represent elements from each of the sets, but different letters may be (temporarily) helpful. A bit of algebraic manipulation allows the representation

$$158 \cdot 33 = \sum_{i=1}^{33} x_i$$

$$162.9 \cdot 27 = \sum_{i=1}^{27} y_i$$

and $$154.1 \cdot 38 = \sum_{i=1}^{38} z_i$$

Adding both sides of these three equalities gives

$$158 \cdot 33 + 162.9 \cdot 27 + 154.1 \cdot 38 = \sum_{i=1}^{33} x_i + \sum_{i=1}^{27} y_i + \sum_{i=1}^{38} z_i$$

so that the sum of the 98 experimental results is

$$15{,}468.1 = \sum_{i=1}^{98} u_i$$

The symbol u_i is used to avoid confusion with previous symbols. Dividing both sides of this equality by 98, the *overall mean* is

$$\bar{x}_4 \cong 157.8 = \sum_{i=1}^{98} \frac{u_i}{98}$$

Still another symbol, $\cong$, has been introduced to convey an idea. As used here, the symbol means *approximately;* and 157.8 has been substituted for the more exact result, 157.8378.

Of course, it remains to be determined if the result $\bar{x}_4$ is the same number that would be obtained if all 98 numbers resulting from the experiments were averaged. Rather than verify that this is the case (in fact, such a verification is impossible unless the elements of the second and third sets are obtainable), a more general result will be established that shows this method of obtaining an overall average to be legitimate.

Formula: **(*Overall mean*)** *If $S_1, S_2, \ldots, S_k$ are k sets of real numbers with $n_1, n_2, \ldots, n_k$ elements, respectively, and with means of $\bar{x}_1, \bar{x}_2, \ldots, \bar{x}_k$, respectively, then the mean of the set S obtained by consideration of the $n_1 + n_2 + \cdots + n_k = n$ elements is the number $n_1\bar{x}_1 + n_2\bar{x}_2 + \cdots + n_k\bar{x}_k$ divided by n. Symbolically, the overall mean is*

$$\bar{x} = \frac{n_1\bar{x}_1 + n_2\bar{x}_2 + \cdots + n_k\bar{x}_k}{n} = \frac{\sum_{i=1}^{k} n_i\bar{x}_i}{n}$$

The argument to support this formula relies only on observing the formulas

$$\bar{x}_1 = \sum_{i=1}^{n_1} \frac{x_i}{n_1}$$

$$\bar{x}_2 = \sum_{i=1}^{n_2} \frac{x_i}{n_2}$$

$$\cdots\cdots$$

$$\bar{x}_k = \sum_{i=1}^{n_k} \frac{x_i}{n_k}$$

or, alternatively, the equivalent statements

$$\bar{x}_1 n_1 = \sum_{i=1}^{n_1} x_i$$

$$\bar{x}_2 n_2 = \sum_{i=1}^{n_2} x_i$$

$$\cdots\cdots$$

$$\bar{x}_k n_k = \sum_{i=1}^{n_k} x_i$$

Addition yields

$$\bar{x}_1 n_1 + \bar{x}_2 n_2 + \cdots + \bar{x}_k n_k = \sum_{i=1}^{n_1+n_2+\cdots+n_k} x_i$$

or, alternatively,

$$\sum_{i=1}^{k} \bar{x}_i n_i = \sum_{i=1}^{n} x_i$$

Division of both sides of this equality by n yields the required relation

$$\sum_{i=1}^{k} \frac{\bar{x}_i n_i}{n} = \sum_{i=1}^{n} \frac{x_i}{n} = \bar{x}$$

Perhaps it will be helpful to state in words the result that the example, and more generally the formula, makes evident. If several sets of data, along with the means of each of the sets, are available, then the mean of the set of data obtained by considering all the elements of the sets simultaneously can be achieved in two ways. The first and most obvious (if possible) way is to determine the average of the set of all of the numbers. The second procedure, which is considerably shorter for sets with a large number of elements, is to use the formula whose validity has just been established. When another measure of central tendency (median) is discussed a little later in the section, it will be shown that this particular property does not hold for all measures of central tendency.

The second property possessed by the mean, to be illustrated for a particular case and then verified for any set of real numbers, is called the *additive property* of the mean. To show why it is sometimes convenient to use such a property, consider again Example 1 of Sec. 4.1.

Example 4. Suppose that after the 33 tests on the strength of the items have been made, recorded, and averaged, it is learned that a malfunction in the gauge on the testing machine has introduced an error in each of the elements in the set of data. Owing to this faulty equipment, each of the numbers has been reported as 17 less than it should have been. This means the corrected data are

$$S = \{60,72,87,102,102, \ldots ,247,313\}$$

A corrected mean can be computed by adding these numbers and dividing once again by 33. But is there an easier method? Intuitively it seems that since the mean first computed was 158 and each element is now 17 larger, the corrected mean should be $158 + 17 = 175$. It is easy to decide whether this process works for this example by a series of elementary computations, and the reader is encouraged to do so. To see another way to resolve the question read on.

For a more general question than that posed in the previous paragraph, suppose that a set of real numbers

$$S_1 = \{x_1, x_2, \ldots, x_n\}$$

is given. Assume, furthermore, that a new set of data is formed from S_1 by adding a number c to each element so that

$$S_2 = \{x_1 + c, x_2 + c, \ldots, x_n + c\}$$

Now both of these sets have means, which will be denoted by $\bar{x}_1$ and $\bar{x}_2$, respectively, given by the formulas

$$\bar{x}_1 = \sum_{i=1}^{n} \frac{x_i}{n}$$

and

$$\bar{x}_2 = \sum_{i=1}^{n} \frac{x_i + c}{n}$$

The question to be resolved is: What relation, if any, exists between these two numbers? Of course, the guess, motivated by the example, is that $\bar{x}_1 + c = \bar{x}_2$, and this will be investigated.

A little algebra will yield the result that is being sought. First, we write

$$\bar{x}_2 = \sum_{i=1}^{n} \frac{x_i + c}{n}$$

as

$$n\bar{x}_2 = \sum_{i=1}^{n} (x_i + c)$$

Using the meaning of the summation symbol, the last expression becomes

$$n\bar{x}_2 = \sum_{i=1}^{n} (x_i + c) = (x_1 + c) + (x_2 + c) + \cdots + (x_n + c)$$

But observe that there are n of the numbers c in this sum, so that regrouping the numbers

$$n\bar{x}_2 = (x_1 + x_2 + \cdots + x_n) + nc$$

Now division of both numbers by n yields

$$\bar{x}_2 = \frac{x_1 + x_2 + \cdots + x_n}{n} + \frac{nc}{n}$$

or

$$\bar{x}_2 = \bar{x}_1 + c$$

This is the expected result, and it is summarized in the following formula.

Formula: (***Additive property of the mean***) *If S_1 is a set of real numbers, $S_1 = \{x_1, x_2, \ldots, x_n\}$, and if the real number c is added to* *each element of S_1 to obtain the set*

$$S_2 = \{x_1 + c, x_2 + c, \ldots, x_n + c\}$$

then the mean of S_1 added to c is the mean of S_2. Symbolically, we write

$$\bar{x}_1 + c = \bar{x}_2$$

There is one aspect of this formula that could cause confusion, so a word of explanation is offered about the phrase *additive property*. Real numbers can be either positive or negative. Addition of a *negative* number c to each element results in a subtraction; that is, when c is negative, each element of the set S_1 is decreased by a real number. The point is that this formula is at the same time also a result about subtraction.

Example 5. To be specific, consider again Example 1 of Sec. 4.1, with the set of 33 numbers, and adjust each number by adding -32. The resulting set of data then has a mean of $158 - 32 = 126$.

The third property, very similar in nature to the second property just established, is the *multiplicative property* of the mean. Before giving an argument to establish the formula, it will again be instructive to consider an example.

Example 6. Refer again to the set of the results of the experiments in Example 1, Sec. 4.1. Assume for this consideration that instead of adding a number to each element of the set, it is found necessary to multiply each number in the set of data by $4/5$. Such an operation presents a new set of data obtained from

$$S = \{43,55,70,85,85, \ldots ,230,296\}$$

which is the set

$$T = \{43 \cdot 4/5,55 \cdot 4/5,70 \cdot 4/5,85 \cdot 4/5,85 \cdot 4/5, \ldots ,230 \cdot 4/5,296 \cdot 4/5\}$$

By forming the indicated products and applying the definition of the mean, a corrected mean is obtained. But clearly if this set of calculations is to result in a number which is after all just $4/5$ of the mean of the original set, then it is much easier to perform one arithmetical operation than $33 + 32 + 1 = 66$ operations, which are necessary if the set T is used. This time, the suggestion is that the reader *not* calculate by using T since we are but a short way from showing a much easier process.

For the purpose of showing how to arrive at a formula for the mean of a second set of data which arises from a first set of numbers by *multi-*

plication, let

$$S_1 = \{x_1,x_2, \ldots ,x_n\}$$

and let c be a real number with

$$S_2 = \{cx_1,cx_2, \ldots ,cx_n\}$$

In the usual notation, the means are

$$\bar{x}_1 = \frac{\sum_{i=1}^{n} x_i}{n}$$

and

$$\bar{x}_2 = \frac{\sum_{i=1}^{n} cx_i}{n}$$

respectively. It will be shown that the relation $c\bar{x}_1 = \bar{x}_2$ holds. The definition of the summation symbol is used to write

$$\bar{x}_2 = \frac{\sum_{i=1}^{n} cx_i}{n} = \frac{cx_1 + cx_2 + \cdots + cx_n}{n}$$

and a property of real numbers (the distributive property) allows

$$\bar{x}_2 = \frac{c(x_1 + x_2 + \cdots + x_n)}{n}$$

which, by the definition of the mean of S_1, is

$$\bar{x}_2 = c\frac{\sum_{i=1}^{n} x_i}{n} = c\bar{x}_1$$

Hence the formula which follows.

Formula: (**Multiplicative property of the mean**) *If S_1 is a set of real numbers, $S_1 = \{x_1,x_2, \ldots ,x_n\}$, and if each element of S_1 is multiplied by the real number c to obtain $S_2 = \{cx_1,cx_2, \ldots ,cx_n\}$, then the mean of S_2 is the mean of S_1 multiplied by c. Symbolically, we write*

$$\bar{x}_1c = \bar{x}_2$$

Note that the formula for multiplication of the mean also includes division of the mean, since multiplication by $1/c$ ($c \neq 0$) is equivalent to division by c. To fully understand this dual role (multiplication or division), reread the additive (subtractive) property of the mean.

It is sometimes necessary (convenient) to use both the additive and multiplicative properties in one problem. There is a formula for this use; its validity is not given here, but aids to support the argument are given in Exercise 12. The formula is a combination of the two previous ones.

Formula: ***(Additive and multiplicative properties of the mean)*** *If* S_1 *is a set of real numbers,* $S_1 = \{x_1, x_2, \ldots, x_n\}$, *and if each element of* S_1 *is multiplied by the real number* c *and then the real number* d *is added to each element to obtain*

$$S_2 = \{cx_1 + d, cx_2 + d, \ldots, cx_n + d\}$$

then the mean of S_2 *is the mean of* S_1 *multiplied by* c, *with* d *added to the product. Symbolically, we write*

$$c\bar{x}_1 + d = \bar{x}_2$$

Example 7. Some of the "nice" properties of the mean as a measure of the average size of a set of numbers have been illustrated, but it would be unfair not to mention one of the disadvantages of the mean. Consider the set of numbers $\{4,6,6,6,8\}$. The mean of this set is

$$\frac{4 + 6 + 6 + 6 + 8}{5} = {}^{30}\!/_5 = 6$$

Suppose, however, that the data are changed to $\{4,6,6,6,73\}$. The mean of this set is

$$\frac{4 + 6 + 6 + 6 + 73}{5} = {}^{95}\!/_5 = 19$$

The intersection of two sets is the set $\{4,6,6,6\}$, so four of the five elements occur in both sets of data. Yet a change (rather radical) in one element changes the mean from 6 (which seems a fair center of the numbers in some sense or another) to 19 (which really does not seem to have much to do with the numbers at all). The point is that if the data include a few very extreme numbers, the mean may not reflect what is expected of it.

The discussion about the three properties of the mean is now concluded, but the mean will be reconsidered in Sec. 4.7 with reference to grouped data. However, there are other measures of central tendencies, and the second to be considered in this section is the *median*. Whereas the mean selects a number which is in the middle, so to speak, by the arithmetical processes of addition and division, the median is often selected by simply choosing the middle number in a set of *ordered numbers*. A definition will put an end to this somewhat vague talk.

Definition: **(*Median*)** *If* S *is a set of real numbers,*

$$S = \{x_1, x_2, \ldots, x_n\}$$

and if the numbers are arranged in ascending order,

$$x_1 \leq x_2 \leq x_3 \leq \cdots \leq x_n$$

then (1) *if n is odd, $n = 2r + 1$, the median of S is x_{r+1}; and* (2) *if n is even, $n = 2s$, the median of S is $(x_s + x_{s+1})/2$.*

The definition shows that if there is an odd number of elements in S, arranged so that each in the sequence is no less than the number which immediately precedes it, then the "middle" number is the median number. If there is an even number of elements in S, then an averaging process is necessary and the number so selected may or may not be in the set S. So that these ideas may be fixed, a number of examples are offered to illustrate the definition. Each of the sets is arranged so that the elements are in ascending order.

	Set	n	r	$r+1$	s	$s+1$	*Median*
Example 8.	{1,2,3,4,5}	5	2	3			3
Example 9.	{2,5,7,9,10}	5	2	3			7
Example 10.	{3,7,8,11,13}	5	2	3			8
Example 11.	{3,6,9,12,15,18,21}	7	3	4			12
Example 12.	{2,4,6,8,10,12,14}	7	3	4			8
Example 13.	{1,2,3,4}	4			2	3	5/2
Example 14.	{2,5,7,9}	4			2	3	6
Example 15.	{5,6,9,14,16,18}	6			3	4	23/2
Example 16.	{1,5,8,11,62,101}	6			3	4	19/2

Another type of chart (a *cumulative frequency* chart), not mentioned in Sec. 4.1, lends itself easily to the discovery of the median. The data of the experiments in Sec. 4.1 will be used to make one of these charts. In this display, Diagram 5, the vertical axis relates the position of the numbers in the *ordered* set and the horizontal axis, as before, contains the range of the numbers.

For this example, since $33 = 2 \cdot 16 + 1$ is odd, the median is in the set S and is, in fact, the number 161. Note that the number (16) of elements in set S that appear to the left of x_{17} is equal to the number of elements of S that appear to the right of x_{17}; hence x_{17} is (in some sense)

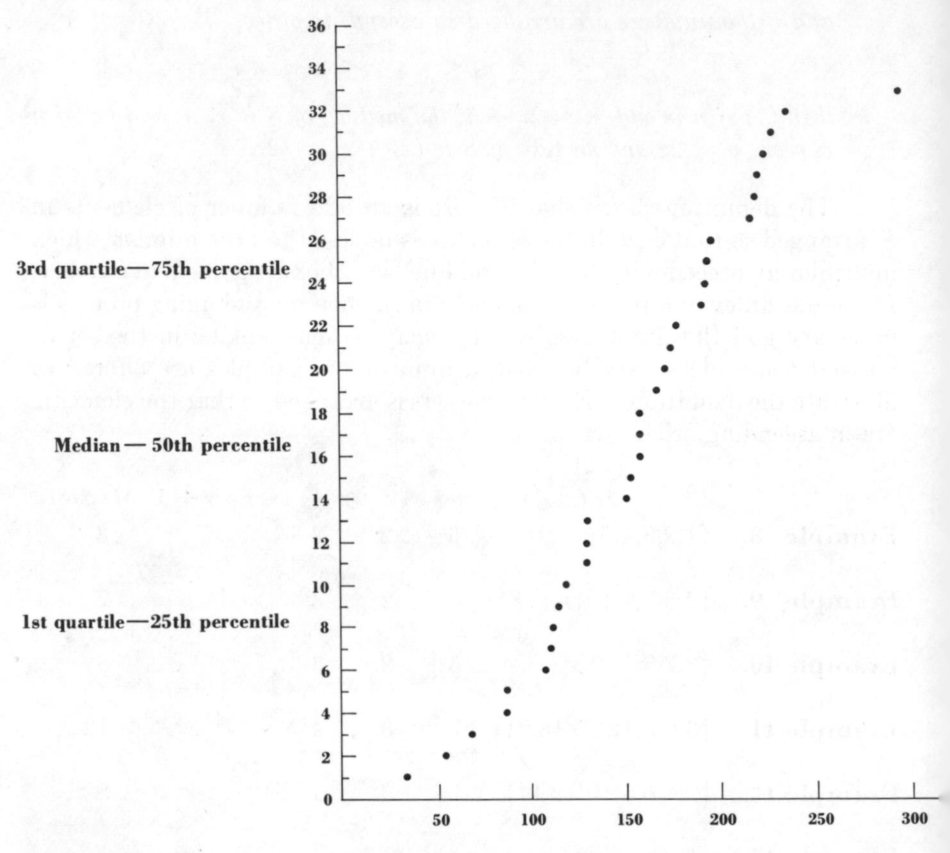

Diagram 5

the center of the numbers. The median partitions the vertical axis into two equal parts. One often reads of first, second, third, and fourth *quartiles* in this connection, and these are achieved by a partition of the vertical axis into four equal portions. These are also indicated in Diagram 5 with the alternate language of 25th percentile, 50th percentile, and 75th percentile. The portion of the set of 33 numbers that is in the first quartile is $\{43,55,70,85,85,102,111,113\}$. The word *decile* is used when the vertical axis is partitioned into 10 equal subsets.

Three properties of the means were advanced as reasons for its importance. Only one of these is investigated for the median, and the acceptance or rejection of the other two properties is left as Exercises 6 to 11. The first property of the mean which was studied was described by the phrase *overall mean.* In imprecise terms, this property says that if k sets of data with $n_1, n_2, \ldots, n_k$ elements and

$$n_1 + n_2 + \cdots + n_k = n$$

are known and if the means $\bar{x}_1, \bar{x}_2, \ldots, \bar{x}_k$ have been computed, then the overall mean is obtained by

$$\sum_{i=1}^{k} \frac{n_i \bar{x}_i}{n}$$

Example 17. One way to phrase this formula is that the overall mean is obtained by averaging the means. Does such a property hold for medians? To see that the answer is *no*, the first two sets in Examples 8 and 9 can be used. They are $\{1,2,3,4,5\}$ and $\{2,5,7,9,10\}$. The medians are 3 and 7, with an overall median of $(3 + 7)/2 = 5$, but the median of the set $\{1,2,2,3,4,5,5,7,9,10\}$ is, by definition,

$$\frac{4+5}{2} = \frac{9}{2} \neq 5$$

The "overall" property for medians does not always hold.

On the other hand, the median (or middle) of qualitative measures is possible, whereas the mean (average) cannot be computed. Refer again to the preference list for colors of towels,

purple, pink, red, blue, white

and observe that if these are arranged in order, the "median" color is red but the mean is undefined.

There is a third measure of central tendency, which is called the *mode*.

Definition: (***Mode***) *If* $S = \{x_1, x_2, \ldots, x_n\}$ *is a set of real numbers, then the mode of S is that element which occurs the largest number of times (if such an element exists).*

Example 18. For example, the mode of $\{1,2,2,3,4\}$ is 2, the mode of $\{2,2,3,3,3,4\}$ is 3, the mode of $\{1,1,2,3\}$ is 1. But what is the mode of the set of numbers in Diagram 5? The number 141 occurs three times, as does the number 161. Hence, the mode is undefined. The concept does generalize to modal intervals, and in Diagram 2, the interval 151 to 200 is the *modal interval* since the frequency is the largest in this interval. Be sure that it is clear why 176 to 200 is the modal interval of Diagram 3. Also see that Diagram 4 does not possess a modal interval. Why? In the preference list of colors, the *modal preference* is that color that more people liked than any other.

Perhaps these discussions have made clear that some judgment must be brought to bear on the problem of selecting the best measure of central tendency. An attempt has been made to show that each of these plays its own role in describing data, and persons who expect to present or read such presentations intelligently should be familiar with the important properties of all three measures.

Exercises

1. Determine the mean, median, and mode of each of the sets of data of Exercise 1 of Sec. 4.1.
2. Use the two sets of data in Exercise 1 of Sec. 4.1, and find the mean, median, and mode of the combined sets.
3. Use the results of the last examination in this course (if they are available), and find the mean, median, and mode of the test results.
4. Let $S = \{6,10,7,9,13,19,22\}$.
   a. Is the median of S the number 9?
   b. If not, what is the median?
5. Let $S = \{2,7,8,13,17,20,24\}$ and $T = \{6,16,18,28,36,42,50\}$.
   a. Compute the mean of S.
   b. Compute the mean of T using only the mean of S. (HINT: Look for a relationship between S and T.)
6. Let $S = \{x_1,x_2,x_3, \ldots ,x_n\}$ and let $T = \{x_1 + c, x_2 + c, x_3 + c, \ldots , x_n + c\}$. Assume S is arranged in ascending order. Assume $n = 2r + 1$; that is, n is odd.
   a. What is the median of S?
   b. What is the median of T?
   c. Is the sum of the median of S and c the median of T?
   d. Does the additive property hold?
7. Rework Exercise 6 for n even.
8. Let S be as in Exercise 6 and $T = \{x_1c,x_2c,x_3c, \ldots ,x_nc\}$, $c > 0$.
   a. What is the median of S?
   b. Is T arranged in ascending order?
   c. What is the median of T?
   d. Is the product of the median of S and c the median of T?
   e. Does the multiplicative property hold?
9. Rework Exercise 8 for n even.
10. Rework Exercise 8 for $c < 0$.
11. Rework Exercise 9 for $c < 0$.
12. Let $S = \{x_1,x_2,x_3, \ldots ,x_n\}$, $T = \{cx_1,cx_2,cx_3, \ldots ,cx_n\}$, and

$$R = \{cx_1 + d, cx_2 + d, cx_3 + d, \ldots , cx_n + d\}$$

Let $\bar{x}_1$ be the mean of S.

a. What is the mean of T?
b. Use the result of part *a* and the additive property of means to write the mean of R.

13. Let $S = \{1,2,3\}$, $T = \{5,6,7\}$, and $R = \{1 + 5, 2 + 6, 3 + 7\} = \{6,8,10\}$.
a. What is the mean of S?
b. What is the mean of T?
c. What is the mean of R?
d. What is the relation between the three means?

14. Rework Exercise 13 for $S = \{x_1,x_2, \ldots ,x_n\}$, $T = \{y_1,y_2, \ldots ,y_n\}$, and $R = \{x_1 + y_1, x_2 + y_2, \ldots , x_n + y_n\}$.

15. If the mean of a set P is $\bar{x}_1$ and the mean of a set Q is $\bar{x}_2$ and if Q has twice as many numbers as P, what is the overall mean?

16. Suppose that a baseball player's batting average for each of the last five years is known. What other information is needed in order to be able to compute his overall average?

Answers to problems

A. $\bar{x} = (86 + 92 + 77)/3 = {}^{255}\!/\!_3 = 85$
$\bar{y} = (9/5) \cdot 85 + 32 = 185$

B. The average is 18.
The average is $3 \cdot 18 = 54$.

C. Accountants use the median because it is less affected by the extremely high values (17.4, for example) and it enables the user to determine at a glance whether his firm is in the top or bottom half of the group.

D. Brand *B*. The mean is based on *total* mileage, while the median is not. Since we are interested in *total* mileage, the mean is much more significant.

4.3 *Variance and deviation*

Problem

A manufacturer of shears finds that the steel from which the blades are made is not always of uniform hardness. If the steel is too soft, the shears will not stay sharp, and if it is too hard, the blades cannot be sharpened with a file. The average hardness desired is specified in the contract with the supplier, but no supplier produces blade metal which is absolutely uniform. How can the manufacturer measure the uniformity of a shipment to determine whether it should be accepted or rejected?

The concepts of variance and deviation are used to describe the uniformity of a set of observations, which in many cases is as important as the measurement of average size. The most important of the measures which describe uniformity or lack of uniformity are the *variance* and its square root, which is called the *standard deviation*. Other words that are sometimes used in this connection may be helpful in indicating what kinds of measures are to be developed, and they are *dispersion* and *variability*. The words *variance* and *deviation* are being used at present in a general sense, but a little later in this section precise meaning will be given to them and formulas will be developed for computational purposes.

Just as there are several ways to measure tendencies toward a center of data, there are also several ways to measure the scatter of a set of data. In this section, we shall discuss several possible measures and point out the obvious weaknesses of some of them before settling on the measure which will occupy most of this section and a large portion of Secs. 4.4, 4.5, and 4.7 as well. Since so much emphasis is to be placed on this measure, it will pay to read with care the preliminary comments about why it is selected from among the list of possible measures.

Since a judgment about the relative values of some measures of dispersion is to be made, it is essential to have several sets of numbers that possess varying degrees of dispersion at hand to analyze. The sets which will serve as examples have been chosen to meet the varying dispersion criteria as well as two other criteria: (1) that the size of the numbers in the sets be small and (2) that the number of elements in each set also be small. The sets are presented for consideration devoid of any practical situation which might possibly give rise to them, such as being the result of some experiment, survey, or other natural occurrence.

Listed below are six sets. Their dot frequency charts are given in Diagram 6.

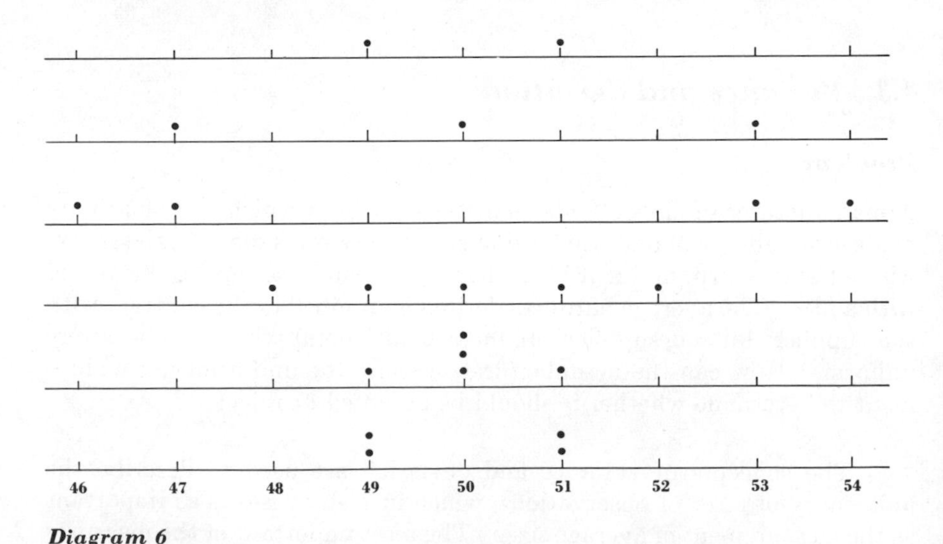

Diagram 6

Example 1. $S_1 = \{49,51\}$

Example 2. $S_2 = \{47,50,53\}$

Example 3. $S_3 = \{46,47,53,54\}$

Example 4. $S_4 = \{48,49,50,51,52\}$

Example 5. $S_5 = \{49,50,50,50,51\}$

Example 6. $S_6 = \{49,49,51,51\}$

Perhaps some of these sets seem to you to be more scattered than others. Since one purpose of this consideration is to select a good measure of dispersion, study each of the charts and try to decide which seems the most scattered and which the least scattered. Before any measures are tried on these sets, it may be helpful to use the other type of diagram, the histogram, as a visual aid in determining the dispersion. Diagram 7 shows a histogram for each of the sets.

Now that a second visual aid is available, make another judgment about which of these sets seems the most scattered. Did your opinion change at all as a result of Diagram 7? It may not be completely clear which of these is the most (least) scattered, but perhaps what follows will aid in deciding.

Since the mean has been demonstrated to be a very important measure of the tendency of the data about a center, it should be observed that the mean of each set is 50. (For review and to verify the previous

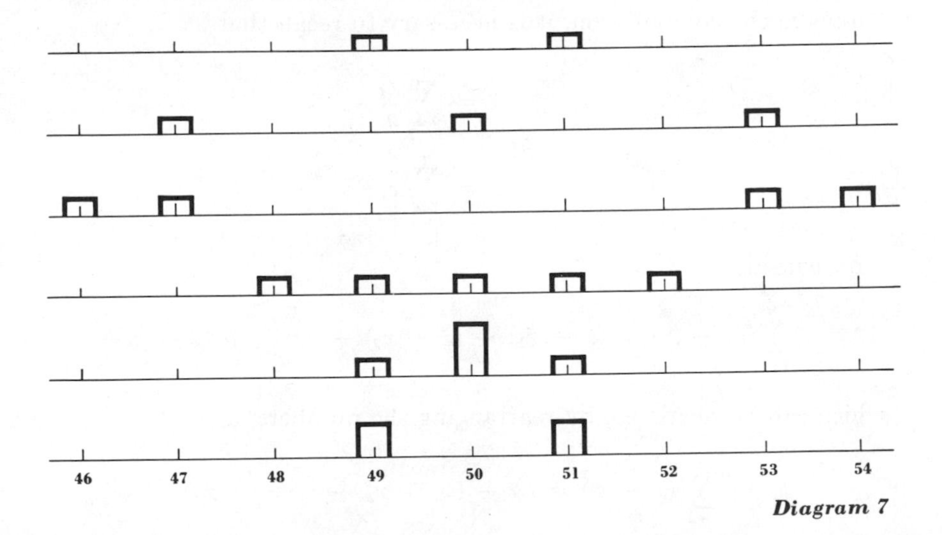

Diagram 7

statement, compute the mean of each set by use of the definition of the mean.) Therefore, the center, as described by the mean, is the same for each of the six sets. The mean portrays the typical size, whereas the variance measures the degree of uniformity (or lack thereof). We are now ready to experiment with some measures of dispersion.

Example 7. The first measure to be tested is the *average scatter* of the elements of each set *from the mean.* Of course, *average scatter* has not been defined, but there is a connotation for the word *scatter* and the average of a set of numbers is obtained by adding the numbers and dividing by the number of elements in the set. Applied to the first set, the element 49 is 1 less (-1) than 50, and the second element is 1 more $(+1)$ than 50. The sum of -1 and $+1$ is zero, and $0/2 = 0$. The corresponding numbers for the second set are -3, 0, $+3$, and their sum is also zero, with $0/3 = 0$. For the third set, the numbers are -4, -3, $+3$, $+4$, with $-4 - 3 + 3 + 4 = 0$ and $0/4 = 0$. It is easy to verify that the sums obtained for the fourth, fifth, and sixth sets are all also zero. Although the sets have been chosen carefully, this particular result is not dependent on the choice of the sets, as the argument of the next paragraph will show.

Suppose that a set $S = \{x_1, x_2, \ldots, x_n\}$ is known and that the mean $\bar{x}$ has been computed. What is the number

$$\sum_{i=1}^{n} (\bar{x} - x_i) = (\bar{x} - x_1) + (\bar{x} - x_2) + \cdots + (\bar{x} - x_n)$$

Remember that for each of the six sets in the six examples, this number is zero. A computation will be made to see if this is the case for any set S. To make the computation, it is necessary to recall that

$$\bar{x} = \sum_{i=1}^{n} \frac{x_i}{n}$$

or

$$n\bar{x} = \sum_{i=1}^{n} x_i$$

Now consider

$$\sum_{i=1}^{n} (\bar{x} - x_i) = (\bar{x} - x_1) + (\bar{x} - x_2) + \cdots + (\bar{x} - x_n)$$

which can be rewritten, by rearranging the numbers, as

$$\sum_{i=1}^{n} (\bar{x} - x_i) = n\bar{x} - (x_1 + x_2 + \cdots + x_n)$$

Replacing $n\bar{x}$ by $\sum_{i=1}^{n} x_i$ yields

$$\sum_{i=1}^{n} (\bar{x} - x_i) = \sum_{i=1}^{n} x_i - (x_1 + x_2 + \cdots + x_n) = \sum_{i=1}^{n} x_i - \sum_{i=1}^{n} x_i$$
$$= 0$$

These computations verify the fact that the average scatter from the mean is always zero. The following formula summarizes the situation.

Formula: **(*Average scatter from the mean*)** *If* $S = \{x_1, x_2, \ldots, x_n\}$ *is a set of real numbers with a mean of* $\bar{x}$, *then the average scatter of* S *from the mean is zero. Symbolically, we write*

$$\sum_{i=1}^{n} \frac{\bar{x} - x_i}{n} = \frac{0}{n} = 0$$

Since the average scatter from the mean is always zero (regardless of the set under consideration), this possible measure of scatter is rejected as uninformative. The average scatter was considered to show that one "natural" measure does not work.

Example 8. The second measure to be tested is the *average scatter with respect to the range.* Certainly the range, which is the difference between the largest element and the smallest element, is important since it tells over how large an interval the data are scattered. If the range is divided by the number of elements in the set, the resulting number is the average length of an interval which contains one member of the set. To proceed with this test, the six averages (with respect to the range) are computed by the use of the formula for the range, $b - a$, and division by n. They are

$$\frac{b-a}{n} = \frac{51 - 49}{2} = 2/2 = 1$$
$$= \frac{53 - 47}{3} = 6/3 = 2$$
$$= \frac{54 - 46}{4} = 8/4 = 2$$
$$= \frac{52 - 48}{5} = 4/5$$
$$= \frac{51 - 49}{5} = 2/5$$
$$= \frac{51 - 49}{4} = 2/4 = 1/2$$

But look at the dot frequency charts in Diagram 6 and compare Examples 1 and 6. From the chart, it would be expected that Examples 1 and 6 should have the same average scatter. They do *not* by this measure since this formula yields 1 for Example 1 and ½ for Example 6. If more convincing evidence of the ineffectiveness of this formula is needed, consider the second and third sets. The charts certainly have a different appearance, and a different variance is anticipated. Yet the formula yields the number 2 for each set. As a result of these observations, the measure "range/n" is not the best because the result is not always the "intuitive" result expected. It is readily admitted that the measure has some merit (as do the median and the mode), but it is not the measure that is found to be consistently the most valuable one.

Example 9. The third measure to be tried is a variant of the first one tested, the average scatter from the mean. Recall that the difficulty encountered with that test was that $\sum_{i=1}^{n} (\bar{x} - x_i)$ is always zero, and while this is a nice property to know, it makes ineffectual the formula $\left[\sum_{i=1}^{n} (\bar{x} - x_i)\right] / n$ since the quotient is always zero. But perhaps this difficulty can be overcome. After all, the trouble lies in the fact that some of the differences are negative (to the right of the mean) while others are positive (to the left of the mean), so that the sum is zero. An obvious correction is to make all the differences positive. But how can this be accomplished? Not all the numbers can be greater than the average! Imagine a sales manager who tells salesmen that every salesman must exceed the average in sales! There is a mathematical function for turning negative numbers into positive, which you may have encountered in previous studies, called the *absolute value*. There is even a well-known symbol for the function, and it consists of two vertical bars, $| \quad |$. The absolute value of zero is zero, the absolute value of x is x, and the absolute value of $-y$ is y. Some examples are

$$|3| = 3 \qquad |5| = 5 \qquad |-4| = 4 \qquad |0| = 0 \qquad |-13| = 13$$

This idea can be applied to the formula $\left[\sum_{i=1}^{n} (\bar{x} - x_i)\right] / n$ so that only positive or zero differences will result. The third trial formula becomes

$$\sum_{i=1}^{n} \frac{|\bar{x} - x_i|}{n}$$

Consider what results are obtained if this formula is applied to each of the six sets of elements under consideration. Make the computations to be sure that the following are correct.

$$\sum_{i=1}^{n} \frac{|\bar{x} - x_i|}{n} = \sum_{i=1}^{2} \frac{|50 - x_i|}{2} = \frac{|50 - 49| + |50 - 51|}{2} = \frac{|1| + |-1|}{2}$$

$$= \frac{1 + 1}{2} = 1$$

$$\sum_{i=1}^{3} \frac{|50 - x_i|}{3} = \frac{|50 - 47| + |50 - 50| + |50 - 53|}{3}$$

$$= \frac{3 + 0 + 3}{3} = 6/3 = 2$$

$$\sum_{i=1}^{4} \frac{|50 - x_i|}{4} = \frac{|50 - 46| + |50 - 47| + |50 - 53| + |50 - 54|}{4}$$

$$= \frac{4 + 3 + 3 + 4}{4} = 7/2$$

$$\sum_{i=1}^{5} \frac{|50 - x_i|}{5} = \frac{2 + 1 + 0 + 1 + 2}{5} = 6/5$$

$$\sum_{i=1}^{5} \frac{|50 - x_i|}{5} = \frac{1 + 0 + 0 + 0 + 1}{5} = 2/5$$

$$\sum_{i=1}^{4} \frac{|50 - x_i|}{4} = \frac{1 + 1 + 1 + 1}{4} = 1$$

These results should be evaluated in terms of the charts in Diagram 6, and it is expected that many, if not all, of the comparisons will agree with what one intuitively expects. For this reason, this is a measure that is frequently used for measuring scatter. But this measure has some shortcomings, which will not be enumerated here, that cause it to be less frequently used than the final measure to be discussed in what follows.

Example 10. The fourth measure of dispersion is also a version of the average scatter from the mean but uses a different method for eliminating the negative differences. Another scheme for turning negative numbers into positive numbers is to square each number. The square of the number x is the product $x \cdot x$, and it is never negative. The formula becomes

$$\sum_{i=1}^{n} \frac{(\bar{x} - x_i)^2}{n}$$

So that this formula can be understood, it is applied to each of the six sets.

$$\sum_{i=1}^{n} \frac{(\bar{x} - x_i)^2}{n} = \sum_{i=1}^{2} \frac{(50 - x_i)^2}{2} = \frac{(50 - 49)^2 + (50 - 51)^2}{2}$$

$$= \frac{1^2 + (-1)^2}{2} = \frac{1 + 1}{2} = 1$$

$$\sum_{i=1}^{3} \frac{(\bar{x} - x_i)^2}{3} = \frac{(50 - 47)^2 + (50 - 50)^2 + (50 - 53)^2}{3}$$

$$= \frac{3^2 + 0^2 + (-3)^2}{3}$$

$$= \frac{9 + 0 + 9}{3} = 18/3 = 6$$

$$\sum_{i=1}^{4} \frac{(\bar{x} - x_i)^2}{4} = \frac{(50 - 46)^2 + (50 - 47)^2 + (50 - 53)^2 + (50 - 54)^2}{4}$$

$$= \frac{4^2 + 3^2 + (-3)^2 + (-4)^2}{4} = \frac{16 + 9 + 9 + 16}{4}$$

$$= 50/4 = 25/2$$

$$\sum_{i=1}^{5} \frac{(\bar{x} - x_i)^2}{5} = \frac{2^2 + 1^2 + 0^2 + (-1)^2 + (-2)^2}{5}$$

$$= \frac{4 + 1 + 0 + 1 + 4}{5} = 10/5 = 2$$

$$\sum_{i=1}^{5} \frac{(\bar{x} - x_i)^2}{5} = \frac{1^2 + 0^2 + 0^2 + 0^2 + (-1)^2}{5} = \frac{1 + 0 + 0 + 1}{5}$$

$$= 2/5$$

$$\sum_{i=1}^{4} \frac{(\bar{x} - x_i)^2}{4} = \frac{1^2 + 1^2 + (-1)^2 + (-1)^2}{4} = \frac{1 + 1 + 1 + 1}{4} = 1$$

It is interesting, and significant, that the results of this test order the sets with respect to variance from the mean in the same way that the third trial formula did. In ascending order, the sets, for this formula and the preceding one, are 5; a tie between 6 and 1; 4; 2; 3. Hence, if it was felt that the third measure was intuitively correct for this example, then this measure should also appear to be "right" for this example.

The fourth measure proves to be the most practical, and it is used to give precise meaning to the word *variance*, which has been used imprecisely in prior discussions. This is not to say that other measures of variance are unimportant. But just as the median and mode have been given secondary roles because of the proved importance of the mean, so the measures discussed in Examples 7 to 9 are relegated to a minor role in favor of the fourth measure considered, which is stated precisely in the following formula.

Formula: **(*Variance*)** *If* $S = \{x_1, x_2, \ldots, x_n\}$ *is a set of real numbers, then the variance of* S *from the mean is the quotient of the sum of the squares of the differences of the mean and each element of* S *by the number* n. *In symbols, we write*

$$v = \sum_{i=1}^{n} \frac{(\bar{x} - x_i)^2}{n}$$

Note that the letter v is to be used to represent variance.

There is a number, which is computed from the variance, that is also used extensively in describing the scatter of a set. The number is the *standard deviation*, and it is explained by the next formula.

Formula: **(*Standard deviation*)** *If* $S = \{x_1, x_2, \ldots, x_n\}$ *is a set of real numbers, then the standard deviation of* S *from the mean is the square root of the variance. Symbolically, we write*

$$s = \sqrt{\sum_{i=1}^{n} \frac{(\bar{x} - x_i)^2}{n}}$$

Note that the formulas for variance and standard deviation are also definitions, adopted as a result of the preceding discussions.

Near the beginning of Sec. 4.2, we observed that the choice of $\bar{x}$ to represent the mean was a temporary decision. It reflects the usual notation for a sample. Similar remarks apply to the use of s to represent standard deviation. Any lengthy discussion of population versus samples is delayed until Sec. 4.9. In the meantime, read s as "standard deviation." The symbol $\sqrt{}$ is read "square root," and it has a special meaning that is reviewed in the following definition.

Definition: **(*Square root*)** *If* x *is a nonnegative real number* $(0 \leq x)$, *then the square root of* x *is* y *if and only if* y *is also a nonnegative number* $(0 \leq y)$ *and the product of* y *and* y *is* x. *Symbolically, we write*

$$\sqrt{x} = |y| \text{ with } y \cdot y = x$$

Some examples of square roots are

$$\sqrt{9} = 3 \qquad \sqrt{0} = 0 \qquad \sqrt{25/16} = 5/4$$
$$\sqrt{4} = 2 \qquad \sqrt{1} = 1 \qquad \sqrt{0.01} = 0.1$$
$$\sqrt{9/4} = 3/2 \qquad \sqrt{16} = 4 \qquad \sqrt{0.25} = 0.5$$

The table of square roots on page 376 is used for some of the results listed in Table 4.

Now that the five measures have been tested, evaluated, and discarded (in the first case) or accepted, a summary is in order. Table 4 lists the results that have been obtained. In the last column, which shows the standard deviation, the numbers in some cases represent an approximation.

Table 4

		1	*2*	*3*	*4*	*5*
Example	$\sum_{i=1}^{n} \frac{\bar{x} - x_i}{n}$	***Range***	$\frac{\textit{Range}}{n}$	$\sum_{i=1}^{n} \frac{\lvert\bar{x} - x_i\rvert}{n}$	$\sum_{i=1}^{n} \frac{(\bar{x} - x_i)^2}{n}$	$s = \sqrt{\sum_{i=1}^{n} \frac{(\bar{x} - x_i)^2}{n}}$
1	0	2	1	1	1	1
2	0	6	2	2	6	2.45
3	0	8	2	7/2	25/2	3.54
4	0	4	4/5	6/5	2	1.414
5	0	2	2/5	2/5	2/5	0.632
6	0	2	1/2	1	1	1

Now that the preliminaries are over and formulas for variance and standard deviation have been selected, compare the results of Table 4 with your original estimate of dispersion to see if they are in agreement.

Example 11. Since a definition for variance has been selected, it will be illustrated with an example. The example is first worked using just the definition, and this is followed with a second method which is somewhat easier. In the next section, still another method will be given, so read the examples to evaluate the various methods of computation. The example will use the set of results of the hypothetical experiments of Sec. 4.1, which yield the set of numbers

$$S = \{43,55,70,85,85,102,111,113,122,136,141,\\ 141,141,150,153,161,161,161,172,177,180,184,\\ 190,192,193,195,210,213,214,217,220,230,296\}$$

with a mean of 158. To determine the variance, 33 differences must be computed, the square of these numbers computed (33 operations), the sum of the squares obtained by addition (32 operations), and the quotient of the sum divided by n computed (1 operation), for a total of 99 operations. It is true that the number of elements in the set is relatively small, as are the elements themselves, but even so, before the computations are very far along, the need for aid such as a desk calculator is apparent. Some of the details will be presented so that the difficulties can be appreciated.

The set of differences,

$$T = \{\bar{x} - x_i | x_i \in S\} = \{158 - x_i | x_i \in S\}$$

is computed as

$$T = \{115,103,88,73,73,56,47,45,36,22,17, 17,17,8,5,-3,-3,-3,-14,-19,-22,-26, -32,-34,-35,-37,-52,-55,-56,-59,-62,-72,-138\}$$

The next step in the process is to form the square of each of these numbers. In symbols, $R = \{y | y = x^2 \text{ and } x \in T\}$. It is suggested that you do *not* verify the elements of the next set, but it is

$$R = \{13{,}225,\ 10{,}609,\ 7{,}744,\ 5{,}329,\ 5{,}329,\ 3{,}136,\ 2{,}209,\ 2{,}025,\ 1{,}296,\ 484,\ 289,\ 289,\ 289,\ 64,\ 25,\ 9,\ 9,\ 9,\ 196,\ 361,\ 484,\ 676,\ 1{,}024,\ 1{,}156,\ 1{,}225,\ 1{,}369,\ 2{,}704,\ 3{,}025,\ 3{,}136,\ 3{,}481,\ 3{,}844,\ 5{,}184,\ 19{,}044\}$$

The next step is to form the sum of these 33 numbers (and again never mind the verification); the result is 99,278. The final operation is to divide by n ($n = 33$ in this example), and the variance is $\cong$ 3,008.424. The square root of the variance is the standard deviation, which is (approximately) 54.85.

There is a second formula for computing the variance that is somewhat easier and with (or without) a desk calculator is more frequently used. To show how this formula is derived from the already known formula, trace through the algebraic manipulation that follows. Variance is given by

$$v = \sum_{i=1}^{n} \frac{(\bar{x} - x_i)^2}{n}$$

or

$$nv = \sum_{i=1}^{n} (\bar{x} - x_i)^2$$

Since $(\bar{x} - x_i)^2 = \bar{x}^2 - 2\bar{x}x_i + x_i^2$ for each element x_i (this is an application of the binomial formula), the right-hand side of the equality involving nv becomes

$$nv = (\bar{x}^2 - 2\bar{x}x_1 + x_1^2) + (\bar{x}^2 - 2\bar{x}x_2 + x_2^2) + \cdots + (\bar{x}^2 - 2\bar{x}x_n + x_n^2)$$

which, upon regrouping the terms, becomes

$$nv = n\bar{x}^2 - 2\bar{x}(x_1 + x_2 + \cdots + x_n) + (x_1^2 + x_2^2 + \cdots + x_n^2)$$

But

$$n\bar{x} = x_1 + x_2 + \cdots + x_n$$

so that

$$\begin{aligned} nv &= n\bar{x}^2 - 2\bar{x}(n\bar{x}) + (x_1^2 + x_2^2 + \cdots + x_n^2) \\ &= n\bar{x}^2 - 2n\bar{x}^2 + (x_1^2 + x_2^2 + \cdots + x_n^2) \\ &= (x_1^2 + x_2^2 + \cdots + x_n^2) - n\bar{x}^2 \end{aligned}$$

Division by n yields the final formula:

$$v = \frac{x_1^2 + x_2^2 + \cdots + x_n^2}{n} - \bar{x}^2 = \frac{\sum_{i=1}^{n} x_i^2}{n} - \bar{x}^2$$

The final result of these computations is contained in the following formula.

Formula: (***Alternate formula for variance***) *If* $S = \{x_1, x_2, \ldots, x_n\}$ *is a set of real numbers, then the variance of S from the mean is the difference of the quotient of the sum of the squares of the elements by n and the square of the mean of S. Symbolically, we write*

$$v = \frac{x_1^2 + x_2^2 + \cdots + x_n^2}{n} - \bar{x}^2 = \frac{\sum_{i=1}^{n} x_i^2}{n} - \bar{x}^2$$

To see how this formula compares with the formula given previously, with reference to Example 1 of Sec. 4.1, note that there are 33 squares to be computed, followed by addition and then a division. Then the mean must be squared (one operation), with the final step a subtraction. Hence there are $33 + 32 + 1 + 1 + 1 = 68$ operations necessary, as compared with 99 operations if the first formula is used. That there are fewer operations to perform is one of the reasons why the second formula for variance is usually preferred over the first.

So that the actual computations can be compared, the second formula will now be used. The set of data is

$$\begin{aligned} S = \{&43,55,70,85,85,102,111,113,122,136,141, \\ &141,141,150,153,161,161,161,172,177,180,184, \\ &190,192,193,195,210,213,214,217,220,230,296\} \end{aligned}$$

The next step is to compute the square of each of these numbers. The set of squares is

$$\begin{aligned} T = \{&1{,}849,\ 3{,}025,\ 4{,}900,\ 7{,}225,\ 7{,}225,\ 10{,}404,\ 12{,}321,\ 12{,}769,\ 14{,}884, \\ &18{,}496,\ 19{,}881,\ 19{,}881,\ 19{,}881,\ 22{,}500,\ 23{,}409,\ 25{,}921, \\ &25{,}921,\ 25{,}921,\ 29{,}584,\ 31{,}329,\ 32{,}400,\ 33{,}856,\ 36{,}100, \\ &36{,}864,\ 37{,}249,\ 38{,}025,\ 44{,}100,\ 45{,}369,\ 45{,}796,\ 47{,}089, \\ &48{,}400,\ 52{,}900,\ 87{,}616\} \end{aligned}$$

The next step is to form the sum of these numbers, which yields 923,090. Division by $n = 33$ gives 27,972.424. The mean, $\bar{x}$, is 158, so $\bar{x}^2$ is $(158)^2$, which is 24,964. The difference of 27,972.424 and 24,964 is 3,008.424, which agrees with the previous result.

Exercises

1. Given the set $S = \{8,12,12,16,18,18\}$.
   a. Compute the average scatter from the mean.
   b. Compute the average scatter in the range.
   c. Compute the average absolute-value scatter from the mean.
   d. Compute the variance by the first formula.
   e. Compute the variance by the second formula.
   f. Compute the standard deviation.
2. Rework Exercise 1 for the set $T = \{12,12,16,18,22\}$.
3. Rework Exercise 1 for the set $R = \{2,6,8,14,18,20\}$.
4. What are the absolute values of the following numbers?
   a. 9 *b.* -7 *c.* $-3/-4$
   d. 12 *e.* -16 *f.* 0
5. Let the standard deviation of a set A be equal to twice the standard deviation of a set B.
   a. What can you say about the scatter of A as compared with the scatter of B?
   b. What can you say about the range?
6. If desk calculators are available, verify the variance and standard deviation of the set in Example 1 of Sec. 4.1 by:
   a. The first formula
   b. The second formula
7. The standard deviation has been defined as the square root of the variance.
   a. If the variance of a set is 2,500, what is the standard deviation? Which is the larger?
   b. If the variance of a set is $\frac{1}{4}$, what is the standard deviation? Which is the larger?
   c. For what numbers is the variance larger than the standard deviation?
   d. For what numbers is the variance smaller than the standard deviation?
8. Explain the following statement: If the variance of a set is zero, then the elements are equal.
9. The coefficient of variance is, by definition, $(s/\bar{x}) \cdot 100$.
   a. Compute the coefficient of variance for Exercise 1.
   b. Compute the coefficient of variance for Exercise 2.
   c. Compute the coefficient of variance for Exercise 3.
   d. Which one is, relatively speaking, less variable?

Answer to problem

The variation in hardness of the blade stock can be measured by the range, the average deviation, the variance, or the standard deviation. Of these, the standard deviation is most widely used.

4.4 *Additive and multiplicative properties of variance and standard deviation*

Problems

A. If $S = \{4,6,10,14,16\}$ and $T = \{12,14,18,22,24\}$, then what is the variance of S? Compute the variance of T by using only the variance of S and the relation between the elements of S and T.

B. Let S be the same as in Prob. A and $T = \{8,12,20,28,32\}$. Compute the variance of T and compare it with the value that you got for the variance of S. Do you see a relationship between the two?

C. A test of five pieces of product A and an equal number of pieces of product B shows the following results:

A		B	
6	$\bar{x}_A = 12$	10	$\bar{x}_B = 16$
10		14	
12		16	
12		16	
20		24	
60		80	

Each observation in set B is exactly 4 greater than the corresponding item in set A.

1. Find the standard deviation for sets A and B and compare them.
2. Would you say that the two sets have the same variance?
3. Assuming that a high score is a desirable quality, would you prefer A or B?
4. Would you agree that, while the two sets have the same variance, set B is *relatively* more uniform than set A?

The effect on the mean of adding a number c to each element of a set of numbers (data) has been established; the mean increases by c. Similarly, the effect on the mean of multiplying each element of a set by a number c has been determined; it multiplies the mean by c. The concern in this section will be with the effect on the variance of addition and multiplication of each element of a set by a number c. Since the standard deviation is the square root of the variance, there will be corresponding results for the standard deviation. Besides the obvious payoff in terms of formulas, there will also be exhibited a third method of computing variance (two were presented in Sec. 4.3), which is often easier to use than the ones already illustrated. The property of addition will be investigated first.

In accordance with the usual practice, an example will be given so that a pattern can be established.

Example 1. Consider the set $S = \{46,50,54\}$, which is chosen so that not many elements are in the set and the numbers are easily manipulated. The mean of S is $\bar{x} = 50$. Average the numbers in S to see that 50 is the mean. Now add 5 to each member of S to obtain $T = \{51,55,59\}$, which has as a mean 55. It has already been established that adding 5 to each element increases the mean by 5. But this example is concerned with determining the effect on the variance of adding 5 to each element. So next the variance of each set is computed. First for S,

$$v = \sum_{i=1}^{n} \frac{(\bar{x} - x_i)^2}{n} = \sum_{i=1}^{3} \frac{(\bar{x} - x_i)^2}{3} = \sum_{i=1}^{3} \frac{(50 - x_i)^2}{3}$$
$$= \frac{(50 - 46)^2 + (50 - 50)^2 + (50 - 54)^2}{3} = \frac{16 + 0 + 16}{3} = 32/3$$

For the variance of T, we have

$$v = \sum_{i=1}^{3} \frac{(\bar{x} - x_i)^2}{3} = \sum_{i=1}^{3} \frac{(55 - x_i)^2}{3}$$
$$= \frac{(55 - 51)^2 + (55 - 55)^2 + (55 - 59)^2}{3}$$
$$= \frac{16 + 0 + 16}{3} = 32/3$$

For this example, the variance of S is the same as the variance of T. It is true that the set S was especially chosen so that the arithmetic would be simple. But the conclusion is not misleading, and the algebraic argument which follows will show the validity of a formula for which this example is but a special case.

To establish a general result, let $S = \{x_1,x_2, \ldots ,x_n\}$ and let the mean of the set S be $\bar{x}$. In keeping with the current trend of thought, let T be a set formed from S by adding a number c to each element of S, so that

$$T = \{x_1 + c,\ x_2 + c,\ \ldots,\ x_n + c\}$$

By a formula of Sec. 4.2, the mean of T is $\bar{x} + c$ (see page 213). This information is sufficient to allow computations which parallel those for

Example 1, and the variance of S is

$$v = \sum_{i=1}^{n} \frac{(\bar{x} - x_i)^2}{n}$$

But how does this compare with the variance of T? Remember that the mean of T is $\bar{x} + c$ and that each element of T is of the form $x_i + c$. These facts (and the definition of variance) lead to the formula for variance of T, which is

$$v = \sum_{i=1}^{n} \frac{[(\bar{x} + c) - (x_i + c)]^2}{n}$$

Since the number $(\bar{x} + c) - (x_i + c)$ is more simply written as

$$(\bar{x} - x_i) + (c - c) = \bar{x} - x_i$$

the variance of T is also

$$v = \sum_{i=1}^{n} \frac{(\bar{x} - x_i)^2}{n}$$

so that the variance of the set T is unchanged. See the next formula for a precise statement.

Formula: (**The additive property of variance**) *If S is a set of real numbers, $S = \{x_1,x_2, \ldots ,x_n\}$, and T is derived from S by adding a real number c to each element of S,*

$$T = \{x_1 + c, x_2 + c, \ldots , x_n + c\}$$

then the variance of S is equal to the variance of T. Symbolically, if v_S is the variance of S and v_T is the variance of T, then

$$v_S = v_T$$

Since standard deviation is the square root of variance, the next formula is an immediate consequence (corollary) of the preceding formula.

Formula: (**The additive property of standard deviation**) *If S is a set of real numbers and T is derived from S by adding the number c to each element of S, then the standard deviation of S is equal to the standard deviation of T. Symbolically, if s_S is the standard deviation of S and s_T is the deviation of T, then*

$$s_S = s_T$$

There is a nice application of these formulas in terms of computing the variance of a set. To appreciate the simplification that this offers, the computations of the next example should be rather closely followed and compared (perhaps even step by step) with the two computations of Sec. 4.3.

Example 2. Consider once again (please) the example offered by the results of the experiment of Sec. 4.1. The set S is

$$S = \{43,55,70,85,85,102,111,113,122,136,141,\\ 141,141,150,153,161,161,161,172,177,180,184,\\ 190,192,193,195,210,213,214,217,220,230,296\}$$

The mean of S has already been computed and is 158. Form a new set from S by adding -158 to each element (of course, this is the same as subtracting 158 from each element). The set T which results is

$$T = \{-115,-103,-88,-73,-73,-56,-47,-45,-36,-22,-17,\\ -17,-17,-8,-5,3,3,3,14,19,22,26,\\ 32,34,35,37,52,55,56,59,62,72,138\}$$

Because the mean of S is 158, the mean of T is zero. Be sure this statement is clear by recalling the additive property of the mean. Since, by the additive property of the variance, the variance of S is the same as that of T, it is possible to compute the variance of S by computing the variance of T. The fact that the mean of T is zero will be crucial in what follows. To compute the variance of T, the formula

$$v = \sum_{i=1}^{n} \frac{(\bar{x} - x_i)^2}{n} = \sum_{i=1}^{n} \frac{(0 - x_i)^2}{n} = \sum_{i=1}^{33} \frac{x_i^2}{n}$$

will be used. Each element of T is squared, the sum of the products is computed (99,278), and the sum is divided by $n = 33$. The result is 3,008.424.

Compare the number of operations of this third procedure with the number of operations of each of the two methods of Sec. 4.3 to see that this third method requires no more operations than the first of the other two. Perhaps more importantly, observe that the numbers in the set T are smaller, hence (?) the operations are more easily performed. At any rate, a new formula is possible; it follows.

Formula: (***Second alternate formula for variance***) *If*

$$S = \{x_1, x_2, \ldots, x_n\}$$

with a mean of $\bar{x}$, then the variance of S is equal to the variance of a set T which is derived from S by adding $(-\bar{x})$ to each element of S.

In all fairness, it should be observed that the mean of the example is a counting number, which makes the arithmetic easier than might be expected. In practice, it is sometimes the case that a number very close to, but not exactly equal to, the mean is used along with the additive (subtractive) property to reduce the size of the numbers for the purposes of computing the variance.

This last formula also makes clear another method of determining the standard deviation, because the square root of the variance is the standard deviation.

So much for the additive property for variance. Now consider the effect of multiplying each element in a set of numbers by c, a real number. A small set to motivate the general result will be used.

Example 3. The set $S = \{8,10,12\}$ has a mean of 10, and the set $T = \{24,30,36\}$, which is formed from S by multiplying each element by 3, has a mean of 30. (Do the arithmetic.) Use of the definition of variance on the set S yields

$$v_S = \sum_{i=1}^{n} \frac{(\bar{x} - x_i)^2}{n} = \sum_{i=1}^{3} \frac{(\bar{x} - x_i)^2}{3}$$

$$= \frac{(10 - 8)^2 + (10 - 10)^2 + (10 - 12)^2}{3}$$

$$= \frac{4 + 0 + 4}{3} = 8/3$$

The same formula used on the set T gives

$$v_T = \sum_{i=1}^{n} \frac{(\bar{x} - x_i)^2}{n} = \sum_{i=1}^{3} \frac{(\bar{x} - x_i)^2}{3}$$

$$= \frac{(30 - 24)^2 + (30 - 30)^2 + (30 - 36)^2}{3}$$

$$= \frac{36 + 0 + 36}{3} = 72/3$$

This time the variances are not equal, so the point of the example is to attempt to guess what relation exists between the important numbers $c = 3$, $v_S = 8/3$, and $v_T = 72/3$. Since $72/3$ is nine times as large as $8/3$ $(8/3 \cdot 9/1 = 72/3)$ and because $c^2 = 3 \cdot 3 = 9$, it can be expected that the two variances are related by a factor of c^2. To see how this guess is correct read the next paragraph.

To establish a formula about variance and muliplication, let

$$S = \{x_1,x_2, \ . \ . \ . \ ,x_n\}$$

with T the set derived from S by multiplying each element of S by the number c,

$$T = \{cx_1,cx_2, \ . \ . \ . \ ,cx_n\}$$

The mean of S is denoted by $\bar{x}$, and the variance of S is given by

$$v_S = \sum_{i=1}^{n} \frac{(\bar{x} - x_i)^2}{n}$$

The mean of the set T is the mean of S multiplied by c, $c\bar{x}$. (See page 215.) Since the elements of T are of the form cx_i, the variance of T is given by

$$v = \sum_{i=1}^{n} \frac{(c\bar{x} - cx_i)^2}{n}$$

Some of the properties of real numbers can be brought into play so that this expression can be reorganized in a fashion that permits easy comparison with the formula for variance of S. The properties used in the computation which follows are the distributive property, the commutative property, and the associative property, but their names are relatively unimportant to this discussion.

$$\begin{aligned}
v &= \sum_{i=1}^{n} \frac{(c\bar{x} - cx_i)^2}{n} \\
&= \frac{(c\bar{x} - cx_1)^2 + (c\bar{x} - cx_2)^2 + \cdots + (c\bar{x} - cx_n)^2}{n} \\
&= \frac{[c(\bar{x} - x_1)]^2 + [c(\bar{x} - x_2)]^2 + \cdots + [c(\bar{x} - x_n)]^2}{n} \\
&= \frac{c^2(x - x_1)^2 + c^2(\bar{x} - x_2)^2 + \cdots + c^2(\bar{x} - x_n)^2}{n} \\
&= \frac{c^2[(\bar{x} - x_1)^2 + (\bar{x} - x_2)^2 + \cdots + (\bar{x} - x_n)^2]}{n} \\
&= c^2 \sum_{i=1}^{n} \frac{(\bar{x} - x_i)^2}{n} \\
&= c^2 v_S
\end{aligned}$$

The important thing about this sequence of observations is the final equality; the variance of the set T is the product of the variance of the set S and c^2. These computations lend credence to the estimated answer suggested by the example and summarized in the next formula.

Formula: **(*The multiplicative property of variance*)** *If S is a set of real numbers, $S = \{x_1, x_2, \ldots, x_n\}$, and T is derived from S by multiplying each element of S by a real number c,*

$$T = \{cx_1, cx_2, \ldots, cx_n\}$$

then the variance of T is the product of the variance of S and c^2. Symbolically, if v_S is the variance of S and v_T is the variance of T, then

$$v_T = v_S c^2$$

Notice that the number c^2 is always nonnegative since the square of a number is always positive (or zero if $c = 0$). But the variance of any set is also nonnegative. Therefore, the product $v_S c^2$ is also nonnegative. This observation is of importance because the next consideration deals with standard deviations, which are expressed as square roots, and square roots of nonnegative numbers are again nonnegative numbers.

The notation v_S and v_T can be extended one more step, and the symbols s_S and s_T will be used to represent the standard deviation of sets S and T, respectively. Because the standard deviation is the square root of variance, it follows that

$$s_T = \sqrt{v_T} = \sqrt{v_S c^2} = \sqrt{v_S}\sqrt{c^2}$$

The square root of c^2 is c or $-c$, depending on which is nonnegative. (See page 229 for the definition of square root.) A notation for indicating the absolute value has already been discussed (see page 226), and the absolute value of c is denoted $|c|$. This means that the standard deviation of the set T is the product of the standard deviation of the set S and the absolute (nonnegative) value of c.

***Formula:* (The multiplicative property of standard deviation)**
If S is a set of real numbers, $S = \{x_1,x_2, \ldots ,x_n\}$, and T is derived from S by multiplying each element of S by a real number c,

$$T = \{cx_1,cx_2, \ldots ,cx_n\}$$

then the standard deviation of T is the product of the standard deviation of S and the absolute value of c. Symbolically, if s_S is the standard deviation of S and s_T is the standard deviation of T, then

$$s_T = \sqrt{v_T} = \sqrt{v_S}\sqrt{c^2} = s_S|c|$$

Example 4. With reference to Example 3 in this section, $c^2 = 9$ and $\sqrt{9} = |3| = 3$. This illustrates the case for c a positive number. To see how the formulas work for the case when multiplication is by a negative number, let $S = \{8,10,12\}$ as before, but let c be chosen as $c = -\frac{1}{4}$. Then the set T becomes $T = \{-2,-\frac{5}{2},-3\}$. The mean of S is 10, and the mean of T is $-\frac{5}{2}$. It will be instructive to ascertain that the mean of T is $-\frac{5}{2}$ both by the method of averaging and by multiplying the mean of S by $-\frac{1}{4}$. The variance of S is $\frac{8}{3}$, and the variance of T can be computed by the formula as

$$\begin{aligned} v_T &= v_S c^2 \\ &= \tfrac{8}{3}(-\tfrac{1}{4})^2 \\ &= \tfrac{8}{3} \cdot \tfrac{1}{16} \\ &= \tfrac{1}{6} \end{aligned}$$

In this example, it is not difficult, nor is it a lengthy process, to verify the result by actually computing the variance using one of the three methods available. But the value of the formula lies in the fact that the variance of T is so easily computed once the variance of S is known. The formula for standard deviation of the set T is now applied and results in

$$\begin{aligned} s_T &= s_S|c| \\ &= \sqrt{8/3}\,|-1/4| \\ &= \sqrt{8/3 \cdot 1/4} \\ &= \sqrt{8/3 \cdot 1/16} \\ &= \sqrt{1/6} = \frac{1}{\sqrt{6}} = \frac{\sqrt{6}}{6} \end{aligned}$$

This is the result obtainable by the definition of standard deviation once the variance v_T is known.

Decimal representations of such numbers as $\sqrt{6}/6$ are often preferable. A table of square roots has been provided in the back of the book for convenience in determining such representations.

Exercises

1. Let $S = \{4,8,10,10,14,16\}$ and $T = \{10,14,16,16,20,22\}$.
   a. Use the first formula of Sec. 4.3 to find v_S.
   b. Compute v_T by use of the fact that $s_i = t_i + 6$ for all $s_i \in S$ and the additive formula of this section.
2. Rework Exercise 1 for $S = \{2,3,4,7,9,11\}$ and $T = \{2{,}502, 2{,}503, 2{,}504, 2{,}507, 2{,}509, 2{,}511\}$.
3. Find v_S of Exercise 1 by the second alternate formula.
4. Find v_T of Exercise 1 by the second alternate formula.
5. Find v_S of Exercise 2 by the second alternate formula.
6. Find v_T of Exercise 2 by the second alternate formula.
7. Let $S = \{8,14,22,64,102\}$ and $T = \{4,7,11,32,51\}$.
   a. Use the second formula of Sec. 4.3 to find v_S.
   b. Compute v_T by use of the fact that $s_i = 2t_i$ for all $s_i \in S$ and the multiplicative formula of this section.
   c. What relation exists between v_S and v_T?
8. Rework Exercise 7 for $S = \{-5,-2,0,4,8\}$ and $T = \{5/3, 2/3, 0, -4/3, -8/3\}$.
9. Find v_S of Exercise 7 by the first formula of Sec. 4.3.
10. Find v_T of Exercise 7 by the first formula of Sec. 4.3.
11. Find v_T of Exercise 8 by the first formula of Sec. 4.3.
12. Find v_S of Exercise 8 by the first formula of Sec. 4.3.

Answers to problems

A. $\bar{x} = 10$, $v_S = (16 + 36 + 100 + 196 + 256)/5 - 10^2 = 120.8 - 100 = 20.8$
$v_S = v_T = 20.8$

B. $v_S = 20.8$, $v_T = 83.2 = 4v_S = 2^2 v_S$. (Note that T is derived from S by multiplying each element of S by 2.)

C. 1. $s_A \cong 4.56$; $s_B \cong 4.56$
2. Yes. Variance is 20.8 for each.
3. *B.* The mean is higher, and the variance equal.
4. Yes

4.5 The number of elements within k standard deviations of the mean

Problems

A. If $S = \{2,8,12,14,18,18\}$, find the mean and the standard deviation. How many elements of S fall within one standard deviation of the mean of S? How many fall within two standard deviations of the mean of S?

B. A manufacturer of parts wishes to bid on a contract to supply a major producer. The contract requires that not over 5 percent of the product fall below a minimum of 460 lb. A test of 100 items shows a mean of 480 and a standard deviation of 4.0 lb. Assuming that these figures are very close to the figures for all the products, can we be reasonably certain that less than 5 percent will lie below the designated minimum?

The mean of a set of numbers indicates a center of the set, while the standard deviation is used to measure the dispersion around the center. But considerations, thus far, have been directed to the compilation of formulas for computation of the scatter rather than to how such measurements can be put to use in practical situations. Here attention will be paid to an important way that the standard deviation can be used as a "measuring stick" to tell how widely the numbers deviate from the center, as represented by the mean.

The application establishes bounds within which a certain fraction of the numbers of any set of data must fall. The bounds for one special case will be determined with certainty, and some comments will be made about the accuracy and usefulness of such bounds.

Example 1. So that some questions that refer to the application can be posed and then answered, refer again to the set of 33 elements given in Example 1 of Sec. 4.1 and used for illustrative purposes in Secs. 4.1 to 4.3. The mean was discovered to be 158, and the standard deviation is (approximately) 54.85. Measuring off one standard deviation to each side of the mean gives the (approximate) bounds 103.15 and 212.85. By observation of the set of numbers, it can be seen that 21 of the 33 numbers are within the bounds.

Example 2. Two standard deviations from the mean yield bounds of 48.3 and 267.7, with 31 of the 33 measurements within the bounds.

These special cases lead to a more general question about such bounds and about the fraction of a set of numbers to be expected within the bounds. A formula is to be given, but it may be helpful to show a special case chosen such that the bounds are determined by three standard deviations.

Suppose that S is a set of n real numbers and that r of the numbers in the set S are such that they are *not* within three standard deviations of the mean of S. Diagram 8 illustrates this situation with a geometric representation of the mean, showing the boundaries to both the left and the right of the mean. This chart indicates the mean of S by $\bar{x}$, with three standard deviations to both the right and the left of the indicated $\bar{x}$. The assumption that is being made for the purpose of the argument is that r of the n numbers in the set S fall to the right of (or on) the line at $\bar{x} + 3s$ or to the left of (or on) the line at $\bar{x} - 3s$. Hence, $n - r$ of the elements of S are within the dashed lines of Diagram 8. Since the preliminary assumptions have now been precisely stated, a question can be posed which is the center of this discussion.

What fraction of the set S lies within the boundaries of $\bar{x} - 3s$ and $\bar{x} + 3s$?

The question has a surprisingly easy answer that is true for all counting numbers n and r; the computations that follow will completely answer the question.

Since subscripts are used only for identification purposes, the simplifying assumption will be made that the elements of S have been designated by subscripts such that $x_1, x_2, \ldots, x_r$ lie outside (or on) the intervals with boundaries $\bar{x} - 3s$ and $\bar{x} + 3s$. It is also assumed that $x_{r+1}, x_{r+2}, \ldots, x_n$ represent the numbers inside the boundaries. With this agreement, the difference of individual elements and $\bar{x}$ can be given in terms of the standard deviation, so Diagram 8 is given again as Dia-

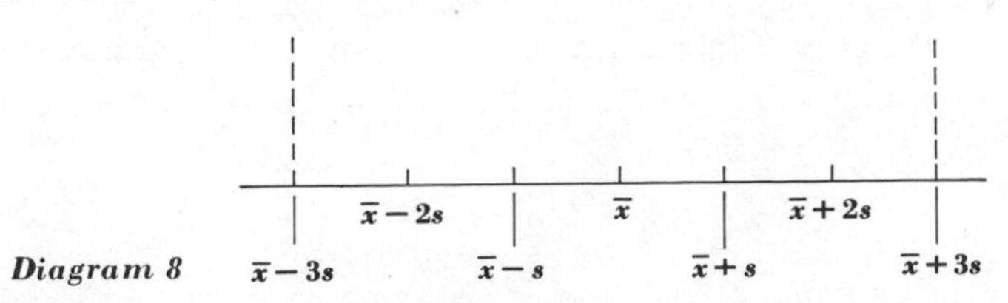

Diagram 8

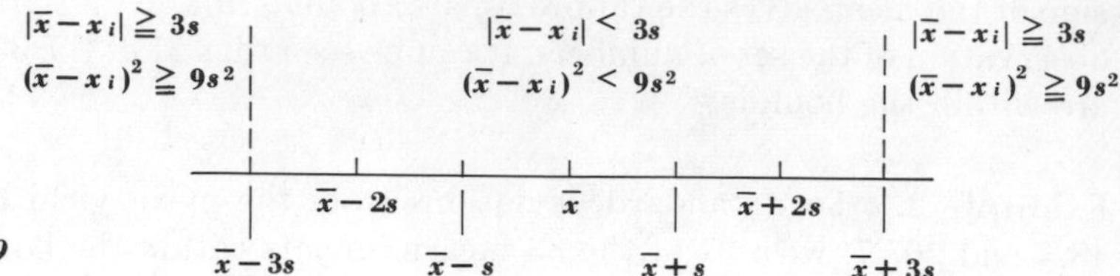

Diagram 9

gram 9. But this time the relation between the number $|\bar{x} - x_i|$ and $3s$ is given.

The relation $(\bar{x} - x_i)^2 \geq 9s^2$ holds for all x_i such that

$$i = 1, 2, \ldots, r$$

while the relation $(\bar{x} - x_i)^2 < 9s^2$ holds for all x_i such that $i = r + 1$, $r + 2, \ldots, n$.

Now the formula for variance, s^2, will be used so that the inequalities in Diagram 9 can be brought into play. Recall that

$$v = s^2 = \sum_{i=1}^{n} \frac{(\bar{x} - x_i)^2}{n}$$

or alternately,

$$s^2 = \frac{1}{n} [(\bar{x} - x_1)^2 + (\bar{x} - x_2)^2 + \cdots + (\bar{x} - x_n)^2]$$

The latter equality can be replaced with another obtained by partitioning the sums into two subsets that correspond to the subsets induced by the boundaries $\bar{x} - 3s$ and $\bar{x} + 3s$. So the formula becomes

$$s^2 = \frac{1}{n} [(\bar{x} - x_1)^2 + (\bar{x} - x_2)^2 + \cdots + (\bar{x} - x_r)^2] + \frac{1}{n} [(\bar{x} - x_{r+1})^2 + (\bar{x} + x_{r+2})^2 + \cdots + (\bar{x} - x_n)^2]$$

The numbers inside the drackets are zero or positive (since the probuct of a number by itself is never negative). Hence, each number is reduced (or left equal) if replaced by a number that is less than or equal to it. In the last equality, those numbers which are outside (or on) the boundaries are replaced by $9s^2$ (see Diagram 9), while for those numbers inside the boundaries, replacement by zero is made. With these replacements, the formula now becomes

$$\begin{aligned} s^2 &\geq \frac{1}{n} (9s^2 + 9s^2 + \cdots + 9s^2) + \frac{1}{n} (0 + 0 + \cdots + 0) \\ &\geq \frac{1}{n} 9s^2 r \\ &\geq \frac{r}{n} 9s^2 \end{aligned}$$

It may happen that $s^2 = 0$, hence $s = 0$, and all the numbers are at the mean, a very uninteresting case. In this case, all the elements of S are within the boundaries of $\bar{x} - 3s$ and $\bar{x} + 3s$. If $s^2 \neq 0$, then $s^2 > 0$, so that division by $9s^2$ yields

$$1/9 \geq \frac{r}{n}$$

This says that *the fraction of elements of S, r/n, in the set S which lie outside (or on) the boundary is less than or equal to* $1/9$.

This is a special result which arises as a consequence of using three standard deviations to the right and left of the mean. The fraction $1/9 = (1/3)^2$ arises because of the choice of three standard deviations. The argument just given can be repeated for k standard deviations, and Exercise 11 gives aid in supplying the argument, with a result that $(1/k)^2 \geq r/n$. This result gives an upper bound on the fractional part of the elements on or outside k standard deviations. The next formula states the same thing but is so worded that it gives a lower bound on the fractional part of the elements within k standard deviations.

***Formula:* (The number of elements within k standard deviations of the mean)** *If S is a set of numbers, $S = \{x_1,x_2, \ldots ,x_n\}$, with a mean of $\bar{x}$ and if s is the standard deviation with k a counting number, $k \geq 1$, then the number of elements in S such that $|\bar{x} - x_i| < ks$ is at least the fraction* $1 - (1/k)^2$.

Example 3. Consider again Example 1 of Sec. 4.1 with $k = 3$ to see that at least $1 - (1/3)^2 = 1 - 1/9 = 8/9$ of the elements of S lie within three standard deviations of the mean. The boundaries are $158 - 3 \cdot 54.85 \cong -7$ and $158 + 3 \cdot 54.85 \cong 323$.

The results claimed in the previous formula are minimal, and such a statement should be explained. With *certainty* it is known that $1 - (1/k)^2$ of the elements of S are within the boundaries $\bar{x} - ks$ and $\bar{x} + ks$. Sometimes it is desirable to estimate the fraction of the number of elements of S within k standard deviations with *almost certainty*. Then the results of the previous formula can be considerably improved. So that some numbers are available, suppose that it is desirable to predict with less than certain accuracy what fraction of the number of the elements of the set S lies within three standard deviations of the mean. It can be shown (and this will be discussed in Secs. 4.12 and 4.13) that under certain circumstances, 99 percent of all measurements are within three standard deviations of the mean. However, this prediction cannot be made with certainty (as was the prediction of the previous formula), so some hedging must be done in the statement of the prediction.

In summary, it is certain that at least 88 percent of a set of numbers will be less than three standard deviations from the mean. It is not certain, but under many circumstances it is expected, that *almost all* (99 percent) of the set will be within three standard deviations.

Exercises

1. Let $S = \{1,1,3,4,5,6,6,8,9,10,14,22,23,25,32,80,83,100\}$.
   a. Find the mean of S.
   b. Find the variance of S (by any of the three formulas).
   c. Find the standard deviation of S.
   d. How many elements of S are within two standard deviations of $\bar{x}$?
   e. How many elements of S are within three standard deviations of $\bar{x}$?
   f. How many elements of S are within four standard deviations of $\bar{x}$?
   g. Compare the results of parts *d* to *f* with the results if the formula $1 - (1/k)^2$ is used.
2. Repeat Exercise 1 for $T = \{4,6,7,7,7,11,13,14,16,18,18,18,18,20,22,32,34,40,52,58\}$.
3. Repeat Exercise 1 for $R = \{-2,-1,-1,0,0,5,6,6,6,7,7,7,8,9,10,11,23,35,46\}$.
4. How many standard deviations is the number 3 of set S from the mean?
5. How many standard deviations is the number 25 of set S from the mean?
6. How many standard deviations is the number 6 of set T from the mean?
7. How many standard deviations is the number 52 of set T from the mean?
8. How many standard deviations is the number -2 of set R from the mean?
9. How many standard deviations is the number 7 of set R from the mean?
10. *a.* What is the least integral value of k such that the interval $|\bar{x} - x_i| < ks$ has at least 70 percent of the elements?
    b. At least 80 percent of the elements?
    c. At least 90 percent of the elements?
    d. At least 95 percent of the elements?
    e. At least 99 percent of the elements?
11. Let the first r of the n elements in $S = \{x_1,x_2, \ldots ,x_r,x_{r+1}, \ldots ,x_n\}$ lie on or outside the boundaries $\bar{x} - ks$ and $\bar{x} + ks$. We will show that $1/k^2 \geq r/n$.
    a. Why does the relation

$$(\bar{x} - x_i)^2 \geq k^2s^2$$

    hold for x_i, $i = 1, 2, \ldots, r$?
    b. Why does the relation

$$(\bar{x} - x_i)^2 < k^2s^2$$

    hold for x_i, $i = r + 1, r + 2, \ldots, n$?

c. Use the formula

$$v = s^2 = \sum_{i=1}^{n} \frac{(\bar{x} - x_i)^2}{n}$$

to explain why

$$s^2 = \frac{1}{n} [(\bar{x} - x_i)^2 + (\bar{x} - x_2)^2 + \cdots + (\bar{x} - x_n)^2]$$

d. Explain why the last formula of (*c*) can be subdivided so that it can be written as

$$s^2 = \frac{1}{n} [(\bar{x} - x_1)^2 + (\bar{x} - x_2)^2 + \cdots + (\bar{x} - x_r)^2]$$
$$+ \frac{1}{n} [(\bar{x} - x_{r+1})^2 + (\bar{x} - x_{r+2})^2 + \cdots + (\bar{x} - x_n)^2]$$

e. Each of the terms of the first sum is replaced with k^2s^2 and each of the terms of the second sum is replaced with 0. Then

$$s^2 \geq \frac{1}{n} (k^2s^2 + k^2s^2 + \cdots + k^2s^2) + \frac{1}{n} (0 + 0 + \cdots + 0)$$

Why is the symbol $\geq$ used instead of $=$?

f. Explain why the formula in (*e*) reduces to

$$s^2 \geq \frac{rk^2s^2}{n}$$

g. If $s \neq 0$, what arithmetical operations permit

$$\frac{1}{k^2} \geq \frac{r}{n}$$

This is the desired formula.

h. If $s = 0$, why are all the elements within k standard deviations?

12. If two sets have the same mean of 300 while their standard deviations are 50 and 20, respectively,
 a. For the first set, what percentage of the elements are between 200 and 400?
 b. For the second set, what percentage of the elements are between 200 and 400?

13. It takes an average of 24 days for delivery of product A after it is ordered. The standard deviation is 2.5 days. If the order is placed 35 days before the stock is needed, at most what percentage of the time will the stock be depleted before the next delivery?

Answers to problems

A. $\bar{x} = 12$, $v = 32$, $s = \sqrt{32} \cong 5.66$. The elements 8, 12, 14, 18 (four elements) fall within one standard deviation of the mean. All the elements in S fall within two standard deviations of the mean.

B. $1 - (1/k)^2$ gives the fraction of the elements within k standard deviations of the mean. Since 460 is five standard deviations below the mean,

$1 - (1/5)^2 = 1.00 - 0.2^2 = 0.96$. If 96 percent are within five standard deviations of the mean, not over 4 percent would be outside the limits and less than 5 percent would be below the minimum requirements.

4.6 Standard scores

Problems

A. For each element of $S = \{4,6,8,12,13,14,15,24\}$, find the number of standard deviations the element is from the mean.

B. The Acme Company manufactures ceramic disks, which may be either too thick or too thin. To reduce the number of rejects, the product is built with a mean thickness halfway between the tolerance limits. Assume that the mean is 0.35 cm, that the standard deviation is 0.02 cm, and that the limits are set at 0.31 and 0.39 cm for minimum and maximum thicknesses.

1. We can be reasonably certain that at least what percentage will be acceptable?
2. If the uniformity of thickness can be increased by more careful workmanship, to what value would the standard deviation have to be reduced to be certain that 95 percent of the disks would be acceptable?

Now the standard deviation will be used in another way as a measuring stick. In the previous section, we saw how the standard deviation could be used as a measuring stick (with the mean as the starting point) to establish intervals in which a fixed percentage of the elements lay. In this section, we shall use the standard deviation to convert a set of numbers into *standard numbers* which by their very nature tell each number's distance (in terms of standard deviation) from the mean. Under this new scheme, the starting point will also be the mean, but the mean of standard scores will always be zero. A formula will be developed to assist in the conversion of measurement. Since one application of this process which has found wide practice has to do with grades (or academic achievements), these numbers are quite frequently called *scores* and *standard scores*, respectively, especially in books devoted to educational measurements.

So that the use of the standard deviation as a device to describe the position of a single number in a set can be appreciated, consider the following example.

Example 1. Suppose that for a set S of numbers the mean is 72 and that one element of S is 83. What true statements can be made about the element 83? Certainly it is above the average, 72, but how much above? Now suppose that it is also known that the standard deviation is 5. Then $\bar{x} + 2s = 72 + 2(5) = 72 + 10 = 82$, and by an earlier formula, at least $1 - (\frac{1}{2})^2 = \frac{3}{4} = 75$ percent of the elements of S are less than 82. Since the particular element under consideration is 83, it follows that 83 is in the upper quartile.

Example 2. But change the assumption that $s = 5$ to $s = 2$. Under this condition, $\bar{x} + 5s = 72 + 5(2) = 72 + 10 = 82$, and at least $1 - (\frac{1}{5})^2 = \frac{24}{25} = 96$ percent of the elements of S are less than 82. That is, it can be concluded with certainty that the element 83 is in the upper 4 percent of the set of elements S.

Clearly the mean is not sufficient information to determine the percentile of a particular element, but the mean and standard deviation (along with the formula of the previous section) can be used to determine the percentile (within bounds) with certainty.

Example 3. So that something concrete can be before us for consideration, reflect again on the set

$$S = \{43,55,70,85,85,102,111,113,122,136,141,\\ 141,141,150,153,161,161,161,172,177,180,184,\\ 190,192,193,195,210,213,214,217,220,230,296\}$$

For each of the elements x_i of S, the new number

$$\frac{x_i - \bar{x}}{s} = \frac{x_i - 158}{54.85}$$

is formed so that a new set T is derived. The set T is given below. The quotients have been rounded off correct to one decimal place.

$$T = \{-2.1,-1.9,-1.6,-1.3,-1.3,-1,-0.9,-0.8,-0.7,\\ -0.4,-0.3,-0.3,-0.3,-0.1,-0.1,0.1,0.1,0.1,0.3,0.3,\\ 0.4,0.5,0.6,0.6,0.6,0.7,0.9,1.0,1.0,1.1,1.1,1.3,2.5\}$$

The numbers in this set give the number of standard deviations that each element is from the mean. The negative numbers indicate that the element is smaller than the mean; the positive numbers indicate that the element is larger than the mean. The mean of the set T is determined by the averaging process, and the sum of the 33 elements is 0.1, which is approximately zero. This implies that the mean is $\cong 0$. The symbol $\cong$ is again used to mean "approximately equal to." The variance can also be computed, and by the formula

$$v = \frac{1}{n}\left(\sum_{i=1}^{n} x_i^2\right) - \bar{x}^2$$

we obtain

$$v \cong \tfrac{1}{33} \cdot 32.71 - 0 \cong 0.99$$

which is approximately 1; hence $s \cong 1$. These two results, $\bar{x} = 0$ and $v = 1$, are not special to this set of numbers; that they apply to any set of numbers will be established in the next few paragraphs. That $\bar{x}$ is not exactly 0 and v not exactly 1 is a result of approximating the numbers in set T.

If S is a set of real numbers, $S = \{x_1, x_2, \ldots, x_n\}$, such that the mean of S is $\bar{x}$ and the standard deviation of S is s and if T is a set of numbers such that $T = \{(x_1 - \bar{x})/s, (x_2 - \bar{x})/s, \ldots, (x_n - \bar{x})/s\}$, then is the mean of T zero? The additive and multiplicative properties of the mean, as established in Sec. 4.2, make it easy to answer the question. It is known that

$$S = \{x_1, x_2, \ldots, x_n\}$$

has a mean of $\bar{x}$. Subtraction of $\bar{x}$ from each element of the previous set yields

$$\{x_1 - \bar{x}, x_2 - \bar{x}, \ldots, x_n - \bar{x}\}$$

which has a mean of $\bar{x} - \bar{x} = 0$. Multiplication of each element by $1/s$ yields the final set

$$\{(x_1 - \bar{x})/s, (x_2 - \bar{x})/s, \ldots, (x_n - \bar{x})/s\}$$

which has a mean of $(1/s)0 = 0$, and this verifies the property stated in the next formula.

Formula: **(*The mean of a set of standard numbers*)** *If S is a set of real numbers, $S = \{x_1, x_2, \ldots, x_n\}$, and if T is a set derived from S by use of the formula $y_i = (x_i - \bar{x})/s$, where $\bar{x}$ is the mean of S and s is the standard deviation of S, such that*

$$T = \{(x_1 - \bar{x})/s, (x_2 - \bar{x})/s, \ldots, (x_n - \bar{x})/s\}$$

then the mean of T is zero. The elements of T are standard numbers.

That the center, as given by the mean, of the new set of data is at zero corresponds geometrically to arranging the data so that approximately half of the set is to the left of the zero and half to the right of the zero.

Under the circumstances of this discussion, what can be said about the standard deviation of T? The example would indicate that the standard deviation of a set of standard numbers is 1. The next argument shows this to be true for all sets of arbitrary numbers.

The additive and multiplicative properties of the variance, as established in Sec. 4.4, will be used in the argument that follows. The set

$$S = \{x_1, x_2, \ldots, x_n\}$$

with a variance of v is the starting point. Next, $\bar{x}$ is subtracted from each element of the set, producing

$$\{x_1 - \bar{x}, x_2 - \bar{x}, \ldots, x_n - \bar{x}\}$$

with an unchanged variance of v. The final step is to multiply each element by $1/s$, and the set

$$\left\{\frac{x_1 - \bar{x}}{s}, \frac{x_2 - \bar{x}}{s}, \ldots, \frac{x_n - \bar{x}}{s}\right\}$$

results. The variance of this set is, by the multiplicative property of the variance, $(1/s)^2v = (1/s^2)v = v/s^2$. But remember that variance and standard deviation are related by $s^2 = v$, so that the variance of the standard scores becomes

$$\frac{v}{s^2} = \frac{v}{v} = 1$$

This is summarized in the next formula.

Formula: **(*The standard deviation of a set of standard numbers*)** *If S is a set of real numbers, $S = \{x_1, x_2, \ldots, x_n\}$, and if T is a set derived from S by use of the formula $y_i = (x_i - \bar{x})/s$, where $\bar{x}$ is the mean of S and s is the standard deviation of S, such that*

$$T = \{(x_1 - \bar{x})/s, (x_2 - \bar{x})/s, \ldots, (x_n - \bar{x})/s\}$$

then the standard deviation of T is 1.

The significance of this last formula has already been commented on for the special example, but it bears repeating as it applies to any set S and the standard set T. Since the mean of T is 0 and the standard deviation is 1, the number of standard deviations of any element of T from the mean is the number itself. For example, if T is a set of standard numbers and -1.9 is an element of T, then the element is 1.9 standard deviations from the mean (to the left). The advantages of such an interpretation should be obvious.

Exercises

1. Write the set of Exercise 1, Sec. 4.5, in standard scores.
2. Write the set of Exercise 2, Sec. 4.5, in standard scores.

3. Write the set of Exercise 3, Sec. 4.5, in standard scores.
4. Determine what percentage of the elements of Exercise 1 are within two standard deviations of the mean.
5. Repeat Exercise 4 for Exercise 2.
6. Repeat Exercise 4 for Exercise 3.
7. A set of numbers has a mean of 25 and a standard deviation of 5.
   a. What percentage of the numbers are between 35 and 15?
   b. Between 40 and 10?
8. Rework Exercise 7 for a standard deviation of $2\frac{1}{2}$.
9. John makes 75 on a test. His class has a mean score of 62, with standard deviation of 6. Ed makes 71. His class has a mean of 69, with a standard deviation of 5. Which boy has the higher percentile score in his own class?
10. In one month Mr. Kole made sales of sheets totaling \$14,000. In the same month Mr. Miller sold \$7,250 worth of comforters. That month the average sale of sheets per salesman was \$12,000 with a standard deviation of \$2,000. The average per salesman for sale of comforters was \$6,000 with a standard deviation of \$500.
    a. Write the standard score for Mr. Kole.
    b. Write the standard score for Mr. Miller.
    c. Which of the two did the best when compared with the salesmen in his own group?

Answers to problems

A. $\{-1.4, -1.0, -0.7, 0, 0.2, 0.3, 0.5, 2.1\}$

B. 1. Since the limits are $0.04/0.02 = 2.0$ standard deviations from the mean, $1 - (1/k)^2$ is equal to $1 - (\frac{1}{2})^2 = 0.75$. So we may be certain that at least 0.75 would be acceptable.

2. If $1 - \left(\frac{1}{k}\right)^2 = 0.95$, then $1 - \left(\frac{1}{0.4/s}\right)^2 = 0.95$ and $\left(\frac{s}{0.04}\right)^2 = 0.05$. $s = 0.04\sqrt{0.05}$; $s \cong 0.0089$ cm.

4.7 *Measures of grouped data*

Problems

A. Compute an approximation to the mean of the set of numbers whose graph is the accompanying histogram. In which interval does the median lie? In which interval does the mode lie?

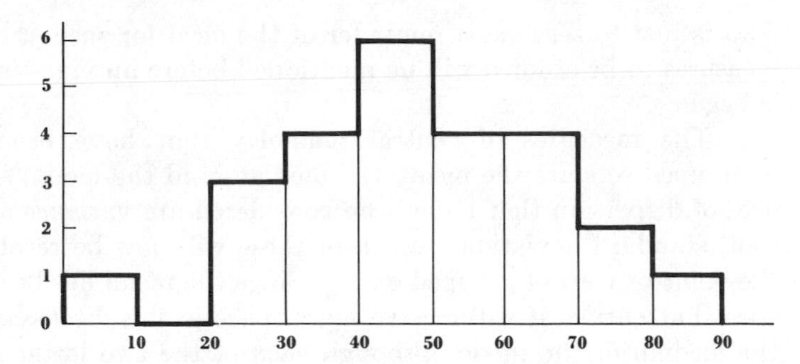

B. For a report to the stockholders on the salaries of the employees of his company, the president of a corporation asks you whether he should use the mean or the median as the average. Assuming the following data, which measure would you use:

1. If you want the average to seem high?
2. If you want the average to seem low?
3. If you want the average to be the one which is most truly representative?
4. Compute the two to check your answers.

Salary	*Number of salaries*	*Total salaries*
$10,000 to 14,999	14	$175,000
15,000 to 19,999	24	420,000
20,000 to 24,999	8	180,000
25,000 to 29,999	3	82,500
30,000 to 34,999	1	32,500
	50	$890,000

Since this section returns to the subject of grouped data, it would be well to recall some of the reasons for partitioning a set of numbers into subsets. The most obvious (and important) reason for treating data in groups rather than as individual elements is that the set has too many elements to study them as individuals. Mention has already been made of such large sets as the result of a national census. To study the set of numbers which represent the ages of the populace, it would be necessary to group the data into convenient subsets with intervals of, say, 5 or 10 years. The number of subsets chosen is dependent on the type of questions to be asked about the data. This is an era of mass production, so a set of data collected from assembly-line production has, by the nature of the operations, a large number of elements. Of course, with the population explosion, census figures also represent mass production.

This is just to serve as a reminder of the need for such studies; now the measures to be studied will be mentioned before an investigation of each is begun.

The measures of central tendency that have been studied for ungrouped sets are the mean, the median, and the mode; and the measures of dispersion that have been considered are variance and its square root, standard deviation. Each of these will now be reconsidered from the point of view of grouped data. Since the mean has been selected for special attention, it will receive more space in this discussion than either the median or the mode, although each of the two latter measures will be discussed briefly as they apply to grouped data.

The mean of a set of numbers has already been defined. It is the average given by $\left(\sum_{i=1}^{n} x_i\right)/n$. The fact that the data have been partitioned into groups does not change the definition of the mean or, of course, the mean itself. There are some adjustments necessary in the process of approximating the mean, and it is those that will occupy our attention here.

Example 1. It is again convenient to have a set of numbers to serve as an example, so the grouping of the set of 33 numbers as given in Table 2 is reproduced here.

Table 2 (repeated)

	Interval 1	*Interval 2*	*Interval 3*	*Interval 4*	*Interval 5*	*Interval 6*
Interval	I_1	I_2	I_3	I_4	I_5	I_6
Range of interval	0–50	51–100	101–150	151–200	201–250	251–300
Frequency	1	4	9	12	6	1

As a first observation, note that the mean of this set has already been computed, and it is 158. If all the information that is known about the set is included in Table 2, then it is *not possible* to compute the mean of the original set S. The table informs us that one element is in the interval 0 to 50, but it does not tell us which number it is; there are four elements in the second interval, but the table gives no information about which numbers they are, etc. We repeat for emphasis that on the basis of the information in the *table, the mean of the original set S cannot be computed.* However, some compromises can be made so that a good approximation to the mean of the set S can be computed. The agreement is that *the elements in each interval are located at the center of the interval.* On the basis of this agreement, the set consists of one element 25, four elements 75, nine elements 125, twelve elements 175, six elements 225, and one element 275. These numbers can be averaged in the usual way, and the average is

$$\frac{1 \cdot 25 + 4 \cdot 75 + 9 \cdot 125 + 12 \cdot 175 + 6 \cdot 225 + 1 \cdot 275}{33}$$

$$= \frac{25 + 300 + 1{,}125 + 2{,}100 + 1{,}350 + 275}{33}$$

$$= \frac{5{,}175}{33} \cong 156.8$$

This average does not agree with the mean of the set S, as might have been anticipated since error is introduced by the agreement made about the location of the numbers within an interval. However, the average of the grouped data that is obtained is quite close to the mean of the original set.

Example 2. Remember that a set of numbers can be partitioned in a number of ways and that the histogram (Diagram 3) used intervals of length 25 rather than 50. Diagram 3 is reproduced here.

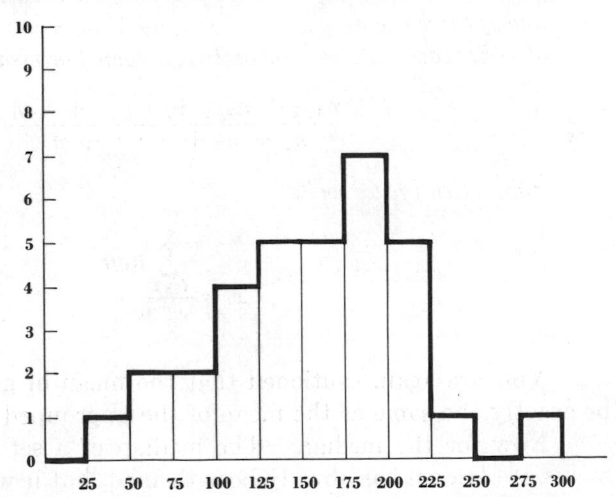

Diagram 3 (*repeated*)

There are 12 intervals in this particular grouping, and the number of elements in each interval is known. The numbers of elements in the intervals are

$n_1 = 0$	$n_4 = 2$	$n_7 = 5$	$n_{10} = 1$
$n_2 = 1$	$n_5 = 4$	$n_8 = 7$	$n_{11} = 0$
$n_3 = 2$	$n_6 = 5$	$n_9 = 5$	$n_{12} = 1$

with n_i representing the number of elements in the ith interval. If the agreement is again made that the numbers within each interval are at the center of the interval, then the mean of the original set can be approximated as

$$\frac{\begin{array}{l}0 \cdot 12.5 + 1 \cdot 37.5 + 2 \cdot 62.5 + 2 \cdot 87.5 + 4 \cdot 112.5 + 5 \cdot 137.5 \\ \quad + 5 \cdot 162.5 + 7 \cdot 187.5 + 5 \cdot 212.5 + 1 \cdot 237.5 + 0 \cdot 262.5 + 1 \cdot 287.5\end{array}}{33}$$

$$= \frac{\begin{array}{l}0 + 37.5 + 125 + 175 + 450 + 687.5 + 812.5 + 1{,}312.5 \\ \qquad\qquad + 1{,}062.5 + 237.5 + 0 + 287.5\end{array}}{33}$$

$$= \frac{5{,}187.5}{33} \cong 157.2$$

The result is again close but not exactly the mean of the original set. The procedure used in these two computations is easily extended to any set which has been partitioned into any number of subsets, and the next formula generalizes the procedure.

Formula: **(*Mean of grouped data*)** *If S is a set of real numbers, $S = \{x_1, x_2, \ldots, x_n\}$, and if a set containing the range of the set S is partitioned into r intervals of equal length with midpoints $y_1, y_2, \ldots, y_r$ and if there are $n_1, n_2, \ldots, n_r$ $(n_1 + n_2 + \cdots + n_r = n)$ elements of S in each interval, respectively, then the mean of the grouped set is*

$$\frac{n_1y_1 + n_2y_2 + \cdots + n_ry_r}{n_1 + n_2 + \cdots + n_r}$$

Symbolically, we write

$$\bar{x} = \frac{\sum_{i=1}^{r} n_iy_i}{n}$$

You are again cautioned that the mean of grouped data need not be exactly the same as the mean of the ungrouped data.

Now for the median. The median of a set whose individual elements are known has already been defined, but it will now be shown how to approximate the median of the ungrouped set by computing the median for grouped data. Recall that the median of a set S is the middle element of S. If the number of elements n in the set S is odd, say $n = 2r + 1$, then the median is in S and is x_{r+1}. If the number of elements n in S is even, then an averaging is made and the number $(x_{n/2} + x_{(n/2)+1})/2$ is by definition the median of S. Remember that in this case the median need not be in the set.

Example 3. Consider again the histogram in Diagram 3 for the purpose of determining the median of the grouped data. If the total information available is contained in the histogram, then the median of the original set *cannot* be determined. This is because the formula for either case, n even or n odd, calls for specific elements of S and they are unknown for grouped data. Nevertheless, the middle element of 33 elements is the seventeenth one, so first we locate the interval which contains the seventeenth element of S, as ordered by size. Counting from the left on this histogram, there are

$$0 + 1 + 2 + 2 + 4 + 5 = 14$$

elements in the first six intervals, so the seventeenth element is in the seventh interval, whose boundaries are 151 and 175. There are five elements in this interval, and the seventeenth is the third of these; so a reasonable approximation to the median of the original set is that it is $3/5$ of the way in the interval, or at

$$151 + 3/5 \cdot 25 \cong 151 + 15 = 166$$

Since for this set the elements are known, this result can be compared with the actual median, which is 161. The approximation is good.

Example 4. If the grouped data in Table 2 are used and a similar procedure is followed, it is discovered that

$$1 + 4 + 9 = 14$$

of the elements are in the first three intervals, so the median is in the fourth interval. There are 12 elements in this interval, and the seventeenth element is the third of these; so a reasonable approximation is that the median is $3/12$ of the way between 151 and 200. A computation yields

$$151 + 3/12 \cdot 50 = 151 + 150/12 = 151 + 12.5 = 163.5$$

which is again close to the median of the ungrouped set.

The mode is the element in the set which occurs the most frequently. Clearly, if the individual elements of the set S are unknown (only grouped data are given), it is impossible to obtain the mode. The corresponding concept for modes is a *modal interval,* which means, as its title indicates, the interval with the largest number of elements of S. In this book, there will be no further need to consider modes of grouped data.

The final topic to be discussed in this section is the variance (hence the standard deviation) of a set of grouped data. The variance of a set of numbers has been defined as

$$v = \sum_{i=1}^{n} \frac{(\bar{x} - x_i)^2}{n}$$

with v representing the variance, $\bar{x}$ the mean, x_i any element of S, and n the number of elements of S. However, there are two other formulas for variance, and the one that proves to be the easiest to use is

$$v = \frac{\sum_{i=1}^{n} x_i^2}{n} - \bar{x}^2$$

Since this equivalent formula is much easier to use, it is the form that will be used to represent the final result. If the total information available is grouped and only information about the groups is known, it is impossible to compute the mean of the original set; but the mean of grouped data has been defined as

$$\bar{x} = \frac{\sum_{i=1}^{n} n_i y_i}{n}$$

with y_i the center of the ith interval, n_i the number of elements in the ith interval, and n the total number of elements. So the mean of grouped data is available. Now look again at the formula

$$v = \sum_{i=1}^{n} \frac{(\bar{x} - x_i)^2}{n}$$

which gives the instructions: Form the difference of the mean and each element, square these differences, add them, and divide by n. Assuming that the mean is known, the difference $\bar{x} - x_i$ for grouped data *cannot* be obtained since the individual elements x_i are unknown. Because of this, the differences of the mean of the grouped data and the midpoint of each interval are substituted for the differences $\bar{x} - x_i$. They are summed as many times as there are elements in each interval. A formula would be

$$v = \sum_{i=1}^{n} \frac{(\bar{x} - y_i)^2 n_i}{n}$$

However, this formula proves to be very impractical to use, and it is replaced temporarily with another (arrived at by algebraic operations that exactly parallel those on page 231), which is

$$v = \frac{\sum_{i=1}^{n} y_i^2 n_i}{n} - \left(\frac{\sum_{i=1}^{n} y_i n_i}{n} \right)^2$$

Example 5. So that the difficulty of even this simplified formula can be appreciated, it is applied to the example in the histogram (Diagram 3). A table of numbers for this example is given next.

Table 5

Interval	*Midpoint,* y_i	*Frequency,* n_i	*Midpoint squared,* y_i^2	$y_i^2 n_i$	$y_i n_i$
0–25	12.5	0	156.25	0	0
26–50	37.5	1	1,406.25	1,406.25	37.5
51–75	62.5	2	3,906.25	7,812.50	225.0
76–100	87.5	2	7,656.25	15,312.50	175.0
101–125	112.5	4	12,656.25	50,625.00	450.0
126–150	137.5	5	18,906.25	94,531.25	687.5
151–175	162.5	5	26,406.25	132,031.25	812.5
176–200	187.5	7	35,156.25	246,093.75	1,312.5
201–225	212.5	5	45,156.25	225,781.25	1,062.5
226–250	237.5	1	56,406.25	56,406.25	237.5
251–275	262.5	0	68,906.25	0	0
276–300	287.5	1	82,656.25	82,656.25	287.5

$$\sum_{i=1}^{12} y_i^2 n_i = 912{,}656.25 \qquad \sum_{i=1}^{12} y_i n_i = 5{,}187.5$$

There are more steps in the computation to complete the use of the formula. But already the undesirable features of this particular formula are apparent. It will, of course, yield an acceptable result. If the arithmetic is completed, the net result is approximately 2,945. This is not the correct variance. The difference arises because of approximating the elements in an interval by the midpoint of the interval. Even so, this is not the formula of the most practical value. There is now a payoff for the computations done in Sec. 4.4 with respect to the effect on the variance of adding a number to each element of the set. Reread Sec. 4.4 to see that the variance is unchanged by the addition of a number to each element of the set.

Example 6. We choose to add -187.5 to each element. This number is chosen because 187.5 is the midpoint of the modal interval. (Usually the midpoint of the modal interval is subtracted.) The substitution $w_i = y_i - 187.5$ will yield a set with the same variance. Multiplication by a number makes the variance change by the square of the number. In this example, we multiply by $1/12.5 = 0.08$. This number is chosen to simplify the arithmetic. A new table with the substitution

$$u_i = 0.08w_i = 0.08(y_i - 187.5) = 0.08y_i - 15$$

is now offered.

Interval	u_i	n_i	u_i^2	$u_i^2n_i$	u_in_i
0–25	−14	0	196	0	0
26–50	−12	1	144	144	−12
51–75	−10	2	100	200	−20
76–100	−8	2	64	128	−16
101–125	−6	4	36	144	−24
126–150	−4	5	16	80	−20
151–175	−2	5	4	20	−10
176–200	0	7	0	0	0
201–225	2	5	4	20	10
226–250	4	1	16	16	4
251–275	6	0	36	0	0
276–300	8	1	64	64	8

$$\sum_{i=1}^{12} u_i^2n_i = 816 \qquad \sum_{i=1}^{12} u_in_i = -80$$

Substituting,

$$\begin{aligned} v &= \frac{\sum_{i=1}^{12} u_i^2n_i}{33} - \left(\frac{\sum_{i=1}^{12} u_in_i}{33}\right)^2 \\ &\cong 24.7273 - 5.8758 \\ &\cong 18.85 \end{aligned}$$

Since the new set was formed by multiplying by the constant $8/100$, the variance is changed by the square of the constant. By the formula derived in Sec. 4.4, the variance of the original set is

$$18.85\left(\frac{100}{8}\right)^2$$

The result of this arithmetic is

$$v \cong 2{,}945$$

This is the same result arrived at earlier by a much more lengthy computation. This gives a standard deviation of 54.27 instead of the exact answer 54.85 determined earlier. The approximation is relatively close.

Exercises

1. Approximate the mean and median of the set represented by the following frequency histogram.

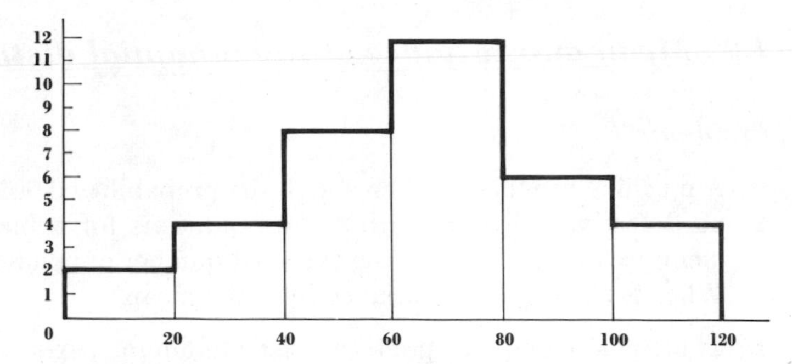

2. The following frequency histogram is derived from the same set as the one in Exercise 1. Approximate the mean and the median of the set represented by it and compare your results with those of Exercise 1.

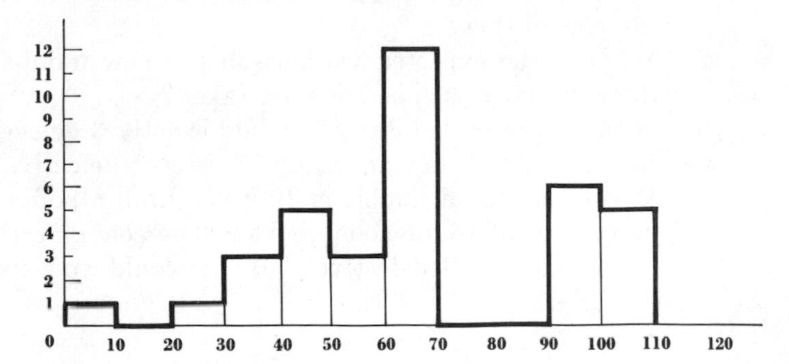

3. *a.* What is the modal interval of Exercise 1?
   b. What is the modal interval of Exercise 2?
4. Approximate the variance of Exercise 1.
5. Approximate the variance of Exercise 2.
6. Rework Exercise 4 with the substitution $u_i = (y_i - 70)/20$.
7. Rework Exercise 5 with the substitution $u_i = (y_i - 65)/10$.
8. A company makes purchases as follows:
   500 units at $3.00 per unit
   200 units at $2.50 per unit
   150 units at $3.10 per unit
   250 units at $1.80 per unit
   What is the average cost per unit?

Answers to problems

A. The mean is 48.2. The median is in the fifth interval. Not answerable.

B 1. Mean
   2. Median
   3. Median
   4. $17,800 and $17,500

4.8 Mean and variance of the binomial distribution

Problems

A. A machine produces 10 items with the probability 0.90 that there are no defectives. (Assume that the conditions for a binomial experiment are met.) What is the expected number of nondefective items? What is the expected variance from the mean?

B. The Acme Company produces cast aluminum parts which have, in the past, been 30 percent defective. A sample of 10 items was selected at random from the production line.

1. What is the expected new number of defective castings in the samples of this size?
2. What is the expected variance in the new number of defective items if many such samples are taken?
3. What is the probability of finding exactly 3 defective items in a sample of 10 if they are really 30 percent defective?
4. If you selected a sample of 10 items from a population which is "guaranteed" to run no more than 30 percent defective, would you expect to find 9 defective? What would you conclude if this occurred?

The mean and variance of sets of numbers have been the subject for many of the examples previously considered. Generally speaking, the sets of numbers were known; i.e., the sets were tabulated. Means of sets of numbers that are generated by experiments with an attached probability will be our concern now. The sets of numbers to be studied will be (except for the first two examples) generated by binomial experiments. Sets of data that arise as the result of binomial experiments are samples taken from a *binomial distribution.*

There have been some previous considerations that are closely related to the subject of this section, so we begin by recalling an example that reviews what is meant by *mathematical expectation.*

Example 1. In Sec. 3.9, mathematical expectation was defined and the expectation for the game of dice was computed. The number of logical possibilities in the game is substantial, and the computation of the expectation somewhat involved. The amounts associated with the events were $+1$ for successful events and -1 for the failures. For the convenience of the reader, the diagram associated with this experiment and the computation of the expectation are reproduced. The diagram is Diagram 10.

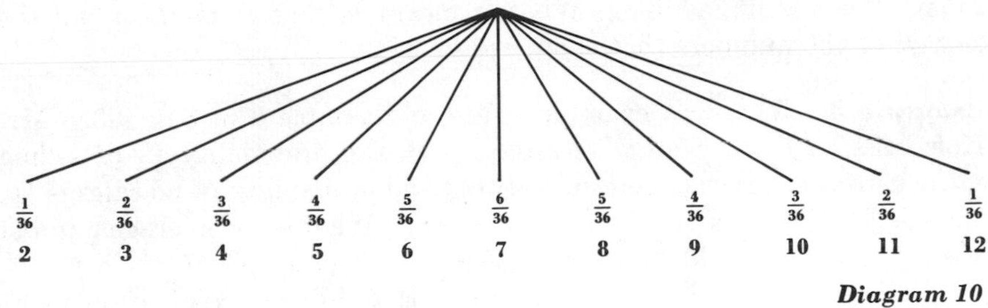

Diagram 10

$$
\begin{aligned}
E = {} & a_2 Pr(E_2) + a_3 Pr(E_3) + a_{4,S} Pr(E_{4,S}) + a_{4,F} Pr(E_{4,F}) + a_{5,S} Pr(E_{5,S}) \\
& + a_{5,F} Pr(E_{5,F}) + a_{6,S} Pr(E_{6,S}) + a_{6,F} Pr(E_{6,F}) + a_7 Pr(E_7) \\
& + a_{8,S} Pr(E_{8,S}) \\
& + a_{8,F} Pr(E_{8,F}) + a_{9,S} Pr(E_{9,S}) + a_{9,F} Pr(E_{9,F}) + a_{10,S} Pr(E_{10,S}) \\
& + a_{10,F} Pr(E_{10,F}) \\
& + a_{11} Pr(E_{11}) + a_{12} Pr(E_{12}) \\
= {} & (-1) \cdot 1/36 + (-1) \cdot 2/36 + (+1) \cdot 3/36 \cdot 3/9 \\
& + (-1) \cdot 3/36 \cdot 6/9 + (+1) \cdot 4/36 \cdot 4/10 + (-1) \cdot 4/36 \cdot 6/10 \\
& + (+1) \cdot 5/36 \cdot 5/11 + (-1) \cdot 5/36 \cdot 6/11 + (+1) \cdot 6/36 \\
& + (+1) \cdot 5/36 \cdot 5/11 + (-1) \cdot 5/36 \cdot 6/11 \\
& + (+1) \cdot 4/36 \cdot 4/10 + (-1) \cdot 4/36 \cdot 6/10 + (+1) \cdot 3/36 \cdot 3/9 \\
& + (-1) \cdot 3/36 \cdot 6/9 + (+1) \cdot 2/36 + (-1) \cdot 1/36 \\
= {} & -7/495
\end{aligned}
$$

The point of this example is that the expectation is an *average*. On a single throw of the dice, one does not expect to lose $7/495$ of the unit wagered. Rather, *in the long run* the expected outcome is an *average* loss for the player of $7/495$ for each unit wagered.

Example 2. For a simpler example, consider the toss of one die. What is the average number that is expected? There are six faces on the die, with the numbers 1 to 6 on them. Each of the six possible events (for an unbiased die) has probability of occurrence of $1/6$. The expectation is, as before, the sum of the products of each amount (the number on the face of the die) and the probability that the number will occur. The amount is 1 if the number that shows is 1, 2 if the number 2 shows, 3 if the number 3 shows, etc. The probability of occurrence of each number is $1/6$. The average expectation is therefore given by the following sum:

$$
\begin{aligned}
1 \cdot 1/6 + 2 \cdot 1/6 + 3 \cdot 1/6 + 4 \cdot 1/6 + 5 \cdot 1/6 + 6 \cdot 1/6 & \\
& = (1 + 2 + 3 + 4 + 5 + 6) \cdot 1/6 \\
& = 21 \cdot 1/6 \\
& = 21/6 = 7/2 = 3\tfrac{1}{2}
\end{aligned}
$$

Of course, it is not expected that on a single throw, the number $3\frac{1}{2}$ will show. That is impossible. What is meant is that *in the long run* the *average* of the numbers that show will be $3\frac{1}{2}$.

Example 3. The next example refers to Example 3 of Sec. 3.3. Mr. Kole calls on three customers a day, with the probability $\frac{1}{2}$ of selling each customer. It was determined that the probability of no sales is $\frac{1}{8}$, one sale $\frac{3}{8}$, two sales $\frac{3}{8}$, and three sales $\frac{1}{8}$. What is the average number of sales that he should expect to make?

This question is recognized as one that has an expectation as an answer. Further, the amounts associated with each event (the number of sales) and the probabilities for the occurrence of the events are known, so that a computation of the expectation is immediate. It is

$$\begin{aligned} 0 \cdot \tfrac{1}{8} + 1 \cdot \tfrac{3}{8} + 2 \cdot \tfrac{3}{8} + 3 \cdot \tfrac{1}{8} &= 0 + \tfrac{3}{8} + \tfrac{6}{8} + \tfrac{3}{8} \\ &= \tfrac{12}{8} = \tfrac{3}{2} = 1\tfrac{1}{2} \end{aligned}$$

Mr. Kole should expect to sell $1\frac{1}{2}$ of his customers per day on the average.

In the previous example, the expectation was computed by the use of the coefficients of the binomial formula $(x + y)^n$, with $x = \frac{1}{2}$ (probability of no sale), $y = \frac{1}{2}$ (probability of a sale), and $n = 3$ (number of trials). For these replacements, the binomial formula is

$$\begin{aligned} (x + y)^n = (\tfrac{1}{2} + \tfrac{1}{2})^3 &= (\tfrac{1}{2})^3 + 3(\tfrac{1}{2})^2 \cdot \tfrac{1}{2} + 3 \cdot \tfrac{1}{2}(\tfrac{1}{2})^2 + (\tfrac{1}{2})^3 \\ &= \tfrac{1}{8} + \tfrac{3}{8} + \tfrac{3}{8} + \tfrac{1}{8} \end{aligned}$$

The next example is patterned after Example 3 except that the probabilities x and y are not specified.

Example 4. In a binomial experiment, the probability of failure is x and the probability of success is $y = 1 - x$. If there are three (independent) trials of the experiment, what is the expected average success? When this question was posed for Example 3, the probabilities x and y were both known to be $\frac{1}{2}$. The answer was the product of the number of trials (3) and the probability of success ($\frac{1}{2}$); $3 \cdot \frac{1}{2} = \frac{3}{2}$. An intelligent guess for this example is that the expected number is the product of the number of trials ($n = 3$) and the probability of success (y): $3y$. A computation verifies that this is indeed the case:

$$\begin{aligned} &0{}_3C_0x^3y^0 + 1{}_3C_1x^2y^1 + 2{}_3C_2x^1y^2 + 3{}_3C_3x^0y^3 \\ &\qquad = 0{}_3C_0x^3 + 1{}_3C_1x^2y + 2{}_3C_2xy^2 + 3{}_3C_3y^3 \\ &\qquad = 0 + 3x^2y + 2 \cdot 3xy^2 + 3y^3 \\ &\qquad = 0 + 3x^2y + 3xy^2 + 3xy^2 + 3y^3 \\ &\qquad = 3xy(x + y) + 3y^2(x + y) \\ &\qquad = 3xy + 3y^2 \\ &\qquad = 3y(x + y) \\ &\qquad = 3y \end{aligned}$$

Some remarks about the algebraic steps in the computation will be made. The first line is a statement of the problem and should be compared with the first line of the computation in Example 3. In the second line, the binomial coefficients have been substituted. The product $2 \cdot 3xy^2$ is equal to $3xy^2 + 3xy^2$, and this replacement is made in line 3. Regrouping of the numbers occurs in line 4, with $3xy(x + y)$ taking the place of $3x^2y + 3xy^2$ and $3y^2(x + y)$ replacing $3xy^2 + 3y^3$. But the sum $x + y$ is 1. It is this fact that permits the simplification of the expression. In Chap. 2, when the binomial formula was first discussed, this important property was stressed. When $x + y$ is replaced by 1, line 5 results. Line 6 is again the result of regrouping, with $3y(x + y)$ replacing $3xy + 3y^2$. The final result is the outcome of replacing $x + y$ by 1.

Examples 3 and 4 point the way toward the formula for the average expectation for events that occur as the result of binomial experiments. They deal with the case $n = 3$. The next example is very similar, but the number of trials is changed.

Example 5. Again suppose a binomial experiment with probability x of failure and $y = 1 - x$ of success. For this example, let $n = 2$. Read the computation below to see that the average expectation is the product of the number of trials (2) and the probability of success (y).

$$\begin{aligned} 0\,{}_2C_0x^2 + 1\,{}_2C_1xy + 2\,{}_2C_2y^2 &= 0 + 2xy + 2y^2 \\ &= 2y(x + y) \\ &= 2y \end{aligned}$$

In this computation, the algebraic steps are much simpler, and the reasons for the steps are not supplied. The reader should use Example 4 as a pattern to supply reasons for this computation.

That the average expectation of a binomial experiment with n trials is ny can be proved. (For binomial distributions, the average expection will be called the *mean*.) But the proof requires the use of induction, and you will recall that we have already rejected (because of difficulty) the use of that tool of mathematics. This was done when the binomial formula was accepted without proof by induction. We state without proof the next formula.

***Formula:* (The mean of the binomial distribution)** *Assume a binomial experiment with n independent trials and probabilities x of failure and $y = 1 - x$ of success. Then the mean of the binomial distribution is $\bar{x} = ny$.*

The reader can verify from Examples 3 and 4 that an alternate formula of computation that yields $\bar{x} = ny$ is

$$\bar{x} = 0{}_nC_0x^n + 1{}_nC_1x^{n-1}y + \cdots + r{}_nC_rx^{n-r}y^r + \cdots + n{}_nC_ny^n$$

This is sometimes written in mathematical shorthand as

$$\bar{x} = E(y) = \sum_{y=0}^{n} yPr(y)$$

where $Pr(y)$ means the probability of y and $E(y)$ means the expected value of y. There will be no occasion in this book to compute the mean (expected value) of probability distributions other than *binomial* distributions, but they are computed in a similar manner.

The reader is to be spared the algebraic manipulations similar to those of Example 4 which motivate the formula for variance of the binomial distribution. For the more mathematically inclined, Exercise 13 derives the formula for the cases $n = 1$ and $n = 2$. Here it is sufficient to say that once the mean is known, the variance is found by a sum which, in notation like that used earlier, is

$$v = E(\bar{x} - x)^2 = \sum_{x=0}^{n} (\bar{x} - x)^2Pr(x)$$

These symbols are further explained in Exercise 13. The formula gives the result.

Formula: (***The variance of the binomial distribution***) *Assume a binomial experiment with n independent trials and probabilities x of failure and $y = 1 - x$ of success. Then the variance is*

$$v = nxy$$

We close with some examples that use the two formulas of the section.

Example 6. A binomial experiment is repeated 10 times with probability of 0.95 of success (and probability 0.05 of failure). What is the expected value of success? This is recognized as an application of the formula for the mean of a binomial distribution. In this example, $n = 10$ and $y = 0.95$, so that the expected value, the mean, is

$$\bar{x} = 10 \cdot 0.95 = 9.5 = 9\tfrac{1}{2}$$

Example 7. In Example 6, what is the variance from the mean? The second formula of the section applies, and

$$v = 10 \cdot 0.95 \cdot 0.05 = 0.475$$

or approximately $\frac{1}{2}$.

It may be helpful to state in words what is meant by the numbers that are given as answers in the last two examples. Under the stated conditions, the experiment is performed 10 times, and the number of successes are recorded. Then the experiment is repeated 10 more times, and the number of successes recorded. This is done over and over. *In the long run,* the average number of successes is expected to be $9\frac{1}{2}$. Furthermore, as the numbers of successes are recorded at the end of each 10 trials, the numbers should vary from $9\frac{1}{2}$ with a variance of approximately $\frac{1}{2}$.

Exercises

1. If a player in Example 1 makes 500 wagers of $1 each, what is his expected gain or loss?
2. There is a probability of $\frac{4}{5}$ that a business venture will net $12,000 and a probability of $\frac{1}{5}$ that it will lose $2,500. What is the expected value of the venture?
3. Assume a binomial experiment of five trials. What is the mean if:
   a. $x = \frac{1}{10}, y = \frac{9}{10}$
   b. $x = \frac{1}{4}, y = \frac{3}{4}$
   c. $x = \frac{1}{3}, y - \frac{2}{3}$
   d. $x = \frac{1}{3}, y = \frac{2}{3}$
   e. $x = \frac{2}{3}, y = \frac{1}{3}$
4. Work Exercise 3 for $n = 6$.
5. Work Exercise 3 for $n = 4$.
6. Compute the variance for each of the parts of Exercise 3.
7. Compute the variance for each of the parts of Exercise 4.
8. Compute the variance for each of the parts of Exercise 5.
9. If Mr. Kole calls on four customers with probability of $\frac{1}{2}$ of making a sale, what is the expected number of sales?
10. Explain how the answers to Exercise 3 can be interpreted as expected numbers of sales if the probabilities refer to the chances of a sale.
11. A salesman claims he sells $\frac{1}{3}$ of his prospects. The sales manager keeps a card file of all prospects. A random sample of 100 prospects shows the salesman sold 30 of the potential customers.
    a. What is the expected number of sales?
    b. What is the standard deviation?
    c. Should the sales manager challenge the salesman?
12. A shaving lotion manufacturer claims that 15 percent of all men use his product. A survey of 100 now reveals that 11 percent use the product. Would you accept the claim with 76 percent certainty?

13. The variance for a binomial experiment for $n = 1$ is computed below. For $n = 1$, the mean is $1y = y$.

$$\begin{aligned} v &= (y - 0)^2 x + (y - 1)^2 y \\ &= y^2 x + x^2 y \\ &= yx(y + x) \\ &= yx \end{aligned}$$

For $n = 2$, the mean is $2y$ and the variance is

$$\begin{aligned} v &= (2y - 0)^2 x^2 + (2y - 1)^2 2xy + (2y - 2)^2 y^2 \\ &= 4y^2 x^2 + (4y^2 - 2y + 1)2xy + 4x^2 y^2 \end{aligned}$$

a. Why does $(2y - 2)^2 = 4x^2$?

b. Do the algebra necessary to reduce the formula to $2xy$. (HINT: Use $x + y = 1$.)

Answers to problems

A. $\bar{x} = 10 \cdot 0.90 = 9$

$v = 10 \cdot 0.90 \cdot 0.10 = 0.9$

B. 1. 3

2. $v = nxy = 10 \cdot 0.7 \cdot 0.3 = 2.1$

3. $\dfrac{10!}{3!7!}(0.3)^3(0.7)^7 \cong 0.2668$

4. $\dfrac{10!}{9!1!}(0.3)^9(0.7)^1 \cong 0.000138$

It is highly unlikely that there would be 9 defective items in a sample of 10 if the population is really only 30 percent defective.

4.9 Statistics and estimation

Problems

A. Is the set {1,2,3,4,5,6,7,8,9,10} a sample or a population?

B. A study of all month-end balances in the accounts receivable of a department store was made. The average balance was found to be \$54, and the standard deviation was \$18.

1. If the purpose of the study were to determine the average balance on that particular date, was this a sample or a population? Were the measures found statistics or parameters?
2. If the purpose of the study were to estimate the average month-end balance which the company could expect, given certain sales and collection policies, is this a sample or a population?
3. In which case would statistical inference be involved?

There has already been occasion (in Sec. 3.0) to call attention to the two aspects of statistics: *descriptive statistics* and *statistical inference.*

But since the remainder of this chapter will be concerned solely with statistical inference, this section offers some general comments about the distinctions between the two aspects of statistics.

The earlier sections of this chapter dealt with descriptive statistics. Because of this, the sets of data used in the examples were treated as samples, although in most cases the sets could have been thought of as a portion of a larger set, the population. However, in *no* case was there any inference made about the population from the sample. We shall begin by reviewing what has been studied previously. Remember that a sample is a subset of a population.

The first section of this chapter dealt with how to organize data so that they can be studied and analyzed efficiently. The next two sections were about measures of the data in a sample. These measures—mean ($\bar{x}$), variance (v), and standard deviation (s)—are examples of statistics. In modern nomenclature, the word *statistic* has come to refer to measures that apply only to the sample. Sections 4.4 to 4.6 were concerned with arithmetical manipulations of the statistics so that a maximum of information could be squeezed from the data. These intermediate sections also provided tools to make the computations of the statistics easier. The properties that were discovered about the statistics, and which were stated as formulas, are the *tools* to which we refer. Section 4.7 was more of the same, except that the elements considered in the examples of data were themselves sets. In Sec. 4.8, we shifted emphasis to sets of data generated by observing events that occur in binomial experiments with an associated probability. However, in that section, as in all the others, the considerations were limited to samples.

Lest this review of topics leads you to believe that there has been an overemphasis on the descriptive aspect of the subject of statistics at the expense of statistical inference, which follows, it should be noted that much of the groundwork for statistical inference was laid in Chap. 3 when the concept of probability was developed. This will be used as a foundation for much of the work on statistical inference.

After this review of what has been accomplished with regard to the study of samples per se, we turn to an example which will show the distinction between descriptive statistics and statistical inference.

Example 1. The annual earnings for a particular year of the employees of a maintenance department of a university are compiled, along with the number of days that each worked, the number of days that each did not work owing to illness, and the number of years that each has been employed. These sets of data, treated as a sample, can be analyzed, and such statistics as the mean, variance, and standard deviation determined. The computation of these numbers belongs in the domain of *descriptive statistics.*

But if the data are used to predict the future earnings of the employees, to estimate the salaries paid to maintenance employees of another university, to predict the average annual number of days of employment that each will miss owing to illness, or to estimate the number of years each employee will remain at the same job, then the work is in the domain of *inference*.

The first example should make clear that the essential difference between the two aspects of statistics is that for descriptive statistics the set of primary concern is the sample, while for inference the set of ultimate concern is the population. The sample is used to make inferences about the population. But with inference comes an aspect of *uncertainty*. When probability is used to show the reliability of these inferences, we have statistical inference. To bring to your attention some of the dangers of incorrect inferences, more examples are offered. In each of the examples, information about a sample is given and a question is posed about the population. Try to find an answer to each question that could be applied to the population as a "correct" inference from the sample.

Example 2. Of the inmates in a certain Federal prison, 70 percent are from homes that could properly be called *poverty homes*. Should it be inferred from this sample (one of a large number of Federal prisons) that about 70 percent of all Federal prisoners come from poverty homes? While you are thinking about the answer to that question, we pose another. It is also a fact that in this same prison, 70 percent of the inmates are from Catholic homes. What inference can be made from the second fact? This example should help you understand the necessity of being careful not to draw "incorrect" inferences. It is also a fact that the population from which prisoners are sentenced to this particular prison is almost totally poor and Catholic, a situation that does not prevail for the national population.

What the example highlights is *caution*. The next three examples are directed to the same point.

Example 3. A study of the number of deaths caused by automobile accidents in New York City in 1930 and in 1960 shows that there were fewer deaths caused by automobile accidents in 1930. Can it be inferred that automobile travel was safer in 1930?

Example 4. If the number of deaths due to tuberculosis in 1945 was greater in Tucson, Arizona, than in any other city of comparable size, can it be inferred that Tucson is an "unhealthy" city as far as tuberculosis is concerned?

Example 5. For families of eight members in a certain city, the average annual income is \$3,600. Is it correct to infer that for families with four members, the average annual income is \$1,800?

As indicated earlier, we shall continue to study samples, but the emphasis will be on the population. The measures mean, variance, and standard deviation will still be of prime importance. In an attempt to lessen any confusion which might arise because of the simultaneous consideration of two sets, some new language and notation will be introduced.

The measures mean, variance, and standard deviation when applied to *populations* are called *parameters*. To aid in distinguishing the frame of reference, the notation for the measures of the population sets of data will be changed. The mean will be denoted by μ (Greek mu), the standard deviation by σ (Greek sigma), and the variance by σ^2. (Remember that the variance is the square of the standard deviation.) In the sections to follow, these symbols will be adopted. A table to assist in keeping these ideas straight follows.

	Sample	*Population*
Mean	$\bar{x}$	μ
Variance	$v = s^2$	σ^2
Standard deviation	s	σ

These comments should make clear the change of emphasis that is about to take place. There is one other caution to be observed. Experience seems to indicate that each person has a reasonably good "built-in" ability to draw inferences. Experience also seems to indicate that most people are prone to overestimate the competence of their "built-in" predictors to make the right inferences. At any rate, you will have a chance to test your own ability to make inferences in the sections that follow, and it is suggested you do so. When a question of inference is posed, try to guess the answer before you study further. This will provide an opportunity to test your own inferences against the *mathematically correct inferences*.

Of course, inferences are only as good as the methods used to make them. So for every inference there should be some test of the reliability of the conclusion reached. This will bring us to two new concepts called *level of significance* and *confidence interval*. These are technical terms to be explained in later sections. But in the meantime, the common sense of having some sort of test of reliability should be obvious. Inferences are usually made by the test of a hypothesis or by the construction of a confidence interval. The *test of a hypothesis* is also a technical phrase. You will be exposed to both methods of making inferences in the sections that follow.

Exercises

1. For each of the sets of data described below, construct a situation where the data would be considered as a population and a different situation where the data would be treated as a sample.
   a. The 1968 withholding taxes from the payroll checks of the employees of the American Laundry Company
   b. The weights of the entering male freshmen at Minnesota University in 1968
   c. The ages of the registered voters in Caldwell County
   d. The ages of the drivers in Alabama who had automobile accidents in 1968
   e. The daily maximum temperature in Philadelphia in January, 1968
   f. The Dow-Jones averages for June, 1968
2. For each of the parts of Exercise 1 where the data might be treated as a sample, construct an inference that might be made about a population.
3. Construct two sets of data from which an "incorrect" inference might be drawn unless the proper caution were used. Pattern your answer after Examples 2 to 5.
4. Relate a personal experience in which you have used a statistic and your "built-in ability to infer" to determine a parameter.
5. Discuss the advisability of the use of the statistic "the mean of the annual income of the set of the employees of General Electric" to predict the parameter "the mean of the annual income of the set of employees in the United States."

Answers to problems

A. Impossible to answer without a frame of reference; it could be either.
B. 1. A population. Parameters.
2. A sample.
3. In case 2.

4.10 Test of hypothesis

Problems

A. The average weight of each orange in a carload shipment of 10,000 oranges is to be estimated. It is known that there is a standard deviation of 1 oz in the weights. A random sample of 64 is selected, and the average weight is 4.6 oz. Should the hypothesis

The average weight is 5.0 *oz.*

be rejected at the 11 percent level of significance?

B. A product which is guaranteed to have a life of 450 hr was sampled to determine whether the product's performance is equal to the manufacturer's claims. If the standard deviation is 10 hr, if the sample mean was 443 hr, and if the sample included 100 items selected at random from the shipment, could we conclude at the 4 percent level that the mean is not as much as 450?

This section deals with inferences about a population which can be drawn from information that comes from a study of some samples of the population. The particular type of method to be used in making these inferences is called the *test of a hypothesis*. The steps involved in this method are the same as those used in a method common to many branches of science. In the framework of experimental science, this method is often called the *scientific method*.

Inference about a population from samples depends upon the ability to choose the samples in a random fashion from the population. But once again, we shall avoid the issue of deciding what is meant by a *random* sample. There are techniques of randomization that can be used to make the selection of each sample truly equally likely. Recall that we have also not attempted to define *equally likely*. In recent days, the widespread use of *random-number tables* has become commonplace in these procedures. But these topics are somewhat beside the point of the major issue to be treated here, which is how to draw conclusions about populations from data collected from samples.

Before we begin the actual discussion about testing hypotheses, we shall study exhaustively two examples. So that the examples can be treated in detail, the numbers in the set are chosen to be small and the number of elements in the set is also small. These compromises will not distract from the main purpose of the study, which is to make clear (by two examples) a more general result (theorem) that will be basic to all the considerations in the remainder of this chapter. Since these examples are for the primary purpose of motivating the more general result that follows, they are presented independent of any business or physical motivation.

Example 1. Let $S = \{8,9,10,11,12\}$ be a population. We propose to study the samples (subsets) of S which have three elements. From Sec. 2.3, the number of subsets of three elements is

$${}_5C_3 = \frac{5!}{3!2!} = \frac{5 \cdot 4}{1 \cdot 2} = 10$$

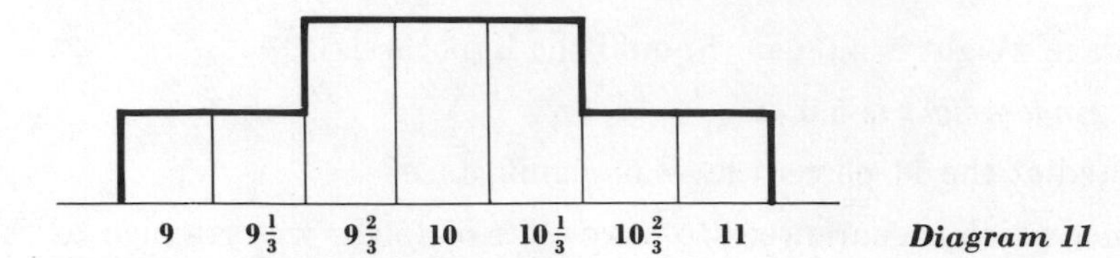

Diagram 11

The 10 subsets are listed below. The subscripts on the letter S in this listing are for identification purposes only.

$$\begin{aligned} S_1 &= \{8,9,10\} & S_6 &= \{8,11,12\} \\ S_2 &= \{8,9,11\} & S_7 &= \{9,10,11\} \\ S_3 &= \{8,9,12\} & S_8 &= \{9,10,12\} \\ S_4 &= \{8,10,11\} & S_9 &= \{9,11,12\} \\ S_5 &= \{8,10,12\} & S_{10} &= \{10,11,12\} \end{aligned}$$

Each of these samples has a mean which is the average of the three numbers. These means are easily computed, and again subscripts are used so that $\bar{x}_1$ is the mean of S_1, $\bar{x}_2$ is the mean of S_2, etc.

$$\begin{aligned} \bar{x}_1 &= 9 & \bar{x}_6 &= 10\tfrac{1}{3} \\ \bar{x}_2 &= 9\tfrac{1}{3} & \bar{x}_7 &= 10 \\ \bar{x}_3 &= 9\tfrac{2}{3} & \bar{x}_8 &= 10\tfrac{1}{3} \\ \bar{x}_4 &= 9\tfrac{2}{3} & \bar{x}_9 &= 10\tfrac{2}{3} \\ \bar{x}_5 &= 10 & \bar{x}_{10} &= 11 \end{aligned}$$

Notice that since these means are for samples, the notation $\bar{x}_i$ rather than μ_i is used for the mean of the set S_i. The set of means (arranged in order of increasing size) can be written as

$$T = \{9,9\tfrac{1}{3},9\tfrac{2}{3},9\tfrac{2}{3},10,10,10\tfrac{1}{3},10\tfrac{1}{3},10\tfrac{2}{3},11\}$$

The set T of means of the samples is displayed by the use of a histogram (Diagram 11). The same information is presented as a graph in Diagram 12. In the latter, the points on the graph have been connected with a smooth curve.

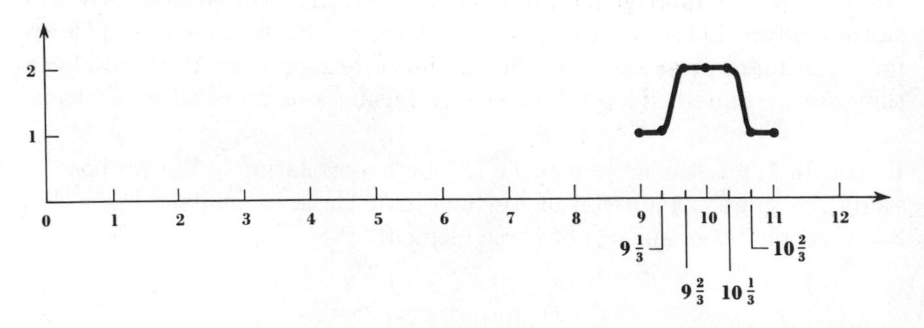

Diagram 12

The set T also has an average which is obtained in the usual fashion. A computation yields that result.

$$\mu_T = \frac{9 + 9\frac{1}{3} + 2 \cdot 9\frac{2}{3} + 2 \cdot 10 + 2 \cdot 10\frac{1}{3} + 10\frac{2}{3} + 11}{10}$$

$$= {}^{100}\!/_{10} = 10$$

The notation μ_T (which will be used in the first two examples) refers to the fact that the average of the elements in T has been computed. This notation is consistent with the use of $\bar{x}_1$ for the mean of S_1. But the average of the set S is also 10. (This should be verified.) The point of all this computation has now been reached. The example deals with samples with three elements each, but the general result is independent of the size of the sample.

The average of all possible sample means for samples of any given size is the same as the average of the population.

For this example, $\mu_T = \mu$, where μ_T is the average of the population of sample means and μ is the average of the population. This is no accident! For all finite sets and with this procedure of sampling, this is always the case. This result, which may come as a surprise to you, can be proved in a rigorous fashion, but such is not to be done in this book. A little later, the italicized result (established for this particular example) will be stated for the general case.

Note that the mean of any one of the samples could have been used as an *estimate* of the mean of the population. The maximum error that could have been made in such an estimate is 1 ($10 - 9 = 1$ and $11 - 10 = 1$). Or, stated differently, in the absence of the "true" mean of the population, any one of the sample means could have been used to infer the mean of S. The discussion that follows will throw some light upon how good such an inference can be expected to be.

Before we summarize, consider another, similar question for the same example. What can be said about the relationship between the variance of the population (S) and the variance of the set T of sample means? First, computations will be made to determine these two numbers. For the set S we have

$$\sigma^2 = \sum_{i=1}^{n} \frac{(\mu - x_i)^2}{n}$$

$$= \sum_{i=1}^{5} \frac{(10 - x_i)^2}{5}$$

$$= \frac{(10-8)^2 + (10-9)^2 + (10-10)^2 + (10-11)^2 + (10-12)^2}{5}$$

$$= \frac{4 + 1 + 0 + 1 + 4}{5}$$

$$= {}^{10}\!/_{5} = 2$$

For the set T we have

$$
\begin{aligned}
(\sigma_T)^2 &= \sum_{i=1}^{n} \frac{(\mu_T - x_i)^2}{n} \\
&= \sum_{i=1}^{10} \frac{(10 - x_i)^2}{10} \\
&= \frac{(10-9)^2 + (10 - 9\tfrac{1}{3})^2 + 2(10 - 9\tfrac{2}{3})^2 + 2(10-10)^2 + 2(10 - 10\tfrac{1}{3})^2 + (10 - 10\tfrac{2}{3})^2 + (10-11)^2}{10} \\
&= \frac{1 + \tfrac{4}{9} + \tfrac{2}{9} + 0 + \tfrac{2}{9} + \tfrac{4}{9} + 1}{10} \\
&= \frac{\tfrac{30}{9}}{10} \\
&= \tfrac{30}{9} \cdot \tfrac{1}{10} = \tfrac{3}{9} = \tfrac{1}{3}
\end{aligned}
$$

Again note that the symbol $(\sigma_T)^2$ (which will be used in the first two examples) is used to make clear that the variance of the set T is being computed. The variances for the sets S and T are not the same! Nor is it likely that the relationship between σ^2 and $(\sigma_T)^2$ could be guessed on the basis of this example (or several other examples, for that matter). The relationship, which is soon to be presented, is usually written for standard deviation rather than variance. Remember that the standard deviation is the square root of the variance. Hence, for this example,

$$\sigma = \sqrt{\sigma^2} = \sqrt{2} \qquad \text{and} \qquad \sigma_T = \sqrt{(\sigma_T)^2} = \sqrt{\tfrac{1}{3}} = \frac{1}{\sqrt{3}}$$

Decimal representations of these numbers will help to make a comparison of their relative sizes. The standard deviation of the population is

$$\sigma = \sqrt{2} \cong 1.4142$$

and the standard deviation of the sample means is

$$\sigma_T = \frac{\sqrt{3}}{3} \cong 0.5774$$

The standard deviation of the sample means is less than that of the population. This means that there is less "scatter," and this property enhances the value of the sample mean as an estimate of the mean of the population.

The formula for the relationship between the standard deviation of the (finite) population and the sample means for finite population is the somewhat formidable expression

$$\sigma_T = \frac{\sigma}{\sqrt{n}} \sqrt{\frac{N-n}{N-1}}$$

In the formula, σ_T is the standard deviation of the set T of sample means, σ is the standard deviation of the population S, n is the number of elements in each sample, and N is the number of elements in the population. For this example, $\sigma = \sqrt{2}$, $\sigma_T = 1/\sqrt{3}$, $n = 3$, and $N = 5$. Replacement in the formula by these numbers yields

$$\begin{aligned}\frac{1}{\sqrt{3}} &= \frac{\sqrt{2}}{\sqrt{3}}\sqrt{\frac{5-3}{5-1}} \\ &= \frac{\sqrt{2}}{\sqrt{3}}\frac{\sqrt{2}}{\sqrt{4}} \\ &= \frac{2}{\sqrt{3}\cdot 2} \\ &= \frac{1}{\sqrt{3}}\end{aligned}$$

Example 2. Another example of the same kind will be given, but since the pattern has been established, less detail will be needed. Let $S = \{3,4,5,6,7,8\}$ be a population, and consider the samples of two elements each. There are ${}_6C_2 = 6!/(2!4!) = (6 \cdot 5)/(1 \cdot 2) = 15$ of these, and they are tabulated as follows:

$$\begin{array}{lll} S_1 = \{3,4\} & S_6 = \{4,5\} & S_{11} = \{5,7\} \\ S_2 = \{3,5\} & S_7 = \{4,6\} & S_{12} = \{5,8\} \\ S_3 = \{3,6\} & S_8 = \{4,7\} & S_{13} = \{6,7\} \\ S_4 = \{3,7\} & S_9 = \{4,8\} & S_{14} = \{6,8\} \\ S_5 = \{3,8\} & S_{10} = \{5,6\} & S_{15} = \{7,8\} \end{array}$$

The means of the samples are computed in the usual way. They are

$$\begin{array}{lll} \bar{x}_1 = 3\frac{1}{2} & \bar{x}_6 = 4\frac{1}{2} & \bar{x}_{11} = 6 \\ \bar{x}_2 = 4 & \bar{x}_7 = 5 & \bar{x}_{12} = 6\frac{1}{2} \\ \bar{x}_3 = 4\frac{1}{2} & \bar{x}_8 = 5\frac{1}{2} & \bar{x}_{13} = 6\frac{1}{2} \\ \bar{x}_4 = 5 & \bar{x}_9 = 6 & \bar{x}_{14} = 7 \\ \bar{x}_5 = 5\frac{1}{2} & \bar{x}_{10} = 5\frac{1}{2} & \bar{x}_{15} = 7\frac{1}{2} \end{array}$$

The means of the samples, arranged in order of increasing size, make up the set T, which is

$$T = \{3\tfrac{1}{2},4,4\tfrac{1}{2},4\tfrac{1}{2},5,5,5\tfrac{1}{2},5\tfrac{1}{2},5\tfrac{1}{2},6,6,6\tfrac{1}{2},6\tfrac{1}{2},7,7\tfrac{1}{2}\}$$

The histogram for T is in Diagram 13, and the graph of T connected by a smooth curve is in Diagram 14.

The means of the set S and the set T are computed in the usual fashion and are

$$\mu = \mu_T = 5\tfrac{1}{2}$$

The standard deviations can also be computed by formulas developed

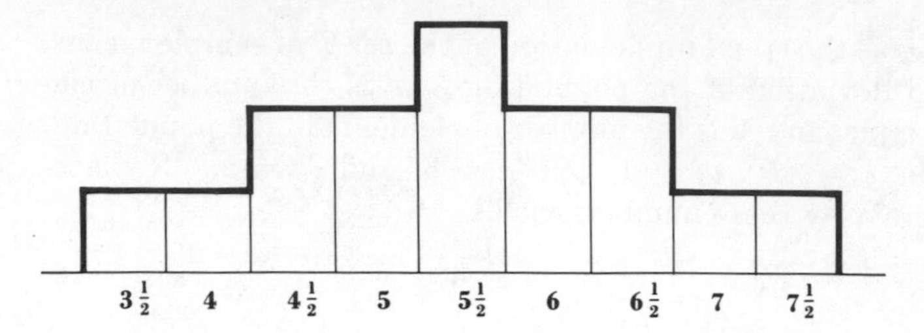

Diagram 13

earlier. They are

$$\sigma = \sqrt{\frac{70/4}{6}} = \sqrt{70/24} = \sqrt{35/12} \cong 1.708$$

$$\sigma_T = \sqrt{\frac{70/4}{15}} = \sqrt{70/60} = \sqrt{7/6} \cong 1.08$$

Note that once again the standard deviation of T is less than the corresponding number for S. The formula which was stated to express the relationship between the standard deviation of the population and the sample means is

$$\sigma_T = \frac{\sigma}{\sqrt{n}} \sqrt{\frac{N-n}{N-1}}$$

For this example, we have

$$\begin{aligned} \sqrt{7/6} &= \frac{\sqrt{35/12}}{\sqrt{2}} \sqrt{\frac{6-2}{6-1}} \\ &= \sqrt{35/24}\,\sqrt{4/5} \\ &= \sqrt{7/6} \end{aligned}$$

These two examples, of course, *prove* nothing. They do, however, verify that the relationship in the formula between the two standard deviations is true for these examples.

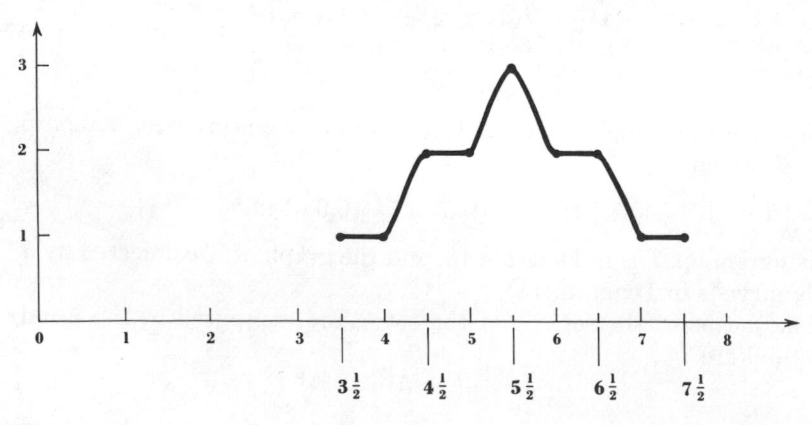

Diagram 14

As we stated earlier, the purpose of these examples is to illustrate a general result which is to be applied in future considerations. This result (theorem) is stated in the formula that follows.

Formula: **(*Relationship between the mean and standard deviation of a finite population and the mean and standard deviation of samples of the population*)** *Let S be a set of real numbers with N elements. Let the mean of S be μ and the standard deviation of S be σ. Consider the set R of all subsets of S with n elements ($n \leq N$). Let T be the set of the means of the set R. If the mean of T is $\mu_{\bar{x}}$ and the standard deviation of T is $\sigma_{\bar{x}}$, then the following relationships hold:*

$$(1) \quad \mu_{\bar{x}} = \mu$$

$$(2) \quad \sigma_{\bar{x}} = \frac{\sigma}{\sqrt{n}} \sqrt{\frac{N - n}{N - 1}}$$

Two comments should be made about this formula. First, a new notation was introduced in the formula. In Examples 1 and 2, the symbol μ_T was used for the average of the sample means. In the formula, the symbol $\mu_{\bar{x}}$ is used instead. This change will make it easier to refer to the average of the sample means when the set T is not available, as will be generally the case. The symbol $\mu_{\bar{x}}$ is also the one in common use in other books of this type. Finally, the subscript $\bar{x}$ on $\mu_{\bar{x}}$ makes clear that the average being considered refers to the means of the samples which are denoted by $\bar{x}$. Similar remarks apply to the use of the symbol $\sigma_{\bar{x}}$, which replaces the symbol σ_T used in the two examples. The notation $\sigma_{\bar{x}}$ will be more convenient than σ_T; it is the standard notation, and it makes clear that the standard deviation refers to the set of sample means which are denoted by $\bar{x}$. The second comment is that the situation to which the formula applies is sometimes called *sampling without replacement*. This expression can be misleading unless fully understood. It refers to the fact that once a sample (subset) is selected, that same sample will not be selected again in the sampling procedure. We shall have occasion to refer to this situation again later.

The next example will show how the relationships in the formula above can be used to test a hypothesis about the mean of a finite set of numbers.

Example 3. The Deluxe Caviar Company packages its product in glass jars from a large container. This process utilizes a machine which is set to stop automatically when the input in the jar reaches 3 oz of net weight. The machine has just filled 10,000 jars with caviar. The machine operates such that there is a standard deviation in the net weights of 0.4 oz.

The company wishes to test the following hypothesis:

The average net weight of caviar per jar is 3 oz.

The company proposes to test the truth of this statement (hypothesis) by selecting a sample of 100 jars and weighing the contents. If the average net weight of the 100 jars in the sample is 3.2 oz, is the hypothesis true?

There is only one word in the total statement of the problem which might be unfamiliar, and that is *hypothesis*. In the context of this problem, *hypothesis* represents a statement which is to be accepted or rejected on the basis of the experiment of selecting a sample of 100 jars and weighing the contents.

First, we identify the various aspects of the problem in terms of the terminology of the formula of this section. The population has 10,000 elements; hence, $N = 10{,}000$. The mean of the population is $\mu = 3$, and the standard deviation of the population is $\sigma = 0.4$. The size of the sample is $n = 100$. The average weight of the elements in the sample, $\bar{x}$, is 3.2 oz.

Before an answer to the question is given (before the hypothesis is tested), it will be helpful to draw two diagrams. First, the number of elements in the population is 10,000 and the individual elements (which are the net weights of the 10,000 jars) are not known nor will they be determined. But if the individual elements were known, it would be expected that they would distribute about the mean $\mu = 3$ with a standard deviation of $\sigma = 0.4$. If these elements were grouped so that a histogram could be drawn and if it, in turn, had the midpoints of each rectangle connected by a *smooth curve*, then the appearance would be as in Diagram 15.

Remember that at least $1 - (1/3)^2 = 1 - 1/9 = 8/9$ of the 10,000 elements *must* be within three standard deviations of the mean. (This does not make any assumption about the distribution of the population.) This is indicated on the graph by showing about 90 percent of the area underneath the graph between $3 - 3 \cdot (0.4) + 3 - 1.2 + 1.8$ and $3 = 3 \cdot (0.4) + 3 = 1.2 + 4.2$. (You may wish to review the results about scatter within three standard deviations on page 245.)

Second, a diagram for the distribution of the means of all possible samples of 100 can be indicated. There are ${}_{10{,}000}C_{100}$ subsets, of 100 elements each. This number is extremely large, and of course, no

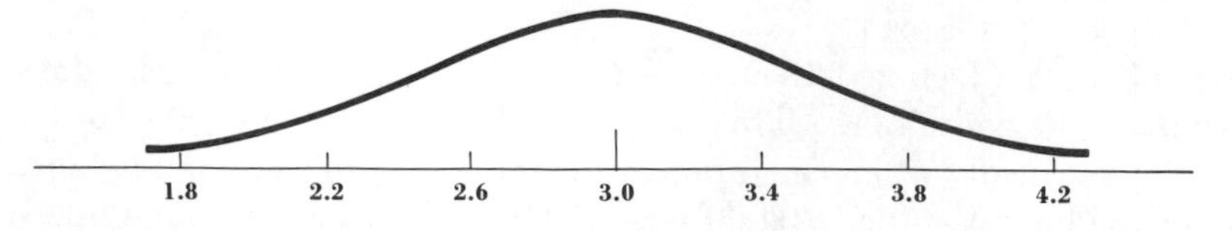

Diagram 15

attempt will be made to compute the mean weights of all the possible samples. This is the set we called R in the formula. If the mean of each sample in the set of all samples of 100 jars, R, were computed, then the set of numbers that would result is the set we called T in the first example. To graph T by computing the mean of each of these samples, as we did for the first example, is absurd. However, remember that the average of the sample means T, which we have denoted by $\mu_T = \mu_{\bar{x}}$, is equal to the mean of the population. Hence, $\mu_{\bar{x}} = \mu = 3$. The standard deviation of the sample means in set T will not be computed either. The task would be tremendous. However, since $\sigma = 0.4$ is known, the number $\sigma_T = \sigma_{\bar{x}}$ can be computed by use of the formula

$$\sigma_{\bar{x}} = \frac{\sigma}{\sqrt{n}}\sqrt{\frac{N-n}{N-1}}$$

But for this example, a simplification of the formula is possible. The population size is large with respect to the sample size, so that

$$\frac{N-n}{N-1} = \frac{10{,}000 - 100}{10{,}000 - 1} = \frac{9{,}900}{9{,}999} \cong 0.99 \cong 1$$

Since this number is approximately 1, so is its square root. In actual practice, this approximation is almost always possible, and the much simpler formula

$$\sigma_{\bar{x}} = \frac{\sigma}{\sqrt{n}}$$

can be used. Replacement by $\sigma = 0.4$ and $n = 100$ yields

$$\sigma_{\bar{x}} = \frac{0.4}{\sqrt{100}} = \frac{0.4}{10} = 0.04$$

Now the second diagram for the distribution of the means of the samples can be indicated. Again the numbers $\mu_{\bar{x}} - 3\sigma_{\bar{x}}$ and $\mu_{\bar{x}} + 3\sigma_{\bar{x}}$ are computed to give

$$3 - 3 \cdot (0.04) = 3 - 0.12 = 2.88$$

and

$$3 + 0.12 = 3.12$$

Diagram 16 shows at least $8/9$ of the area clustered between 2.88 and 3.12.

Now we can make a statement about the truth of the hypothesis on the basis of the outcome of the experiment. The experiment resulted in a sample mean of 3.2. But the probability of such an event is less than $1/9$ because at least $8/9$ of the sample means are between 2.88 and 3.12. This is true with no assumption about the distribution of the population.

In fact, a much stronger statement is possible. Since

$$\mu_{\bar{x}} - 5\sigma_{\bar{x}} = 3 - 5 \cdot 0.04 = 3 - 0.2 = 2.8$$

and

$$\mu_{\bar{x}} + 5\sigma_{\bar{x}} = 3.2$$

and since $1 - (1/5)^2 = 1 - 1/25 = 24/25$, the probability of selecting a random sample of mean weight greater than 3.2 or less than 2.8 is less than $1/25$. The probability of the selection of the sample is less than 4 percent. Since this event is highly unlikely, the hypothesis is *rejected.*

Note that no positive assertions have been made about the "true" mean of the population. The only remark that is made as the result of this experiment is that it is highly improbable that the mean of the population is 3.

Example 4. The next example is a variation of the preceding one. Suppose that the same company knows that government regulations are such that they must be sure that the net weight of their containers averages at least 3 oz. In this case, they might decide to test the hypothesis

The average net weight of caviar per jar is less than 3 *oz.*

Suppose that the experiment with the sample of 100 jars yields the same results as before: $\bar{x} = 3.2$.

The hypothesis to be tested in this example permits the possibility of $\mu_{\bar{x}} < 3$. But if such were the case, the probability of a sample with $\bar{x} = 3.2$ would be even more unlikely. Consult Diagram 16 to see why such a hypothesis would decrease the probability of the selection of a sample with a mean of 3.2.

Again the hypothesis is rejected. It should be noted that the statement of the hypothesis is somewhat negative. The company really wants to ensure that their product meets government regulations. Rejec-

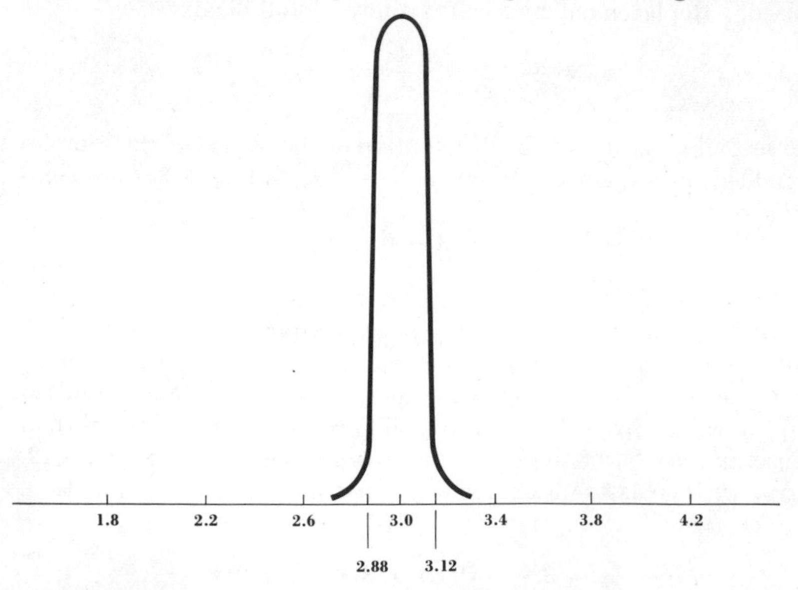

Diagram 16

tion of this hypothesis is based on the fact that the probability is "very great" that the mean is 3 oz or more than 3 oz, and they can cease worrying that government regulations have been violated. The rejected hypothesis is sometimes called a *null hypothesis.* The hypothesis that the population mean is 3 or more than 3 can be accepted, and it is sometimes called an *alternate hypothesis.*

Example 5. The Roller Steel Company wishes to estimate the average time between the date of completion of a product and its delivery date. The standard deviation in average time elapsed between these two dates is four days. They wish to accept or reject the hypothesis

The average time between production and delivery is 34.2 *days.*

The company has records on 10,000 shipments and selects (at random) the records on 50 shipments. The average time is computed for the sample, and it is 30.8 days. What conclusions can be reached about the hypothesis?

With reference to the symbols in the formula of this section, the number of elements in the population is $N = 10{,}000$; the number of elements in the sample is $n = 50$. The standard deviation of the population is $\sigma = 4$. The standard deviation of the sample can be computed by

$$\sigma_{\bar{x}} = \frac{\sigma}{\sqrt{n}}\sqrt{\frac{N-n}{N-1}}$$

Again, because the population size is large with respect to the sample size, the number

$$\sqrt{\frac{10{,}000-50}{10{,}000-1}} = \sqrt{\frac{9{,}950}{9{,}999}} \cong \sqrt{0.995} \cong 1$$

is very close to 1 and can be ignored in the computation. So

$$\sigma_{\bar{x}} \cong \frac{4}{\sqrt{50}} \cong 0.57$$

Again it may be helpful to see graphic displays. The first of these (Diagram 17) uses the hypothetical mean for the population $\mu = 34.2$ and shows that portion of the curve which lies within four standard deviations. It is known that at least $1 - (\frac{1}{4})^2 = \frac{15}{16}$ of the elements of the popula-

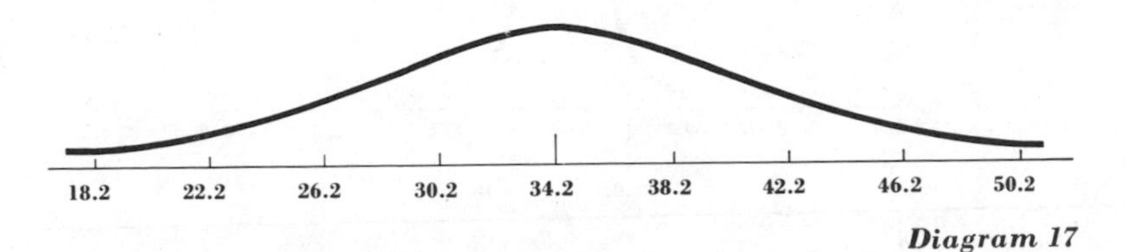

Diagram 17

tion must be in the interval whose *end points* are

$$\mu - 4\sigma = 34.2 - 4 \cdot 4 = 18.2$$
$$\mu + 4\sigma = 34.2 + 4 \cdot 4 = 50.2$$

Notice that a *smooth curve* has been used to approximate the graph of the distribution, which actually has a finite number of points. In Diagram 18, the mean is also 34.2, but since the standard deviation is only 0.57, the number of points in the sample distribution that are near the mean is much greater. Again the end points have been chosen to represent four standard deviations from the mean. These numbers are

$$\mu_{\bar{x}} - 4\sigma_{\bar{x}} = 34.2 - 4 \cdot (0.57) = 31.92$$
$$\mu_{\bar{x}} + 4\sigma_{\bar{x}} = 34.2 + 4 \cdot (0.57) = 36.48$$

Since 30.8, which is the mean of the random sample selected, lies to the left of the end point 31.92, it can be stated with probability of at least $^{15}\!/\!_{16}$ that the estimated mean is incorrect. It can be stated that the probability of $\mu = 34.2$ is less than $^{1}\!/\!_{16}$. The hypothesis is rejected.

The section is concluded with several general remarks.

First, the percentage to be used as the test for acceptance or rejection of the hypothesis should be decided before the sample is selected. In Example 3, the first observation was that the mean of the sample selected was more than three standard deviations from the mean. Another observation was that the mean of the sample selected was also more than five standard deviations from the mean of the population.

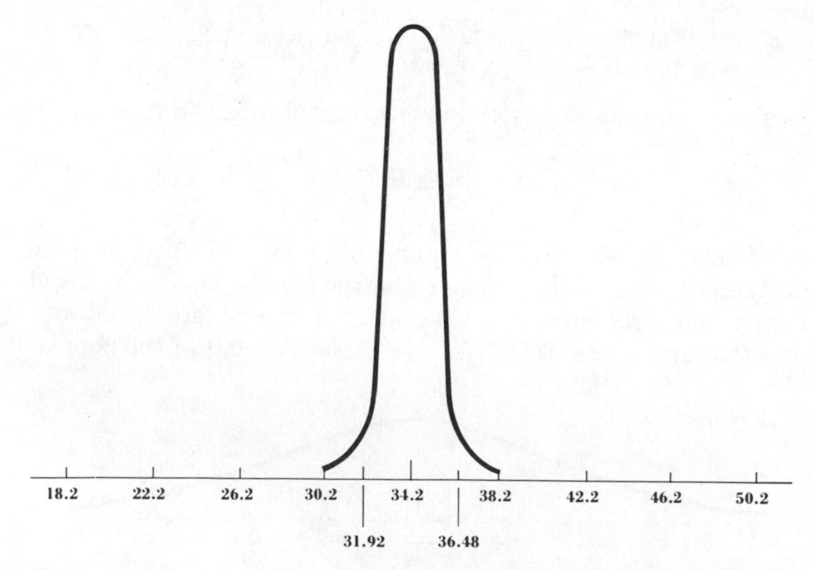

Diagram 18

The first rejection is at the 11 percent *level of significance;* the second rejection is at the 4 percent *level of significance.* In some books these are called the 89 percent and 96 percent levels of significance.

For Example 5, the rejection is at the 7 percent $(1 - 15/16 = 1/16)$ level of significance (or, alternately, at the 93 percent level of significance). These levels of significance should be selected before the experiment is made. The level of significance selected depends upon the degree of certainty of prediction required.

Second, the size of the sample will clearly affect the outcome. To see this, remember that the standard deviation of the sample means $\sigma_{\bar{x}}$ is determined by dividing the standard deviation of the population σ by $\sqrt{n}$ (when N is large with respect to n). If n is larger, then $\sigma_{\bar{x}}$ is smaller and the sample mean curve is not so dispersed. We shall not try in this book to justify the choice of a proper size of sample, but there are well-established procedures for the choice of a sample size.

Third, the use of a sample to test a hypothesis can lead to a "bad" decision. The pitfalls of making bad decisions have already been discussed briefly. But for this situation, the dangers are clear and are easily specified. The hypothesis is either true or false. The hypothesis may be true and accepted, which is a "good" decision. The hypothesis may be false and rejected, and this is also a good decision. However, the hypothesis may be true and rejected. This is a bad decision. In the nomenclature of statistics, this is called an *error of type I.* The remaining case is that the hypothesis is false and accepted. This is also a bad decision and is often called an *error of type II.* A proper discussion of the relative importance of avoiding each type of error and of the procedures adopted to assist in avoiding these errors would be much too lengthy to be included here. There are numerous books that include such discussions. A related class of problems called *decision problems* is also discussed at length in other books.

Finally, it must be clear that the examples were constructed so that rejection of the hypothesis was mandatory. In other problems, the decision to accept the hypothesis is also evident. A little thought will show that for others a clear-cut decision is not possible and some further action such as more sampling or increasing the size of the sample is necessary.

Exercises

1. Refer to Example 1 to answer each of the following:
   a. How many subsets of S are there with two elements?
   b. Write the subsets of S with two elements.
   c. Compute the means of the sets in part *b*.

d. Display the results of part *c* on a histogram.
e. Display the results of part *c* on a graph, as in Diagram 12.
f. Compute the average of the set determined in part *c*.
g. Compute the standard deviation of the set determined in part *c*.
h. Verify that the relationship between the result in part *g* and the standard deviation of the set S satisfy the formula on page 279.

2. Work Exercise 1 again for subsets of four elements.

3. A random sample of 64 is chosen from a population of 10,000. The mean of the sample is 14.5. The standard deviation of the population is 4. For each of the following, reject the hypothesis if the probability is $\frac{1}{4}$ or less of the selection of each sample.
   a. The mean of the population is 15.0.
   b. The mean of the population is 16.1.
   c. The mean of the population is 16.6.
   d. The mean of the population is 13.5.
   e. The mean of the population is 12.9.
   f. The mean of the population is 12.4.

4. Rework Exercise 3 for probability $\frac{1}{9}$.

5. Rework Exercise 3 for probability $\frac{1}{16}$.

6. Assume that you are buying a product which is described as having a mean strength of 650 lb. Assume that the standard deviation is 25 lb. Test the hypothesis that the mean is 650 lb. Reject the hypothesis at the four percent level. Assume a sample of 100 items with a mean of 632. Would you reject the hypothesis?

7. Investigating an alleged unfair trade practice, the Federal Trade Commission studied a sample of fifty "12-oz" bottles from a large shipment. The sample mean is $\bar{x} = 11.94$. The population has a standard deviation of $\sigma = 0.14$ oz. Test whether this constitutes evidence on which to base a finding of unfair practice at the
   a. 75 percent level of confidence.
   b. 89 percent level of confidence.
   c. 96 percent level of confidence.

Answers to problems

A. For this example, N is the number of oranges in the carload, which is very large with respect to the number $n = 64$. So the factor

$$\frac{N - n}{N - 1}$$

is approximately 1 and is ignored. It is known that $\sigma = 1$, so that

$$\sigma_{\bar{x}} = \frac{1}{\sqrt{64}} = \frac{1}{8} = 0.125$$

The hypothesis to be tested is that $\mu = 5$. But

$$5 - 3 \cdot (0.125) = 5 - 0.375 = 4.625$$

which is larger than 4.6. The hypothesis is rejected at the 11 percent level of significance.

B. Ignoring the factor

$$\frac{N - n}{N - 1}$$

the hypothesis to be tested is $\mu = 450$. But $\sigma = 10$ implies that

$$\sigma_{\bar{x}} = \frac{10}{\sqrt{100}} = 1$$

Then

$$450 - 5 \cdot 1 = 445$$

At the 4 percent level of significance, the mean is not as much as 450.

4.11 Confidence intervals

Problem

A lot of 5,000 loaves of bread is delivered to a supermarket. A random sample of 80 loaves is selected and weighed. The average weight is 7 oz, with a standard deviation of 0.1 oz. Give a point estimate of the average weight of the total lot. Construct a confidence interval for the average weight of the entire lot with 75 percent certainty.

Estimation of a parameter of a population by the use of the statistics for a random sample is, as it was in Sec. 4.10, the subject of this section. The outcome of the test of a hypothesis was its ultimate rejection or acceptance on the basis of the outcome of a sampling event. This rejection (acceptance) was, of necessity, tempered by a statement about the probability of being wrong. The method used in this section is very closely related to the test of a hypothesis, but the eventual outcome is the construction of a *confidence interval* rather than a probability statement about a hypothesis. The reader will no doubt observe several similarities between the reasoning used in this second method and that used in the previous section. On the other hand, no new formulas are needed for this new attack on the same problem.

There is one other departure from the technique employed in Sec. 4.10, and this involves the standard deviation of the population. Since this change in method is best introduced by example, a detailed explanation of it will be delayed until later.

Example 1. The reading ability of an incoming class of 5,000 freshmen at a large state university is to be estimated by using the examination scores for a sample of 50 members of the class (chosen at random). The exam is given to the 50 students, and the scores on the exam result in a mean score of 76, with a standard deviation of 6. What statements could be made about the mean score on the test if all 5,000 of the scores were considered rather than the sample of 50?

One estimate of the population mean μ is that it is equal to the sample mean, $\bar{x} = 76$. Such an estimate is called a *point estimate*. But another technique will yield an *interval estimate*, and the meaning of this expression will become clear as we proceed.

Neither the mean μ nor the standard deviation σ of the population is known. It is also desirable *not* to compute these numbers, for the obvious reason that this would be a very large task. At this stage of the investigation for the examples in Sec. 4.10, the standard deviation of the population was given and the mean of the population was estimated by the statement of a hypothesis. In practice, the standard deviation of the population is rarely known, and it is also usually impractical to compute it. You should review the previous section to see that this is true for the particular examples given there. Try to formulate the reasons why the standard deviation of the population for those examples would not be known or easily computable. Because of these reasons some *estimator* of the number (the standard deviation of the population) is needed, and an obvious choice is the standard deviation of the sample. For this example, the standard deviation of the sample is $s = 6$, and this will be used to estimate σ. The use of the standard deviation of the sample as an estimator of the standard deviation of the population clearly has some shortcomings, but in practice this compromise is often necessary.

The number of elements in the population is $N = 5{,}000$, and the number of elements in the sample is $n = 50$, so that once again the factor

$$\sqrt{\frac{N-n}{N-1}} = \sqrt{\frac{5{,}000-50}{5{,}000-1}} = \sqrt{\frac{4{,}950}{4{,}999}} \cong \sqrt{1} = 1$$

which occurs in the relationship between σ and $\sigma_{\bar{x}}$ can be ignored in the computation.

Now that an estimate has been made of the standard deviation of the population σ, it can be used to compute the standard deviation of the means of all possible samples of the same size, $\sigma_{\bar{x}}$. By the formula of the previous section, we have

$$\begin{aligned} \sigma_{\bar{x}} &= \frac{\sigma}{\sqrt{n}} \\ &= \frac{6}{\sqrt{50}} \cong 0.85 \end{aligned}$$

The mean of the sample is used as an estimator of the average of the sample means, so that $\mu_{\bar{x}} = \mu = 76$. With these estimates, the knowledge about the fraction of the numbers in the population that are within two standard deviations of the average can be used to construct an *interval of confidence*, which will include the mean of the population. We choose this example to be 75 percent "sure." That is, we predict with at least $1 - (\frac{1}{2})^2 = 1 - \frac{1}{4} = \frac{3}{4} = 75$ percent certainty that the mean of the population is within two standard deviations of the sample mean 76. With these agreements, it can be stated that the *end points* of the interval of confidence are

$$\mu_{\bar{x}} - 2\sigma_{\bar{x}} = 76 - 2 \cdot 0.85 = 76 - 1.7 = 74.3$$

and

$$\mu_{\bar{x}} + 2\sigma_{\bar{x}} = 76 + 2 \cdot 0.85 = 76 + 1.7 = 77.7$$

The ultimate conclusion is, as before, a statement with a probability attached. The mean of the population μ is such that

$$74.3 < \mu < 77.7$$

with a probability of at least 0.75.

In this first example, the confidence interval was constructed on the premise that the mean of the population would not be more than two standard deviations removed from the mean of the sample selected at random. This resulted in the necessity of assigning the probability of at least 0.75 to the end points. For some purposes, this degree of certainty might not be good enough. The desirability of a "goodness" test was discussed earlier. Suppose that *more than* 75 percent certainty is desired in the final statement of the estimate of the population mean. This increase in certainty is obtainable, but the price which must be paid in obtaining it is an increase in the length of the confidence interval.

For Example 1, the mean of the population was located (with at least 75 percent certainty) in an interval of length

$$77.7 - 74.3 = 3.4$$

Example 2. For this second example, suppose that at least $1 - (\frac{1}{3})^2 = 1 - \frac{1}{9} = \frac{8}{9} \cong 89$ percent certainty is desired. Then the number of standard deviations to be used in the construction of the confidence interval is three, and the end points of the interval are computed using the formulas $\mu_{\bar{x}} - 3\sigma_{\bar{x}}$ and $\mu_{\bar{x}} + 3\sigma_{\bar{x}}$, with the resulting numbers

$$76 - 3 \cdot (0.85) = 76 - 2.55 = 73.45$$
$$76 + 3 \cdot (0.85) = 76 + 2.55 = 78.55$$

In summary, the mean of the exam scores is such that

$$73.45 < \mu < 78.55$$

and this statement is made with at least 89 percent certainty.

The certainty of the final statement for this example has increased, but so has the length of the confidence interval, which is now

$$78.55 - 73.45 = 5.1$$

Example 3. The third example deals with the estimation of the parameter of a binomial distribution.

A city is to hold an election for mayor, with 100,000 persons expected to vote. A random sample shows that of 500 voters questioned, 280 favor the incumbent. An estimate of the percentage of the total set of voters favoring the current mayor is desired. Further, a confidence interval for this estimate is to be constructed.

Note that this set of voters is a binomial population, with individual voters for or against the incumbent. If we denote by $N = 100{,}000$ the number who will vote and by w the number who will vote for the present mayor, then the true proportion of those for him, $w/N = w/100{,}000$, is the parameter to be estimated. The proportion of the sample for him, $^{280}\!/_{500} = 0.56 = 56$ percent $= x$, is a point estimate of the parameter. The first question raised, namely, what is a point estimate of the percentage of total voters favoring the incumbent, has been answered: It is 56 percent.

For this example, neither the standard deviation of the population (voters) nor the standard deviation of the sample is known. But now the results of Sec. 4.8 can be used. In that section, it was established that the variance of a set which distributes by the binomial distribution is given by

$$v = nxy$$

For this example, the variance is

$$v = 500 \cdot (0.56) \cdot 0.44$$

The number y is given by $y = 1 - x = 1 - 0.56 = 0.44$. A computation shows that

$$v = 123.2$$

and since $s = \sqrt{v}$,

$$s = \sqrt{123.2} \cong 11.1$$

This is the standard-deviation estimate. It is stated in terms of the number of voters in a sample of 500. So first we construct a confidence interval of 75 percent certainty for the true number in a sample of 500 favoring the incumbent. It has a lower bound of

$$280 - 2 \cdot (11.1) = 257.8$$

and an upper bound of

$$280 + 2 \cdot (11.1) = 302.2$$

These numbers are easily converted to percentages, and the 75 percent confidence interval for the true proportion of voters favoring the incumbent is

$$\frac{257.8}{500} < \frac{w}{N} < \frac{302.2}{500}$$

or

$$51.56\% < \frac{w}{N} < 60.44\%$$

Note that once again the factor $\sqrt{(N - n)/(N - 1)}$ has been ignored because of the relative sizes of N and n.

Example 4. The National Woolen Goods Company manufactures men's ties. In a lot of 100,000 ties, there are a number of ties with minor defects. The company selects a sample of 250 ties, which are inspected with the result that 200 are found to be free of defects. Estimate the percentage of the population (100,000 ties) that are defective and construct a confidence interval for the percentage of nondefective items with 75 percent certainty.

The estimator of the percentage of nondefective items is the quotient of nondefective ties and the number in the sample, which is

$$x = {}^{200}\!/_{250} = 80\%$$

The estimated defective percentage is $1 - 0.80 = 0.20 = 20$ percent. The variance of the population, assuming a binomial distribution, is given by

$$\begin{aligned}\sigma^2 &= n \cdot x \cdot y \\ &= 250 \cdot (0.80) \cdot (0.20) \\ &= 40\end{aligned}$$

The number 40 is used as an estimator of the variance of the population. The estimated standard deviation of samples of 250 each is then

$$\sigma = \sqrt{40} \cong 6.33$$

This is the number of ties and is not a percentage of the sample.

The question calls for the construction of an interval of 75 percent confidence. But we know that the end points are to be constructed with two standard deviations from the point estimate. First we construct an interval for the *number* of defective ties in a sample of 250. It is

$$\begin{aligned}50 - 2 \cdot 6.33 < \text{number of defectives} < 50 + 2 \cdot 6.33 \\ 37.44 < \text{number of defectives} < 62.66\end{aligned}$$

In terms of percent, the interval is approximately

$$\begin{aligned}\frac{37.44}{250} < \text{percent of defectives} < \frac{62.66}{250} \\ 15\% < \text{percent of defectives} < 25\%\end{aligned}$$

In summary, the percentage of defectives, with 75 percent confidence, is between 15 and 25.

Exercises

1. Ten thousand students take the same mathematics examination. A random sample of $n = 36$ papers has an average score of $\bar{x} = 81$, with a standard deviation of $s = 5$. Use s as an estimator of the standard deviation σ of the population and construct a 75 percent confidence interval for the mean of the population.
2. Rework Exercise 1 for $\bar{x} = 75$, $s = 7$.
3. Rework Exercise 1 for $\bar{x} = 92$, $s = 3$.
4. Rework Exercise 1 for an 89 percent confidence interval.
5. Rework Exercise 2 for an 89 percent confidence interval.
6. Rework Exercise 3 for an 89 percent confidence interval.
7. For a forthcoming election involving 50,000 voters, a poll shows that 60 members of a sample of 100 voters favor candidate A. Assume a binomial distribution. Construct a 75 percent confidence interval for the percentage who "truly" favor candidate A.
8. Rework Exercise 7 for $N = 100{,}000$.
9. Rework Exercise 7 for 65 voters favoring A.
10. Rework Exercise 9 for $N = 100{,}000$.
11. Rework Example 4 for an 89 percent level of confidence.

Answer to problem

The point estimate $\mu_{\bar{x}}$ is 7 oz. The sample standard deviation of 0.1 oz is used to estimate σ: $\sigma = 0.1$. The standard deviation of the sample means is

$$\sigma_{\bar{x}} = \frac{\sigma}{\sqrt{n}} = \frac{0.1}{\sqrt{80}} \cong \frac{0.1}{9} \cong 0.0111$$

The numbers

$$\mu_{\bar{x}} - 2\sigma_{\bar{x}} = 7 - 2 \cdot 0.0111 = 6.9778$$

and

$$\mu_{\bar{x}} + 2\sigma_{\bar{x}} = 7 + 2 \cdot 0.0111 = 7.0222$$

are the end points of the confidence interval.

4.12 *The normal distribution and estimation*

Problems

A. If an unbiased coin is tossed 100 times, what is the probability of at least 43 heads?

B. Able and Baker work on a production line. The number of units produced per day and the standard deviations are 180 and 12 for Able and 175 and 20 for Baker. To encourage production, the men are offered a bonus for any day on which their individual production rate exceeds 190 units.

1. How often will each man receive a bonus if the number of units produced daily follows a normal distribution in each case?
2. If the results in part 1 indicate that Baker, who is the slower worker, gets a bonus more often than Able, what quota would be required to give them bonuses an equal percentage of the time?

This section and the next one closely parallel the previous two sections, so a review of Secs. 4.10 and 4.11 will be the starting point of this discussion. The theme has been (and will continue to be) the estimation of a parameter of a population by the use of a statistic of a sample. The mean has received all the attention here, but similar discussions of the variance of a sample is possible. In Sec. 4.10, a formula with two parts was developed that could be used as an aid in making a "good" estimate of the parameter from the statistic. However, even with the help of the formulas, the estimate had to be made with an accompanying probability statement—which made clear the reliability of the estimate. But this is the way things are, and this section and the next are aimed at improving the reliability of the statement that qualifies the estimator. We shall see that an improvement in the probability statement made by use of the formula in Sec. 4.10 is often possible. The circumstances under which this improvement is possible are the major concern of the early portion of this section. The formulas developed in Sec. 4.10 were also used in Sec. 4.11. However, in Sec. 4.11 there was a change in the approach used, and the eventual outcome of the analysis was a confidence interval rather than the test of a hypothesis.

Now we turn to what is to appear in this section and the next. The mathematical tool to be developed is the *normal distribution*. The development of this aid will be accomplished relatively early in this section, and it will be used for problems in the later portion of this section and also in Sec. 4.13.

Some apology to the more mathematically minded reader may be appropriate since the development of the technique to be used is, of necessity, not very rigorous. This lack of rigor in the mathematical derivation is a necessity because this book does not extend to the mathematical depths necessary to prove many of the assertions that are to be made about the normal distribution. The reader is asked to accept these assertions without proof. Of course, accepting facts about a technique without supporting proof does not detract from the accuracy of the results obtained by use of the technique. The point is that some statements, which are rather difficult to establish mathematically, must be accepted at their face value. In short, the emphasis is to be on *practice rather than theory.*

On the bright side of the picture, much of the vocabulary that is needed for the work in the last two sections of this chapter has already been developed in Secs. 4.10 and 4.11. However, one concept that will be needed in the solution of many of the examples is *standard scores.* This idea was also developed earlier, and it will prove helpful to review Sec. 4.6, where the concept of standard scores was discussed.

The main idea to be developed in this section is that of a *normal distribution.* But we begin with a question whose answer can be computed using only two familiar concepts: *probability* and a *binomial distribution.* Hopefully, this example will point up the need for some better techniques than those that have been developed so far.

Example 1. A machine is used to sort cantaloupes after they are picked and delivered to the sorting shed to be prepared for the market. Because of such factors as oversize, undersize, unripeness, and overripeness, *in the long run* the machine rejects 40 percent of those picked as being unsuitable for marketing. These rejections must be made according to a set of predetermined rules in a contract between the farmer and the wholesaler who is purchasing the crop. In a sample of 1,000, what is the probability that 370 or more will be rejected?

The mathematical tools to solve the problem are available, and some of the preliminary steps toward the answer will be taken. Note first that this operation meets the criteria established for a binomial experiment (see page 171). Hence the formulas developed for the mean and standard deviation for a binomial distribution apply. Note that if, in a sample of size n, x is the percentage defective, then

$$\mu = nx = 1{,}000 \cdot (0.40) = 400$$

and $$\sigma = \sqrt{nxy} = \sqrt{1{,}000 \cdot (0.40) \cdot (0.60)} = \sqrt{240} \cong 15.49$$

The expected number of rejects out of 1,000 is 400, with a standard deviation of approximately $15\frac{1}{2}$.

But the question that has been posed asks for the probability of 370 rejects *or* 371 rejects *or* 372 rejects *or* $\cdot\cdot\cdot$ *or* 1,000 rejects. From Sec. 3.5, it is known that since these are mutually exclusive events, the answer to the question is the sum of the probabilities of all the events. Furthermore, formulas are available to write these probabilities, so that the answer is

$$\begin{aligned} Pr(E_{370}) + Pr(E_{371}) + \cdots + Pr(E_{1,000}) = {}& {}_{1,000}C_{370} 0.4^{370} \cdot 0.6^{630} \\ &+ {}_{1,000}C_{371} 0.4^{371} \cdot 0.6^{629} + \cdots \\ &+ {}_{1,000}C_{1,000} 0.6^{1,000} \end{aligned}$$

Surely the complexity of this expression is sufficient to demonstrate the immenseness of the computational task. Even with the aid of tables, adding machines, and desk calculators, it would take a *very* long time to make the computations necessary to answer the question. If the reader has any doubts about how enormous this computational task is, he is encouraged to *start* the computation.

This example illustrates that some other means of solution is desirable, and the normal distribution to be discussed next provides that method. This example will be returned to a little later when the alternate method of solution has been developed.

In prior considerations, the sets studied have been *finite*. This simply means that the sets can be counted and that the process terminates with a counting number. In recent examples, the number of elements in a population set has been denoted by N. For a finite set, we are able to make statements such as "The set has 17 elements," or "The set has 1,027 elements," or "The set has 69 elements." When *finite* sets are graphed, the graph is a set of points which appear as dots on the paper. (Consult the charts in Sec. 4.1 to review this idea.) The *normal distribution* is an *infinite* set. Therefore, the set of points in a normal distribution cannot be counted. When a normal-distribution set is graphed, the picture shows a smooth curve. It has already been hinted that often the graphs of finite distributions are approximated by a smooth curve, and some of the diagrams of previous sections were drawn using this fact.

The normal-distribution curve is completely determined by the mean μ and the standard deviation σ. But a little reflection will convince you that once again some mathematical justifications are being omitted. What is the mean of an infinite set? How can the standard deviation be computed if an infinite sum must be computed? It should be clear that some new methods will have to be used if the mean and the standard deviation are to be determined for an infinite set. These are two of the points that were in mind when the apology for the mathematical "sloppiness" of this presentation was offered. It is sufficient to say that the mean for infinite sets is still an average and that standard deviation for infinite sets still refers to the scatter, or dispersion, of the points. In Diagram 19, a normal distribution is illustrated. On the horizontal line, some points are marked. The center point is the mean μ, and the other

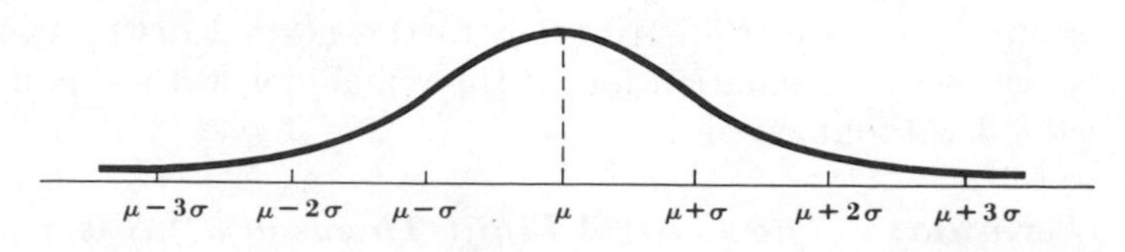

Diagram 19

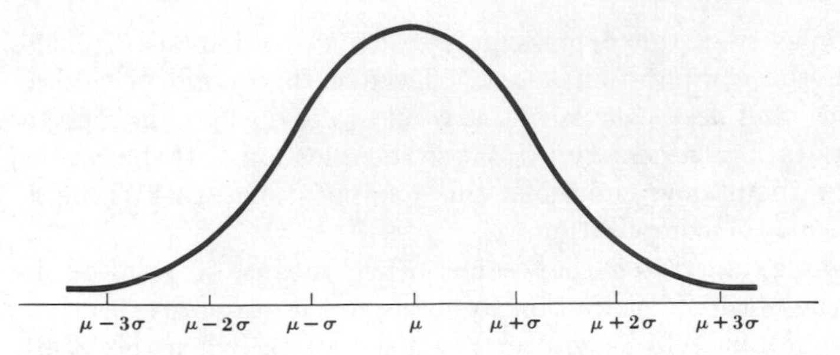

Diagram 20

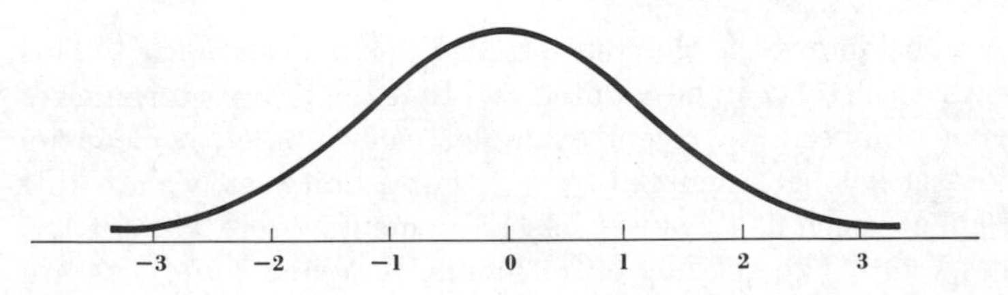

Diagram 21

points show units of standard deviation away from the mean. Lest one diagram lead to a misconception about the shape of the *normal curve,* another normal distribution with a smaller standard deviation is displayed in Diagram 20. The point is that there are many normal distributions and their graphic appearance will vary as the mean and the standard deviation are changed. As the standard deviation is decreased, the curve becomes more "peaked" if the scale remains the same.

It was mentioned earlier that it is often advantageous to use standard scores when normal distributions are studied. If standard scores are used, then the mean of the distribution is 0 and the standard deviation is 1. These facts were proved for finite distributions in Sec. 4.6. They also hold for infinite normal distributions. This is another of those facts that you are asked to accept without proof. In Diagram 21, a normal-distribution curve has been sketched with standard scores shown along the horizontal axis. When the variable is the standard score, the curve is said to be in *standard form.*

The standard form of the normal curve has been introduced because of its usefulness in answering questions about the distribution of sample means. There is a general statement (theorem) which relates the set of sample means of a distribution to the normal distribution, and it will serve as the foundation for all the remainder of this chapter. It also is stated without proof.

Formula: **(The Central Limit Theorem)** *If S is a population with finite mean μ and finite standard deviation σ and if samples of size n*

are drawn at random from S, then if n is "large," the distribution of sample means T is approximately a normal distribution, with a mean of

$$\mu_{\bar{x}} = \mu$$

and a standard deviation of

$$\sigma_{\bar{x}} = \frac{\sigma}{\sqrt{n}}$$

Since we propose to use these two formulas to answer questions about particular examples, there are several points which should be mentioned.

First, note the word *approximately* in the formula. How well the normal distribution approximates the actual distribution of the sample means depends on several factors. For one thing, the sample size must be large. Precisely what is meant by *large* will not be explained, but tests are known. Another factor of the "goodness" of the approximation is the size of S. Generally speaking, the normal distribution is used when S is either infinite or sufficiently large to be considered infinite. But perhaps a more practical statement is that the approximation is good if the size of S is large with respect to n. Still another factor is the distribution of the underlying population S. The closer the distribution of S is to a normal distribution, the better is the approximation of the normal curve to the actual distribution of the sample means. These are all vague statements, made despite the emphasis placed on precision in language elsewhere in this book. But the truth of the matter is that there is a great deal of uncertainty in prediction and estimation, and the best is being made of a bad situation with a sincere attempt at honesty. The difficulty lies in the fact that the mathematical theory necessary to make more precise statements is not available.

The predictions and estimates that are to be made about the parameter of a population from a statistic of a sample will be arrived at by the use of properties of the normal-distribution curve (normal curve) and the Central Limit Theorem. But as in the previous two sections, the estimates will have to be combined with statements about probability. But what has probability to do with the normal curve? So far, there has been no mention of how to use a normal-distribution curve to compute the probability of an event. Once again, the patience of the reader is tested. You are again requested to accept a fact that is provable but which will not be proved here. *If the conditions of the Central Limit Theorem apply and if a sample mean $\bar{x}$ is determined and written as a standard score, then the probability of its occurrence or nonoccurrence is computable from probabilities associated with the normal curve.* In Diagram 22, the area underneath a normal curve has been partitioned into subsets. In each subset is written a percentage which can be correctly assigned to the occurrence of the event. For example, the probability of a sample mean with a standard score between 0 and 1 is approximately 0.34, between 1 and 2 is 0.136, and between -3 and -2 is 0.021.

As illustrated in Diagram 22, approximately 68.2 percent of the area under the curve lies within two standard deviations of the mean. The end points for this interval, in standard scores, are -1 and 1. In Sec. 4.5, the percentages of the elements within certain standard deviations of the mean were computed, but the formula was of no assistance for $k = 1$. This keeps us from making a comparison between the probabilities for the interval of length two standard deviations. However, from Diagram 22 there is an approximate probability of 95 percent that a sample mean will be within two standard deviations of the true mean. Compare this with the formula in Sec. 4.5. The strongest permissible statement was that about 75 percent of the data were clustered within two standard deviations of the mean. Recall also that in Sec. 4.5 you were told that under certain circumstances the 75 percent probability statement about an element being within two standard deviations of the mean could be considerably improved. The comparison between the 95 and 75 percent was what was referred to. Diagram 22 also shows that approximately 99.6 percent of the area under the normal curve is within three standard deviations of the mean. The formula of Sec. 4.5 could only guarantee that approximately 89 percent of the data would be clustered within three standard deviations of the mean.

In order to use the normal curve, a table that gives the probability between the mean and a standard score z will be needed. There is a rather complete table in the back of the book. However, since a smaller table will serve for most of the examples, a portion of the complete table is reproduced as Table 6. Not all the numbers are significant, but they are close enough for practical purposes. Since the normal curve is *symmetric* about the mean, it is sufficient to give the numbers for positive values only. We shall return to this subject later in the examples.

The first column in the table gives the number of standard deviations that the sample mean $\bar{x}$, written as a standard score z, is from the mean μ. The second column gives the probability of a sample event falling between the population mean μ and the sample mean written as a standard score. The figures in the third column are twice those in the

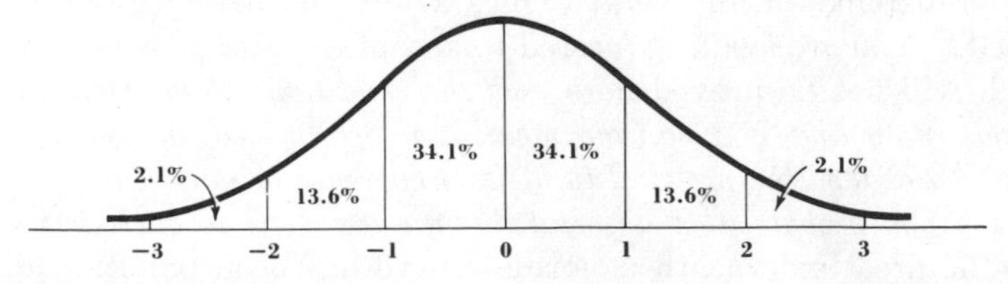

Diagram 22

Table 6

z	$Pr(z)$	$2Pr(z)$
0	0.0000	0.0000
0.1	0.04	0.08
0.25	0.10	0.20
0.4	0.16	0.32
0.5	0.19	0.38
0.675	0.25	0.50
0.8	0.29	0.58
1.0	0.34	0.68
1.5	0.43	0.86
1.96	0.475	0.95
2.0	0.477	0.954
2.5	0.494	0.988
2.58	0.495	0.99
3.0	0.499	0.998

second. Hence, the third column gives the probability that a sample mean $\bar{x}$, written as a standard score, will fall between $\mu - \bar{x}$ and $\mu + \bar{x}$.

The ultimate goal of this section is to use the concepts developed so far to test hypotheses. But some preliminary elementary examples will now be given to help you become acquainted with how to use the diagram for a normal curve to visualize the problem. These examples will also serve to instruct you in the use of the table.

Example 2. Assume a set S such that the Central Limit Theorem holds. What is the probability that a sample mean, written as a standard score, will lie between 0 and 0.5? A diagram will assist in understanding the question. In Diagram 23, the area under the curve between 0 and 0.5 is shaded. The percentage of the total area is the probability that the sample mean will be between 0 and 0.5. From the table, the probability is 0.19.

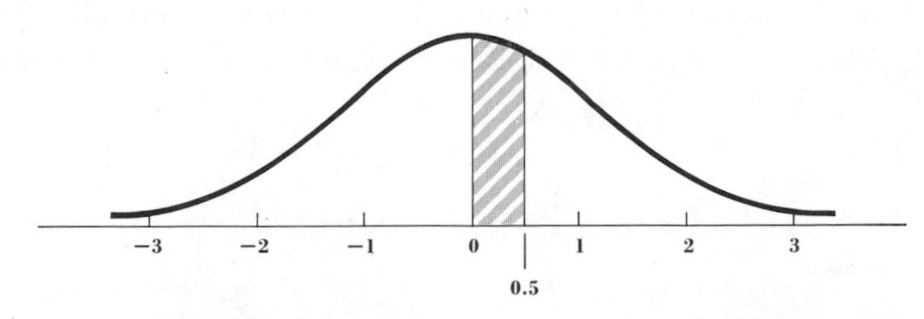

Diagram 23

Example 3. Assume a set S such that the Central Limit Theorem holds. What is the probability that a sample mean, written as a standard score, will be between 0 and 1.96? From the table, the answer is 47.5 percent. It will be instructive to construct a diagram similar to Diagram 23 with the proper area shaded.

Example 4. Assume a set S such that the Central Limit Theorem holds. What is the probability that a sample mean, written as a standard score, will be between 0 and -1.0? From Diagram 22, the answer is approximately 0.34. Note that the table could also be used because of the symmetry property mentioned earlier. The probability of a sample mean between 0 and -1 is the same as that between 0 and 1.

Example 5. Assume a set S such that the Central Limit Theorem holds. What is the probability that a sample mean, written as a standard score, will be between -0.5 and 1.0? For this question, Diagram 22 cannot be used since the areas are not divided into the proper subdivisions so that the answer is apparent. So the table will be used. But first an equivalent question will make clear how to proceed: What is the probability that z will be between -0.5 and 0 or between 0 and 1? The symmetry of the curve and the table give the first probability as approximately 19 percent and the second as approximately 34 percent. Since these are mutually exclusive events, the probabilities are added:

$$0.19 + 0.34 = 0.53$$

The probability of the occurrence of the event is approximately 0.53.

Example 6. Example 5 of Sec. 4.10 will now be reconsidered. Recall that the Roller Steel Company wishes to estimate the average time between date of completion of a product and its delivery date. The standard deviation in time elapsed between the two dates is four days. The hypothesis to be tested is

> *The average time that elapses between the production date and the delivery date is* 34.2 *days.*

A sample of 50 items selected yields a mean of 30.8. Assume that the Central Limit Theorem holds and test the hypothesis. Since $\sigma = 4$ is given, the formula of the theorem can be used to compute $\sigma_{\bar{x}}$:

$$\sigma_{\bar{x}} = \frac{4}{\sqrt{50}} \cong 4/7$$

The standard score is determined by the use of the formula on page 250. It is

$$z = \frac{30.8 - 34.2}{4/7} = \frac{-34/10}{4/7} = \frac{-34}{10} \cdot \frac{7}{4} = \frac{238}{40} \cong -6$$

But the probability of an event deviating from the mean by this standard deviation is less than 1 percent. For this example, the deviation is by six standard deviations. This occurrence is so unlikely that it is usually not even included in the tables. It is virtually certain that if the hypothesis is true, the sample event would not occur. Hence, the hypothesis is false and is rejected.

Example 7. We return now to Example 1. Recall that this example described a binomial experiment in which the probability of the occurrence of an event (a particular cantaloupe is discarded) is 0.4. It is desired to compute the probability that in a sample of 1,000, 370 or more will be discarded. A normal distribution will be used to find the probability. But for what reason may the binomial distribution be replaced by a normal curve? The reason is that for certain circumstances, such a replacement is permissible. The Central Limit Theorem as applied to a binomial distribution is the justification of the replacement. However, two tests can be used as "rules of thumb" for determining when the replacement is permissible. First, we give a statement of the replacement principle:

The binomial distribution can be closely approximated by a normal distribution with mean $\mu = nx$ *and standard deviation* $\sigma = \sqrt{nxy}$.

The above statement can usually be used if $nx > 5$ and $ny > 5$. This is one of the rules of thumb. The other test is that $\mu + 2\sigma$ and $\mu - 2\sigma$ should each lie between 0 and n.

Before the example is worked, the tests should be made in order to see that the normal distribution can be used. Two computations give μ and σ:

$$\mu = nx = 1{,}000 \cdot (0.4) = 400$$
$$\sigma = \sqrt{nxy} = \sqrt{1{,}000 \cdot 0.4 \cdot 0.6} = \sqrt{240} \cong 15.49$$

Now $nx = 400 > 5$ and $ny = 600 > 5$, so the first test is satisfied. Further,

$$\mu + 2\sigma \cong 400 + 30.98 = 430.98$$

and

$$\mu - 2\sigma \cong 400 - 30.98 = 369.02$$

are both between 0 and $n = 1{,}000$, so the second test is satisfied.

To use the table, the numbers are converted to standard scores:

$$z = \frac{370 - 400}{15.49} = \frac{-30}{15.49} \cong -1.94$$

The number 370 is approximately 1.93 standard deviations from the mean 400. Since 340 is less than 400, it is to the left of the mean. Hence, the point on the standard score line that corresponds to 340 is -1.94. The question calls for the probability of the event that more than

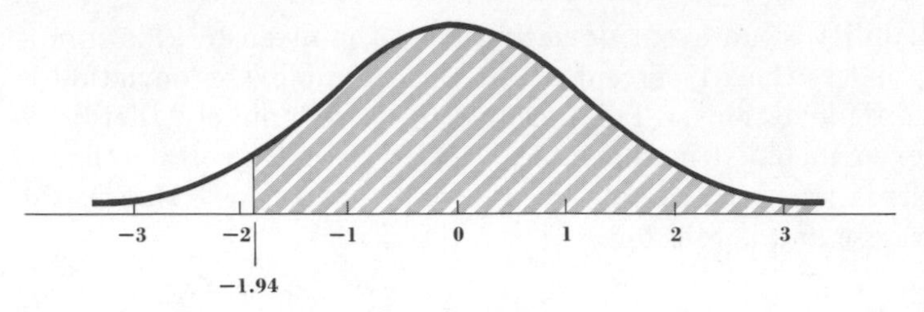

Diagram 24

370 of the cantaloupes will be rejected, so the area under the normal curve that corresponds to the probability is to the right of the standard score −1.94. In Diagram 24, the area is shaded.

Use of the table in the back of the book shows there is a probability of 0.97 that the event described in the example will occur.

Example 8. The reliability of a manufactured item is the probability that it will function under the conditions for which it has been designed. A company has a contract to produce an item with 0.98 reliability. They wish to test the hypothesis that the probability of a defective product is 0.02. They will do this by testing a sample of 1,000. A random sample is selected, and the test is made. It is determined that 29 of the items in the sample do not perform as they are designed to do. Should the hypothesis be accepted?

The normal curve will be used to answer the question. The expected number of defectives is

$$\mu = nx = 1{,}000 \cdot 0.02 = 20$$

with a standard deviation of

$$\sigma = \sqrt{1{,}000 \cdot 0.02 \cdot 0.98} \cong 4.4$$

The standard-score value corresponding to 29 is

$$z = \frac{29 - 20}{4.4} = \frac{9}{4.4} \cong 2.05$$

The number 29 is larger than the number $\mu = 20$, so the point corresponding to 29 is to the right of the point corresponding to 20. In fact, it is over two standard deviations to the right. Since the probability of observing 29 or more defectives is the area under the curve to the right of 2.05, the probability of the sampling event is less than 0.02. The null hypothesis is rejected with 98 percent confidence. This example should be compared with those of Sec. 4.10.

Example 9. In a large factory that uses a lot of equipment, many of the tasks are routine. The jobs can be handled by a person with an

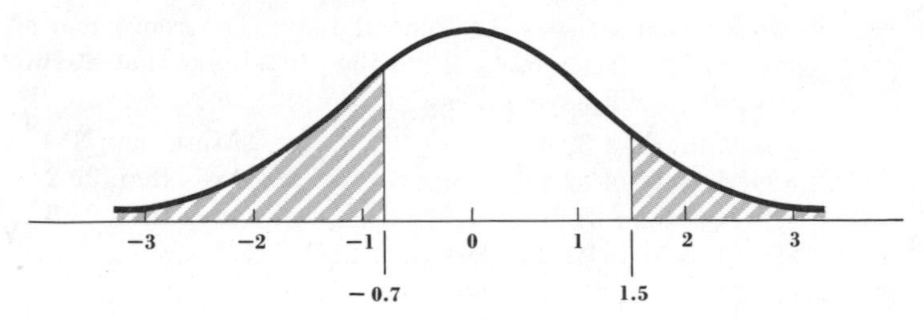

Diagram 25

IQ of 93 or more, but persons with IQs of over 115 become bored and are not suitable for the work. For the 5,000 employees, the IQs are normally distributed, with a mean IQ of 100 and with a standard deviation of 10. What percentage of the employees are not suited for their jobs?

There are two scores that have to be computed to answer the question. They are

$$z = \frac{93 - 100}{10} = \frac{-7}{10} = -0.7$$

and

$$z = \frac{115 - 100}{10} = \frac{15}{10} = 1.5$$

In Diagram 25, the shaded areas show those persons not suited for their jobs.

The area on the left represents those with less than the satisfactory IQ, and the area on the right represents those whose high IQ will cause them to be bored. From the tables in the back of this book, we find that the corresponding probabilities are $0.5 - 0.258 = 0.242$ and $0.5 - 0.433 = 0.067$. Since these are mutually exclusive areas, the total probability is $0.242 + 0.067 = 0.309 \cong 31$ percent.

Exercises

1. For each of the following, assume that the Central Limit Theorem holds. What is the probability that a sample mean, written as a standard score, will lie between the following numbers?

   a. 0 and 0.25 *b.* 0 and 1.5
   c. 0 and 2.5 *d.* 0 and −0.4
   e. 0 and −1.5 *f.* −0.25 and 1.0
   g. −1 96 and 1.96 *h.* −2.0 and 2.0
   i. −3.0 and 3.0 *j.* −1.5 and 0.675

2. Sketch a normal curve and shade the area associated with each of the regions in Exercise 1.

3. A set S which satisfies the Central Limit Theorem has a mean of 30 and a standard deviation of 4. Find the probability that a sample of 25 selected at random will have a mean of:

   a. More than 30.8
   b. More than 31.6
   c. More than 32.4
   d. Less than 29.2
   e. Less than 28.4
   f. Less than 27.6
   g. More than 31.6 or less than 28.4
   h. More than 32.4 or less than 27.6
   i. More than 31.6 or less than 29.2
   j. Between 30.8 and 29.2
   k. Between 31.6 and 28.4
   l. Between 31.6 and 27.6

4. Sketch a normal curve and shade the area associated with each of the regions of Exercise 3.

5. A binomial experiment is conducted with probability 0.97 of success. If a sample of 1,000 is selected, what is the probability that it will contain each of the following numbers of failures?

   a. More than 35
   b. More than 41
   c. More than 46
   d. Less than 30
   e. Less than 30
   f. Less than 19
   g. Less than 14
   h. Between 25 and 35
   i. Between 19 and 41
   j. Between 14 and 46
   k. Between 19 and 30
   l. Between 30 and 41

6. Sketch a normal curve and shade the area associated with each of the regions of Exercise 5.

7. Rework Example 9 for IQs of 95 and 112.

8. Rework Example 9 for IQs of 92 and 117.

9. The average cost of 100,000 hospital confinements is \$325 with a standard deviation of \$75.

   a. If no information is known about the distribution, what is the minimum percentage that cost between \$175 and \$475?
   b. If the costs distribute normally, what percentage of the confinements cost between \$175 and \$475?

10. The strength of the product produced by a company has a mean of 100 with a standard deviation of 20. Use the table for normal distributions to answer the following questions:

    a. What percentage of the product has strength between 100 and 120?
    b. What percentage of the product has strength greater than 120?
    c. What percentage of the product has strength between 60 and 140?
    d. What percentage of the product has strength greater than 60?
    e. What percentage of the product has strength between 40 and 160?
    f. What percentage of the product has strength between 110 and 130?
    g. Ninety percent of the product has strength above ______.
    h. Ninety-five percent of the product has strength below ______.

i. The middle ninety-five percent has strength between what values?
j. What percentage of the product has strength of 50 or more?
k. What percentage of the product has strength less than 145?
l. What percentage of the product has strength between 110 and 160?

Answers to problems

A. This is a binomial experiment, with $x = y = \frac{1}{2}$. The desired probability is the sum

$$Pr(E_{43}) + Pr(E_{44}) + \cdots + Pr(E_{100}) = {}_{100}C_{43}(\tfrac{1}{2})^{43}(\tfrac{1}{2})^{57} + {}_{100}C_{44}(\tfrac{1}{2})^{44}(\tfrac{1}{2})^{56} + \cdots + {}_{100}C_{100}(\tfrac{1}{2})^{100}$$

Owing to the difficulty in computing these numbers, we use the normal distribution to approximate the binomial distribution. Now $\mu = 100 \cdot \frac{1}{2} = 50$ and $\sigma = \sqrt{100 \cdot \frac{1}{2} \cdot \frac{1}{2}} = \sqrt{25} = 5$. The standard score is

$$\frac{42.5 - 50}{5} = \frac{-7.5}{5} = -1.5$$

The graph is

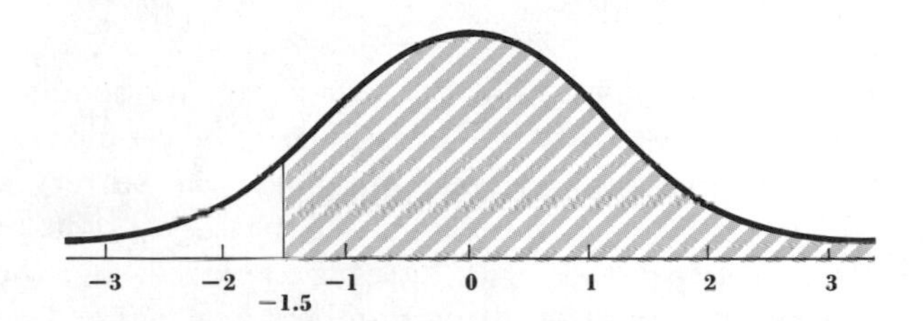

From the table, the probability is $0.4332 + 0.5 = 0.9332$.

B. 1. For Able: $z = \dfrac{190 - 180}{12} = 0.83$; $0.50 - 0.2967 \cong 20\%$

For Baker: $z = \dfrac{190 - 175}{20} = 0.75$; $0.50 - 0.2734 \cong 23\%$

2. $$\frac{x - 180}{12} = \frac{x - 175}{20}$$
$$12(x - 175) = 20(x - 180)$$
$$12x - 2{,}100 = 20x - 3{,}600$$
$$8x = 1{,}500$$
$$x = 187.5$$

4.13 The normal distribution and intervals of confidence

Problem

A random sample of 50 of the employees of a large company have an average of 13.2 years of formal education, with a standard deviation of

0.8 year. Assume that the sample means distribute normally, and construct a 99 percent confidence interval for the true average number of years of formal education of all employees.

For certain population sets S, it can be assumed, without any serious loss of accuracy, that the random-sample statistic will be distributed in such a way that it may be approximated by a normal distribution. This statement is often written more simply as: The population set S is a normally distributed random-sample statistic. For the examples to be discussed in this section, this assumption will be made. The question asked about means of populations will be answered using the normal-distribution curve in its standard form, as illustrated in Diagram 22.

This section deals with examples. The facts necessary to answer questions were advanced in the previous section. The examples all deal with the construction of confidence intervals. The reader will find it helpful to reread Sec. 4.11, where the concept of confidence intervals was first developed.

Example 1. The Blue Battery Company conducts a study of 100 of their batteries with the assistance of a retailer. The retailer records the date that the customer purchases the battery and also the date on which the customer replaces the battery. The number of days that elapse between these two dates is averaged, and the standard deviation is computed, with resulting numbers $\bar{x} = 800$ and $s = 9$. The company wishes to decide how long to guarantee their batteries on the basis of this experiment.

First, all the assumptions are made that are necessary to use the techniques at hand. In particular, assume that the sample is random and that the sample means distribute normally. A point estimate is that $\mu = \bar{x} = 800$ days. But a better statement can be made. Of course, the standard deviation of the number of days of use for every battery produced (the population) is not known, nor can it be computed. As in Sec. 4.11, the sample standard deviation is used as an estimate for the population standard deviation, so that $\sigma = 9$. Now the formula

$$\sigma_{\bar{x}} = \frac{\sigma}{\sqrt{n}}$$

will give the sample-mean standard deviation, which is

$$\sigma_{\bar{x}} = \frac{9}{\sqrt{100}} = 0.9$$

We construct a confidence interval at the level of significance of 99 percent. Since the normal curve is to be used for this construction, 99 percent confidence is achieved with a standard deviation of approximately 3. For a normal distribution, the end points are $\mu - 3\sigma$ and $\mu + 3\sigma$. For this example, we have

$$800 - 3 \cdot 0.9 = 800 - 2.7 = 797.3$$

and

$$800 + 3 \cdot 0.9 = 800 + 2.7 = 802.7$$

It is asserted with 99 percent confidence that the population mean of all the batteries is such that

$$797.3 < \mu < 802.7$$

Example 2. At some government defense bases, cash incentive awards are given to a few employees who make suggestions that lead to more economical operation of the base. One year at a particular base, the average incentive payment for 50 awards was \$240, with a standard deviation of \$25. What can be said, with probability 0.95, about the true average of the total awards?

The normalizing assumptions are made, and it is accepted that a point estimate of μ is \$240 and that σ = \$25. To see the importance of the assumption that the sample is random, read Example 2 in Sec. 4.9 again to see how one could be misled. The standard deviation is computed with

$$\sigma_{\bar{x}} = \frac{25}{\sqrt{50}} = \frac{25}{5\sqrt{2}} = \frac{5}{\sqrt{2}} \cong 3.536$$

The number of standard deviations chosen is two because Diagram 22 shows that approximately 95 percent of the area under the normal curve lies within approximately two standard deviations of the mean. The end points of the confidence interval are

$$240 - 2 \cdot 3.54 = 240 - 7.08 = 232.92$$

and

$$240 + 2 \cdot 3.54 = 240 + 7.08 = 247.08$$

With 0.95 probability, the true mean of the awards is such that

$$232.92 < \mu < 247.08$$

Example 3. The reliability of a small steel spring to function under the conditions for which it was designed is to be tested. The experiment is made on 1,000 springs, and 25 of them fail to perform adequately. What statement can be made with 99 percent confidence about the true percentage of good springs?

The experiment is a binomial one, and the techniques of Sec. 4.11 could be employed. But the normal distribution can also be used.

Recall that the "rule-of-thumb" test

$$nx > 5$$

and

$$ny > 5$$

was suggested as a criterion to decide if the normal distribution can be used. For this example,

$$nx = 1{,}000 \frac{25}{1{,}000} = 25 > 5$$

and

$$ny = 1{,}000 \frac{975}{1{,}000} = 975 > 5$$

The expected number of successes (reliable springs) is

$$\mu = ny = 1{,}000 \cdot 0.975 = 975$$

and the standard deviation is

$$\sigma = \sqrt{nxy} = \sqrt{1{,}000 \cdot 0.975 \cdot 0.025} = \sqrt{24.375} \cong 5$$

The end points of the 99 percent confidence interval are

$$975 - 3 \cdot 5 = 975 - 15 = 960$$

and

$$975 + 3 \cdot 5 = 975 + 15 = 990$$

The expected percentage of reliable springs is, with 99 percent probability, such that

$$96\% < \mu < 99\%$$

Exercises

1. The sample means of a population distribute normally. A random sample of 64 elements is selected, and the mean and standard deviation of the sample are computed. Construct a 68 percent confidence interval for the mean of the population for each of the following:
   a. $\bar{x} = 50, s = 3$
   b. $\bar{x} = 120, s = 7$
   c. $\bar{x} = 80, s = 5$
   d. $\bar{x} = 1{,}240, s = 12$
   e. $\bar{x} = 46, s = 2$
2. Work Exercise 1 for a 95 percent confidence interval.
3. Work Exercise 1 for a 99.74 percent confidence interval.
4. Work Exercise 1 for $n = 100$.
5. Work Exercise 4 for a 95 percent confidence interval.
6. Work Exercise 4 for a 99.74 percent confidence interval.
7. Work Exercise 1 for $n = 25$.
8. Work Exercise 7 for a 95 percent confidence interval.

9. Work Exercise 7 for a 99.74 percent confidence interval.

10. A random sample of 81 secretaries' wages is averaged with the result of $\bar{x}$ = $96 and a standard deviation of $6. Assume a normal distribution.
    a. Determine the 96 percent confidence limits.
    b. If the data distribute in any manner whatsoever, what are the 96 percent confidence intervals?

Answer to problem

Since $\sigma = 0.8$,

$$\sigma_{\bar{x}} = \frac{0.8}{\sqrt{50}} = \frac{0.8}{5\sqrt{2}} \cong 0.113$$

The end points of the confidence interval are

$$13.2 - 3 \cdot 0.113 = 13.2 - 0.339 = 12.861$$

and

$$13.2 + 3 \cdot 0.113 = 13.2 + 0.339 = 13.539$$

5

SYSTEMS OF LINEAR EQUALITIES AND INEQUALITIES

5.0 Introduction

Since the concepts to be presented in this chapter are a bit different from those in the preceding chapters, we begin by observing some similarities. First, the main concern continues to be the solution by the use of mathematical tools of problems that arise in business situations. New ideas will again be introduced by means of examples. Second, a *pattern of solution* developed in previous work will again be used in this chapter. This pattern can be described by the following steps: (1) *the statement of a problem in a business or other situation,* (2) *restatement of the problem in the language of mathematics,* (3) *solution of the mathematical problem, and* (4) *interpretation of the mathematical solution to the physical problem.*

In this chapter, the statement of the problems will lead to a restatement in mathematics that uses *linear equations* or *linear inequalities*. Consideration of several linear equations simultaneously results in the study of a *system* of linear equations. In the first three sections, the simplest case, a system of two equations in two unknowns, will be studied. Techniques for the solution of systems of equations will be developed. In Sec. 5.4, systems of three equations in three unknowns will be studied. From the solution techniques discussed earlier, the technique that uses matrices, and row operations on matrices, will be selected as the most suitable. In Sec. 5.5, some systems in which the number of equations is different from the number of variables will be studied. Also, a mathematical justification of the matrix method of solution will be given. The remaining two sections are about systems of *linear inequalities.* One section is devoted to problems in which a profit is to be made a *maximum,* and the other section contains problems in which a cost is to be made a *minimum.* This class of problem is often referred to as *linear-programming* problems. The graphic method of solution will be the one used.

Systems of equations such as those to be presented in this chapter are probably already familiar to the reader since they are frequently studied in high school algebra. The treatment here assumes no previous knowledge, and while an involved treatment is beyond the scope of this book, the method of solution may be different from that which you have seen before.

5.1 *Systems of two linear equations in two unknowns: algebraic solution*

Problems

A. A drugstore makes a profit of 2 cents on each bottle of brand *X* aspirin that it sells and a profit of 6 cents on each bottle of brand *Y* aspirin that it sells. How many bottles of each brand must the druggist sell if he is to make a total profit of $8 from the sales of these two brands of aspirin?

B. Refer to Prob. *A* and decide how many bottles of each brand the druggist must sell if the total number of bottles sold is 280.

C. Refer to Prob. *A* and decide how many bottles of each brand the druggist should buy if the number of bottles of brand *X* purchased (and sold) is 450?

D. Is it possible to make a profit of 1 cent per bottle of brand *X* and 3 cents per bottle of brand *Y* for a total profit of $4 from the sale of these two brands that is a common solution with the condition of Prob. *A*?

E. A retail paint store has a gross profit on sales of brand *A* paint of $39,000 during the preceding year. The manager is considering adding a new line of paint, brand *B*, which will sell for $5.10 a gallon, as compared with $7.80 for brand *A*. It is believed that the addition of the lower-priced brand will cause some switching but will also attract some new customers who were unwilling to pay $7.80 a gallon for house paint. If the paint costs the store $5.20 for *A* and $3 for *B*:

1. How many gallons of brand A were sold to earn a gross profit of \$39,000?
2. If the addition of the lower-priced brand causes 25 percent of the customers to buy brand B, how many gallons of each would have to be sold to earn the same gross profit as the preceding year?
3. If the addition of the lower-priced line will increase the total gallons sold by 25 percent, what percentage of the customers sold could buy the lower-priced brand and still allow the firm to earn a gross profit of \$40,000?

As indicated in the introductory section of this chapter, the type of mathematics to be studied here is rather different from that of the preceding chapters. However, the motivation will continue to be furnished by problems that occur in business situations. For that purpose, we return again to the hypothetical business firm created in Chap. 1.

Example 1. At a meeting of the board of directors of CCC, the production and financial goals of the next fiscal year are discussed. The directors have many factors such as cost of materials and market potential to consider, but after much deliberation, they decide that they should be able to make a gross profit of \$240,000 on the two products, sheets and blankets. They would like to determine the number of each of these products that must be sold so that this gross profit can be attained. After a careful study of their anticipated production costs, they estimate that the sales prices can be such that they can make \$0.12 gross profit on each sale of a sheet and \$0.60 on each sale of a blanket. How many of each of the two products must be sold in order to make a profit of \$240,000?

There are two numbers to be determined: the number of sheets to be sold and the number of blankets to be sold. These unknown numbers will be represented (in the usual notation for such problems) by the symbols x_1 and x_2. The symbol x_1 is read "x subscript 1" or, more briefly, "x sub 1." Similarly, x_2 is read "x subscript 2" or "x sub 2." The mathematical conditions that state the problem must now be written. It is estimated, and to be treated as a fact in this problem, that the gross profit that can be made on the sale of each sheet is \$0.12 and the gross profit that can be made on the sale of each blanket is \$0.60. The profit that can be made on the sale of x_1 sheets is $0.12x_1$, and similarly, the profit that can be made on the sale of x_2 blankets is $0.60x_2$. The profit is the product of the profit on each item and the number of items. The total profit to be made is \$240,000, and the sum of the two profits is the total profit. That is,

$$(1) \quad 0.12x_1 + 0.60x_2 = 240{,}000$$

is a *condition* that is to be met if the problem is to be solved. Expressions of this type are called *linear equations* (in two unknowns). Both the word *condition* and the word *equation* will be used for such expressions.

Since linear equations are to be studied for some time, a few remarks about nomenclature are in order. The symbols x_1 and x_2 are examples of *variables*, the word signifying that they represent quantities that vary. This is in contrast to such numbers as 240,000, which are always fixed and hence called *constants*. The constants which precede the variables in equations, 0.12 and 0.60 for this example, are *coefficients* of the variables. The reason for the adjective *linear* is best explained geometrically and is deferred until the next section. The *solution* to Eq. (1) is an *ordered* pair of numbers (a,b) which, when used as replacements for x_1 and x_2, respectively, satisfy the condition. If a number is chosen as a replacement for x_1, then x_2 is determined; x_2 is therefore dependent on the replacement chosen for x_1. To make this last statement more easily seen, Eq. (1) can be rewritten as

$$(1) \quad x_2 = \frac{240{,}000 - 0.12x_1}{0.60}$$

This is sometimes expressed in mathematical shorthand as

$$x_2 = f(x_1)$$

which is read "x_2 is a function of x_1." The set of ordered pairs of replacements is an example of a *function*. However, we shall have little or no occasion to write the equation in this form. The language introduced here will make it easier to discuss the equation. You will be reminded as we proceed of how to use the terms which have just been introduced.

The numbers x_1 and x_2 are to be determined in such a way that the condition expressed in Eq. (1) is satisfied. By observation, one ordered pair of numbers that satisfies the condition is (2,000,000, 0). The symbol (2,000,000, 0) means that the variable x_1 is to be replaced by the number 2,000,000 and x_2 is to be replaced by the number 0. In other words, if 2 million sheets are sold and no blankets are sold, then the desired profit of $240,000 is made. To verify this statement, x_1 is replaced in Eq. (1) by 2,000,000 and x_2 is replaced by 0. The equation then becomes

$$0.12 \cdot 2{,}000{,}000 + 0.60 \cdot 0 = 240{,}000 + 0 = 240{,}000$$

This shows that (2,000,000, 0) is a solution of the equation.

Another pair of numbers that is a solution to the condition expressed in Eq. (1) is (0, 400,000). The physical interpretation of this ordered pair is that no sheets are sold and 400,000 blankets are sold. Again to verify that this is a solution, the replacements are made:

$$0.12 \cdot 0 + 0.60 \cdot 400{,}000 = 0 + 240{,}000 = 240{,}000$$

Thus far, two solutions to the equation have been found. However, the company does not wish to discontinue manufacturing and selling either of the two products. The first solution requires that they sell no blankets, and the second solution requires that they sell no sheets. Therefore, both of these solutions are rejected and a "better" solution is sought.

One other ordered pair of numbers that satisfies the condition expressed in Eq. (1) is (1,000,000, 200,000). If these quantities are sold at the profits that have been estimated, the desired total profit is obtained. Verification of this statement is by replacement in Eq. (1), and

$$0.12 \cdot 1{,}000{,}000 + 0.60 \cdot 200{,}000 = 120{,}000 + 120{,}000 = 240{,}000$$

establishes that the condition is satisfied.

A review of the results obtained in this example with reference to the four steps in the solution of a problem outlined in Sec. 5.0 will be helpful. The statement of a problem was given, and it was restated in the language of mathematics [Eq. (1)]. The mathematics problem was solved; in fact, it was solved three times. The first two solutions were rejected because, when the solutions were interpreted with reference to the physical problem, they implied an undesirable change of company policy. Now the third solution must be accepted or rejected. But there are many other solutions, and some of them may be more desirable from the company's point of view than the third one.

From a strictly mathematical viewpoint, there are actually an unlimited number of solutions to Eq. (1), and Exercise 1 requires that others be found. However, this discussion should make clear that more information is needed to choose that *particular solution* best suited to the needs of the company. In the next example, we shall see one way that the firm might choose to generate this information, and at the same time, we shall lead to the main topic of the first five sections of the chapter: systems of linear equations.

Example 2. The board decides to ask some of the vice-presidents (and their staffs) to submit estimates on sales of sheets and blankets along with estimated gross profits. Vice-presidents Gumm, Hand, and Ivan are selected to furnish more estimates. After consultation with his staff, Vice-president Gumm reports that he and his staff concur on an estimated profit of $0.08 on the sale of each sheet and $1.20 on the sale of each blanket. A condition that expresses this estimate is

$$(2) \quad 0.08x_1 + 1.20x_2 = 240{,}000$$

This equation is derived in the same fashion as Eq. (1). The number $0.08x_1$ represents estimated profit on the sale of sheets; $1.20x_2$ represents estimated profit on the sale of blankets; the sum $0.08x_1 + 1.20x_2$ is equal to the total profit, 240,000. Some particular solutions of Eq. (2) are (3,000,000, 0), (0, 200,000), and (1,500,000, 100,000). The reader should verify that these ordered pairs are solutions by using them as replacements in Eq. (2). It should be clear that the first two of these would be rejected for the reasons stated earlier. Similarly, more information is needed to evaluate the third solution. This leads to the next example.

Example 3. Two equations which represent two points of view are available for comparison. The board would like to know if there are solutions to Eq. (1) that are also solutions to Eq. (2), and vice versa. In other words, does the system of linear equations

$$\begin{aligned} (1)\quad 0.12x_1 + 0.60x_2 &= 240{,}000 \\ (2)\quad 0.08x_1 + 1.20x_2 &= 240{,}000 \end{aligned}$$

have a *simultaneous solution?* The pair of estimates are to be considered together, and a common solution sought.

The solution will be obtained by algebraic means. So that the procedures will be clear, the arithmetical manipulations are written on the left and the reasons for the steps are written on the right.

$$\begin{aligned} (1)\quad 0.12x_1 + 0.60x_2 &= 240{,}000 \\ (2)\quad 0.08x_1 + 1.20x_2 &= 240{,}000 \end{aligned}$$ Statement of problem

$$\begin{aligned} (1)\quad x_1 + 5x_2 &= 2{,}000{,}000 \\ (2)\quad 8x_1 + 120x_2 &= 24{,}000{,}000 \end{aligned}$$ Multiply (1) by $1/0.12 = {}^{100}\!/_{12}$; Multiply (2) by 100

$$\begin{aligned} (1)\quad x_1 + 5x_2 &= 2{,}000{,}000 \\ (2)\quad 80x_2 &= 8{,}000{,}000 \end{aligned}$$ Multiply (1) by -8 and add to (2)

$$\begin{aligned} (1)\quad x_1 + 5x_2 &= 2{,}000{,}000 \\ (2)\quad x_2 &= 100{,}000 \end{aligned}$$ Multiply (2) by $\frac{1}{80}$

$$\begin{aligned} (1)\quad x_1 &= 1{,}500{,}000 \\ (2)\quad x_2 &= 100{,}000 \end{aligned}$$ Multiply (2) by -5 and add to (1)

The last pair of equations shows that if the company sells 1,500,000 sheets and 100,000 blankets, then the profit that they have established as their goal will be achieved. Furthermore, this profit can be made by either a profit of $0.12 on each sheet and $0.60 on each blanket or a profit of $0.08 on each sheet and $1.20 on each blanket.

Equations (1) and (2), when considered simultaneously, are an example of a *system of two linear equations in two unknowns.* We shall refer to such a system as a 2×2 system. This example has a *unique (one) solution* (1,500,000, 100,000).

Some remarks about the techniques used to arrive at the solution are in order. The procedures used were multiplication of an equation by a (well-chosen) constant and addition of the product to the other equation. The justification that these procedures yield an *equivalent system* will be deferred to Sec. 5.5, where the correctness of such operations will be verified for any system. In the meantime, the techniques will continue to be used, with the understanding that they will eventually be justified.

Example 4. As another example, suppose Mr. Hand submits an estimate of \$0.18 profit on sheets and \$0.90 profit on blankets. The equation that represents this possibility is

$$(3) \quad 0.18x_1 + 0.90x_2 = 240{,}000$$

and is derived in a now familiar way. Does the system

$$\begin{aligned} (1) \quad 0.12x_1 + 0.60x_2 &= 240{,}000 \\ (3) \quad 0.18x_1 + 0.90x_2 &= 240{,}000 \end{aligned}$$

have a solution? Again the algebraic procedures previously used will be exploited.

If Eq. (1) is multiplied by $1/0.12 = {}^{100}\!/_{12}$, then the system becomes

$$\begin{aligned} (1) \quad x_1 + 5x_2 &= 2{,}000{,}000 \\ \text{and} \qquad (3) \quad 0.18x_1 + 0.90x_2 &= 240{,}000 \end{aligned}$$

If Eq. (1) is multiplied by -0.18 and added to Eq. (3), then the following are obtained:

$$\begin{aligned} (1) \quad x_1 + 5x_2 &= 2{,}000{,}000 \\ \text{and} \qquad (2) \quad 0x_1 + 0x_2 &= -120{,}000 \end{aligned}$$

But this system of linear conditions is not satisfied for (x_1,x_2) regardless of the numbers chosen for x_1 and x_2, since the second condition will always state $0 = -120{,}000$. This means that regardless of the amount of sheets and blankets sold, it is not possible to make the total profit \$240,000 with the profits as estimated by both the board and Mr. Hand. This system of equations has *no solution*.

Example 5. Finally, Mr. Ivan submits an estimate whose equation is

$$(4) \quad 0.24x_1 + 1.20x_2 = 480{,}000$$

Note that this estimate rejects the estimated profit of \$240,000 and projects that \$480,000 is an obtainable profit. Do the estimates of the board and Mr. Ivan have a common solution?

The conditions which represent the estimates given by the board and by Mr. Ivan are

$$(1)\quad 0.12x_1 + 0.60x_2 = 240{,}000$$
$$(4)\quad 0.24x_1 + 1.20x_2 = 480{,}000$$

If Eq. (1) is multiplied by -2 and added to Eq. (4), then the following system results:

$$(1)\quad 0.12x_1 + 0.60x_2 = 240{,}000$$
$$(4)\quad 0x_1 + 0x_2 = 0$$

Regardless of the values chosen for x_1 and x_2, Eq. (4) is satisfied. Why? This means that common solutions of the two equations are precisely those solutions of the first equation. Stated differently, any solution of Eq. (1) is also a solution of Eq. (4), and vice versa. Particular solutions of the first equation have already been listed in Example 1.

Although Mr. Ivan and his staff submitted an estimate which appeared to be different from the estimate given by the board, these considerations show that the solutions are the same. This system of linear equations has an infinite number of solutions, although many would be rejected for the reasons mentioned earlier.

Before this section is concluded, there are several general remarks that should be made. If you suspect that the four equations were rather carefully chosen in order to represent all three possible types of solutions for two equations in two unknowns, you are correct. Given two linear equations in two unknowns, there will be either:

1. A unique solution
2. No solution
3. An infinite number of solutions

The examples were chosen to illustrate each of these cases.

You may already be familiar with methods of solving two equations in two unknowns, but the method that was used in this section may be new to you. However, there are some very good reasons why this particular method of solution of systems of equations is often preferred over other methods, and this subject will be discussed further in Sec. 5.5.

Exercises

1. Find x_1 such that (x_1,x_2) is a solution of

$$0.12x_1 + 0.60x_2 = 240{,}000$$

when:

a. $x_2 = 100{,}000$ *b.* $x_2 = 200{,}000$
c. $x_2 = 300{,}000$ *d.* $x_2 = 400{,}000$
e. $x_2 = 500{,}000$

2. Rework Exercise 1 for $0.08x_1 + 1.20x_2 = 240{,}000$.
3. Rework Exercise 1 for $0.18x_1 + 0.90x_2 = 240{,}000$.
4. Rework Exercise 1 for $0.24x_1 + 1.20x_2 = 480{,}000$.
5. Given $2x_1 + 3x_2 = 6$ and $ax_1 + 4x_2 = 8$. Choose a such that for the system of equations, there is:

   a. A unique solution
   b. No solution
   c. An infinite number of solutions

6. Rework Exercise 5 for the system

$$\begin{aligned} 3x_1 + ax_2 &= 12 \\ 5x_1 + 5x_2 &= 3 \end{aligned}$$

7. Solve the following systems:

   a. $\begin{aligned} 3x_1 + 2x_2 &= 5 \\ x_1 + 3x_2 &= 4 \end{aligned}$ *b.* $\begin{aligned} 2x_1 + 4x_2 &= 16 \\ x_1 + x_2 &= 6 \end{aligned}$

   c. $\begin{aligned} 3x_1 + 15x_2 &= 18 \\ x_1 + 5x_2 &= 6 \end{aligned}$ *d.* $\begin{aligned} 2x_1 + 6x_2 &= 8 \\ x_1 + 3x_2 &= 5 \end{aligned}$

8. A butcher is to buy some chickens and ducks. He will sell the chickens at a profit of 2 cents per lb and will lose 1 cent per lb on the ducks. He does not wish to make a profit since these items will be used to attract customers. Nor does he intend to lose money on the sales. The supplier will sell to him at attractive prices only if he buys the same number of pounds of ducks as of chickens. How many pounds of each should the butcher buy in order to meet these conditions?

9. The same butcher decides to make a profit of $20 on the sale of chickens and ducks. The market has changed, and now he must buy 100 lb more of ducks than of chickens. How many pounds of each should he buy?

10. Suppose that the butcher now makes 2 cents per lb on chickens and loses 1 cent per lb on ducks. He wishes to make $10 for handling costs. The supplier now requires that he buy in such a way that the pounds of chickens purchased exceed by 500 lb one-half the number of pounds of ducks.

    a. How many pounds of ducks should he buy?
    b. Chickens?
    c. If he buys 300 lb of chickens, can be achieve his goal of $10?
    d. If he buys 150 lb of ducks, can he achieve his goal?

11. In studying the relationship between hardness and strength of a company's product, an analyst needs the values for a and b to make the required estimates. Solve the two equations to obtain the necessary values:

$$\begin{aligned} 10a + 110b &= 2{,}224 \\ 110a + 1{,}320b &= 26{,}664 \end{aligned}$$

12. A mail-order firm uses the following equation to estimate the number of dollars of sales for the day from the weight of first-class mail received by

7 A.M.:

$$\text{Sales} = a + b(x) \qquad \text{where } x \text{ is the weight of the mail}$$

If the number of shipping clerks needed is one per \$12,000 of sales, find the number of clerks needed on a day when the mail weighs 300 lb, given:

$$\begin{aligned} 11a + 115b &= 49{,}620 \\ 115a + 1{,}595b &= 683{,}700 \end{aligned}$$

Answers to problems

A. The equation is $2x_1 + 6x_2 = 800$. Particular solutions are (400, 0), (100, 100), and (0, $133\frac{1}{3}$).

B. The system of equations is

$$\begin{aligned} 2x_1 + 6x_2 &= 800 \\ x_1 + x_2 &= 280 \end{aligned}$$

The solution is (220, 60).

C. The system of equations is

$$\begin{aligned} 2x_1 + 6x_2 &= 800 \\ x_1 &= 450 \end{aligned}$$

No solution is possible.

D. The system of equations is

$$\begin{aligned} 2x_1 + 6x_2 &= 800 \\ x_1 + 3x_2 &= 400 \end{aligned}$$

This system of equations has an infinite number of solutions. Almost all would be rejected because of reasons discussed in the text.

E. 1. $39{,}000/(7.80 - 5.20) = 15{,}000$ gal

2. $x_1 = 3x_2$
   $2.60x_1 + 2.10x_2 = 39{,}000$
   $x_1 \cong 11{,}820$ gal of A
   $x_2 \cong 3{,}940$ gal of B

3. $2.60x_1 + 2.10x_2 = 40{,}000$
   $x_1 + x_2 = 1.25 \cdot 15{,}000$
   $x_2 = 17{,}500$
   Percent $x_2 = 17{,}500/18{,}750 = 93.3\%$

5.2 *Systems of two linear equations in two unknowns: geometric solution*

Problems

A. Describe the set

$$\{(x_1,x_2)|2x_1 + x_2 = 5\} \cap \{(x_1,x_2)|x_1 + 2x_2 = 5\}$$

B. A company which has made a study of the relationship between its sales and the spendable income in each of the counties which it serves arrived at the following (x_1,x_2) coordinates for the line which best summarizes this relationship: ($0,$5) and ($50,000,$2,000).

1. If spendable income is on the horizontal (x_1) axis and company sales are on the vertical (x_2) axis, prepare a graph which shows this relationship.
2. In what quadrant are all the values shown?
3. Use your graph to estimate the sales which could be expected in county *A*, where the spendable income is equal to $35,000.
4. If you were the sales manager for the firm and found that county *A* sales were actually $1,200, would you feel that the county *A* salesman had done well or poorly?

Geometry has often been used in this book to illustrate ideas. Diagrams, histograms, and decision trees are examples of geometric devices that have been used for illustrative purposes. In this section, a small portion of that branch of mathematics called *analytic geometry* will be used to picture the three types of possible solutions of a system of two equations in two unknowns that were discussed in Sec. 5.1. Before the examples of the previous section are reinvestigated from a geometric viewpoint, some language and assumptions will be introduced.

Associated with a linear equation in two unknowns is a set of points in a plane called the *graph* of the equation. Graphs are made with the aid of a *coordinate system.* A coordinate system (of the type needed to illustrate the graph of a linear equation in two unknowns) consists of two perpendicular lines in a plane, called *axes,* the *orientation* of the axes, and a *unit of measurement.* Diagram 1 exhibits such a set of lines along with some of the points on each of the axes and their *coordinates* on the lines.

Some comments about notation and conventions are in order. The axes are labeled x_1 and x_2 in accordance with the notation adopted in the previous section for the variables. By convention, x_1 is the horizontal axis and x_2 is the vertical axis. Also by agreement, the positive portion of the x_1 axis is to the right and the positive portion of the x_2 axis is the top half. These agreements are what was referred to as the *orientation* of the axes. There is a unit of measurement, and once it is established, it is used to locate the points associated with the integers shown in the diagram. The point of intersection of the axes is the *origin,* and the four portions of the plane are called *quadrants.* The quadrants are labeled first, second, third, and fourth in Diagram 1; this is just a convention made for convenience of reference.

There is a one-to-one correspondence between the real numbers and a line in a plane.

Part of this correspondence has been shown in Diagram 1. It follows that

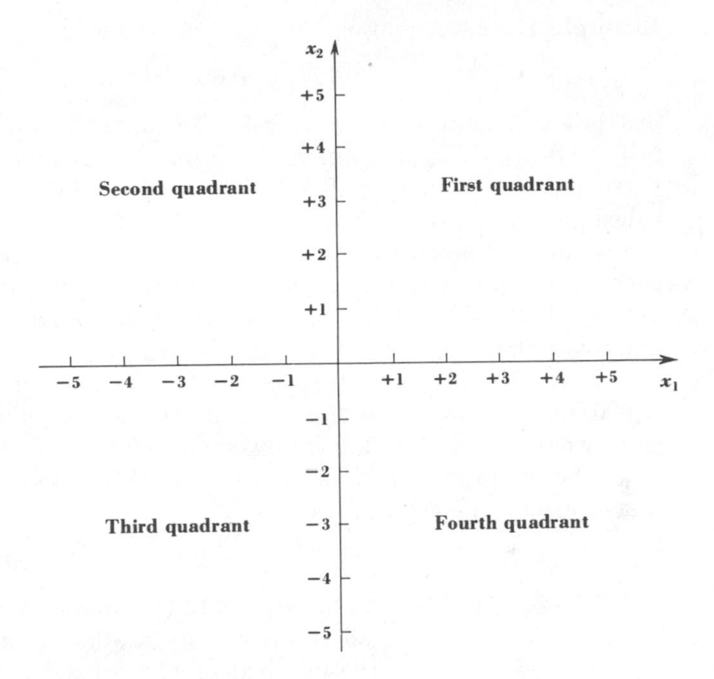

Diagram 1

There is a one-to-one correspondence between the set of ordered pairs of real numbers (x_1,x_2) and the points of the plane.

The ordered pairs are called the *coordinates* of the points.

The purpose of this mathematical machinery is to draw the graphs of some linear equations. But before the graphs can be drawn we need one more agreement. Although it will be accepted here without supporting argument, this agreement is one of the main reasons that equations of the type $Ax_1 + Bx_2 = C$ are called *linear* equations.

If A, B, C are real numbers, then the graph of the equation

$$Ax_1 + Bx_2 = C$$

is the set of ordered pairs of real numbers (u,v) such that

$$Au + Bv = C$$

The points in a plane that correspond to (u,v) are on a line, and every point on the line has coordinates which satisfy the equation

$$Ax_1 + Bx_2 = C$$

Such an equation is called linear.

The preliminaries are over, and now the graphs of linear equations can be studied. The examples of this section will refer to the examples of Sec. 5.1.

Example 1. The equation

$$(1) \quad 0.12x_1 + 0.60x_2 = 240{,}000$$

has the solutions observed previously: (2,000,000, 0), (0, 400,000). These *two points are sufficient to determine the line* which is the graph of Eq. (1). (Two points determine a line.) The first of these, (2,000,000, 0), is called the x_1 *intercept* because it is the point where the line intercepts the x_1 axis. Similarly, (0, 400,000) is the x_2 intercept. Along with these intercepts, Diagram 2 shows the other particular solution noted before, (1,000,000, 200,000). The *unit of measurement* chosen for this graph is 250,000.

Note that (0, 400,000) corresponds to a point on the x_2 axis, that (2,000,000, 0) corresponds to a point on the x_1 axis, and finally, that the point whose coordinates are (1,000,000, 200,000) is in the first quadrant.

The set notation adopted in Chap. 1 can be used to write the equation as a set of points described by

$$(1) \quad \{(x_1,x_2)|0.12x_1 + 0.60x_2 = 240{,}000\}$$

The points of the line correspond to the solution set of the equation, but it is clear that not all the points on the line would be acceptable results when related to the physical problem that gave rise to Eq. (1). For example, one point on the line has coordinates (3,000,000, −200,000). (You should verify that this is indeed a solution.) But this solution applied to the problem means the company must have a sale of 3 million sheets and at the same time have a "negative" sale of 200,000 blankets—an impossibility. Solutions such that either x_1 or x_2 is negative are therefore unacceptable for this particular problem. This means that only points in the first quadrant are meaningful and feasible.

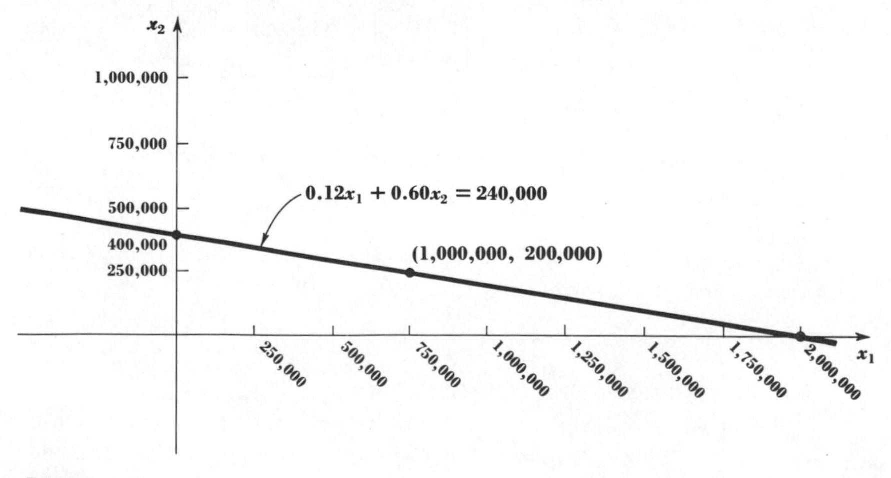

Diagram 2

Another point on the line has coordinates $(1{,}000{,}000\frac{1}{2}, 199{,}999\frac{9}{10})$. [Verify that this is a solution. This can be done by replacement in Eq. (1).] This solution is also not *acceptable* since it is not feasible to sell a part of a sheet or a part of a blanket.

It will be helpful to reread that portion of Sec. 5.0 which describes the solution of a problem again. See how, for the problem under consideration, a mathematical statement of the problem was made and a number of solutions determined. In fact, every solution to Eq. (1) is associated with a point on the line in Diagram 2, and every point on the line has an ordered pair of coordinates which, when used as replacements for x_1 and x_2, satisfy Eq. (1). Nevertheless, not all solutions to the mathematical statement of the problem are acceptable solutions to the problem which gave rise to the equation.

Example 2. Now consider the equation

$$(2) \quad 0.08x_1 + 1.20x_2 = 240{,}000$$

The graph of Eq. (2) is also a line, and the graph of the line is shown in Diagram 3. The unit of measurement is again chosen to be 250,000.

From the graph, it can be seen that (0, 200,000) and (3,000,000, 0) are each solutions to Eq. (2). They are, of course, the *intercepts*. Each of these solutions, when applied to the physical problem, requires that the CCC discontinue a product; because of this, these two solutions were rejected. There are many solutions (every point on the line), and among the solutions are (1,500,000, 100,000) and (750,000, 150,000). These points are indicated on the graph. Verify that these two points are solutions by using them as replacements in Eq. (2).

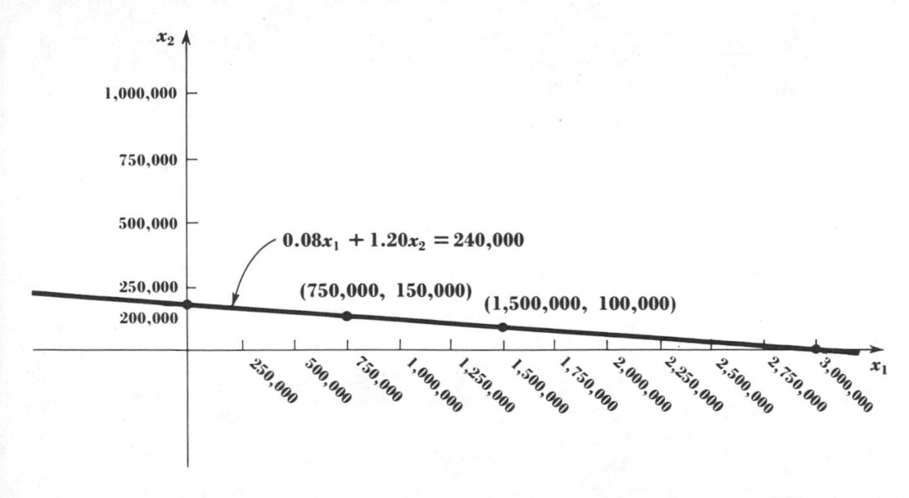

Diagram 3

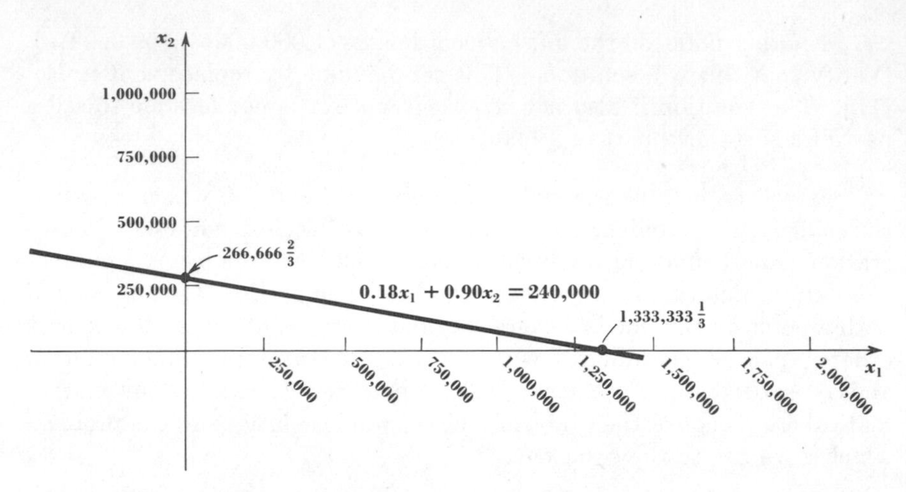

Diagram 4

Example 3. The third equation of the previous section is

$$(3) \quad 0.18x_1 + 0.90x_2 = 240,000$$

It also has a line as a graph, and the line and the intercepts (1,333,333⅓, 0) and (0, 266,666⅔) are shown in Diagram 4. What is the unit of measurement?

Example 4. The graph of

$$0.24x_1 + 1.20x_2 = 480,000$$

is shown in Diagram 5. Verify that the intercepts are as indicated.

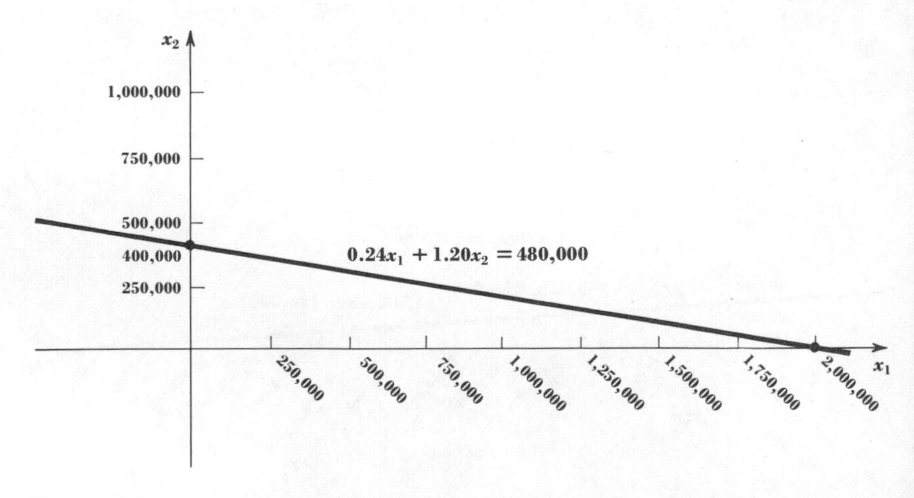

Diagram 5

Example 5. The graphs of the equations of Sec. 5.1 have all been shown. But we saw that it is sometimes essential to consider two equations simultaneously, i.e., as a system of equations. Previously, we considered the system

$$\begin{aligned} (1)\quad 0.12x_1 + 0.60x_2 &= 240{,}000 \\ (2)\quad 0.08x_1 + 1.20x_2 &= 240{,}000 \end{aligned}$$

These equations lend themselves to expression by set notation, and using S for the set described in Eq. (1) and T for the second set, we have

$$\begin{aligned} S &= \{(x_1,x_2)|0.12x_1 + 0.60x_2 = 240{,}000\} \\ T &= \{(x_1,x_2)|0.08x_1 + 1.20x_2 = 240{,}000\} \end{aligned}$$

The set to be determined is $S \cap T$. The lines which represent sets S and T were drawn in Diagrams 2 and 3, respectively. In Diagram 6, the graphs of these two equations are shown with the same coordinate system so that the equations can be considered simultaneously. From Diagram 6, an estimate of the coordinates of the point which is common to the two lines can be made. (This can also be determined by solving the two equations simultaneously, as was done in Sec. 5.1.) Try to estimate the solution from the graph.

You can test your estimate by remembering that the algebraic solution yielded the ordered pair (1,500,000, 100,000). In set notation, $S \cap T = \{(1{,}500{,}000,\ 100{,}000)\}$.

Example 6. Recall that the system of equations

$$\begin{aligned} (1)\quad 0.12x_1 + 0.60x_2 &= 240{,}000 \\ (3)\quad 0.18x_1 + 0.90x_2 &= 240{,}000 \end{aligned}$$

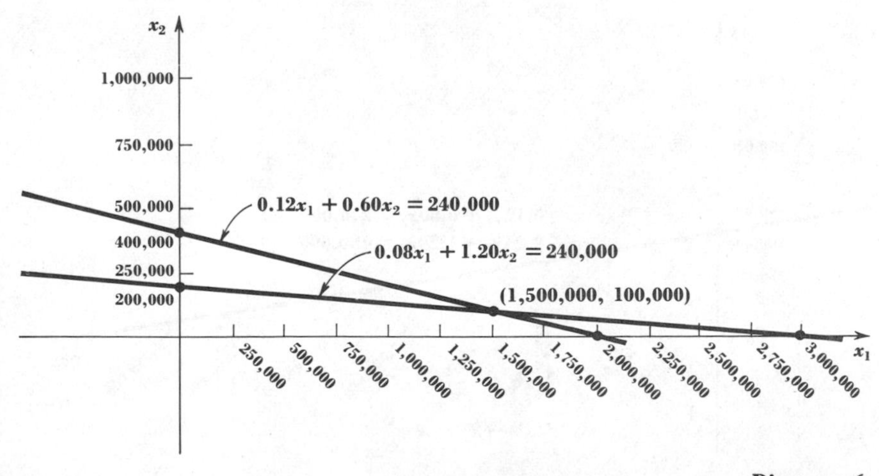

Diagram 6

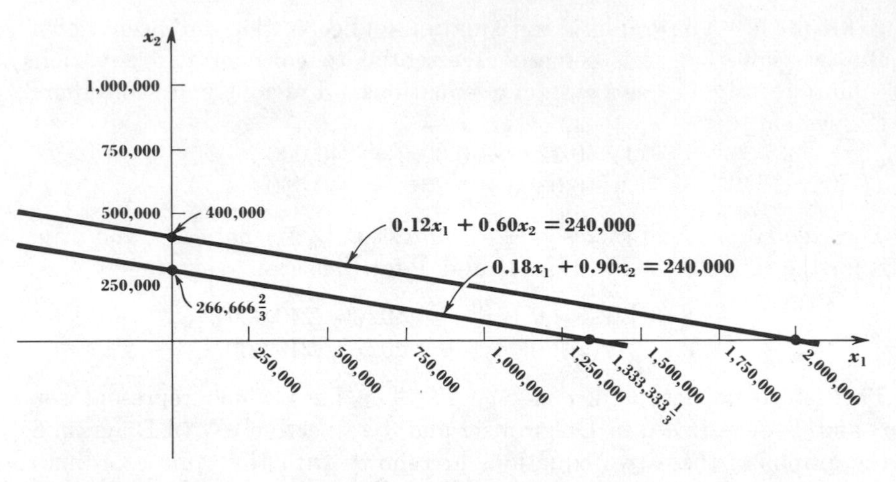

Diagram 7

has no solution. To see what this means geometrically, Diagrams 2 and 4 are reproduced simultaneously in Diagram 7.

The lines which are the graphs of the two equations are *parallel;* they have no point in common.

Example 7. Finally, the system

$$\begin{aligned} (1) \quad 0.12x_1 + 0.60x_2 &= 240{,}000 \\ (4) \quad 0.24x_1 + 1.20x_2 &= 480{,}000 \end{aligned}$$

has an infinite number of solutions, which are the points of the graph of either Eq. (1) or Eq. (4) since the graph of both equations is the same.

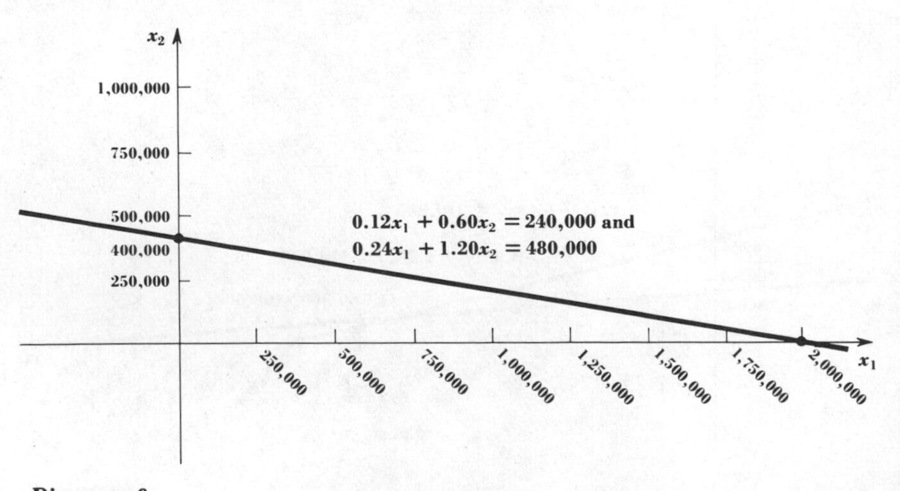

Diagram 8

Diagram 8 shows this graph. Geometrically, the two lines (the graphs) are the same.

One point of this presentation is that systems of two linear equations in two unknowns are easily graphed. The graphs not only offer an alternate (but perhaps less accurate) method of solving the system but at the same time make possible a visual understanding of the problem. In terms of sets of points, the solution is either:

1. A single point
2. No points
3. A line

Exercises

1. For each of the following systems, find the intercepts and draw the graphs. Estimate the solution from the graph.

   a. $x_1 + x_2 = 7$, $2x_1 + x_2 = 10$
   b. $2x_1 + 4x_2 = 16$, $x_1 + x_2 = 6$
   c. $3x_1 + 5x_2 = 4$, $x_1 + 10x_2 = 3$
   d. $x_1 + 2x_2 = 5$, $2x_1 + 4x_2 = 7$
   e. $3x_1 + 12x_2 = 27$, $x_1 + 4x_2 = 9$

2. Check your graphical estimate for each of the systems in Exercise 1 by solving algebraically.

3. Draw the line which represents the set of solutions of each of the following equations. In each case, determine x_1 and x_2 such that $(x_1,0)$ and $(0,x_2)$ are solutions.

   a. $x_1 + 3x_2 = 13$
   b. $x_1 + 2x_2 = 7$
   c. $x_1 + 8x_2 = 13$
   d. $2x_1 + 6x_2 = 11$
   e. $2x_1 - x_2 = 5$
   f. $x_1 + 2x_2 = 10$
   g. $2x_1 + x_2 = 4$
   h. $4x_1 + 2x_2 = 6$
   i. $x_1 + 5x_2 = 12$
   j. $2x_1 + 10x_2 = 14$
   k. $x_1 + 3x_2 = 6$
   l. $2x_1 + 6x_2 = 12$
   m. $x_1 + x_2 = 3$
   n. $3x_1 + 3x_2 = 9$

4. Find the solution (if possible) of the systems of equations which result by considering simultaneously the following pairs of the equations in Exercise 3:

   a. *a* and *b*
   b. *c* and *d*
   c. *e* and *f*
   d. *g* and *h*
   e. *i* and *j*
   f. *k* and *l*
   g. *m* and *n*

5. Draw the graphs of your answer to Exercise 5, Sec. 5.1.

6. Draw the graphs of Exercise 9, Sec. 5.1.

7. Draw the graphs of Exercise 10, Sec. 5.1.
8. Which of the following equations are linear?
   a. $x_1 + 17x_2 = 43$
   b. $3x_1 + 2x_2 + x_3 = 22$
   c. $1.5x_1^2 + x_2 = 17$
   d. $x_1 - 0.172x_2 = 3.946$
   e. $22a + 44b = 79$
   f. $4.3x_1 - 2.1x_2 - 8.9x_3 + x_4 + 2x_5 = 136.3$
   g. $0.7x_1 + \sqrt{x_2} = 23$
   h. $x_1x_2 = 7$
   i. $x_1/x_2 = 6$
   j. $x_1 = 9$

Answers to problems

A. It is a point whose coordinates are $(5/3, 5/3)$.

B. 1.

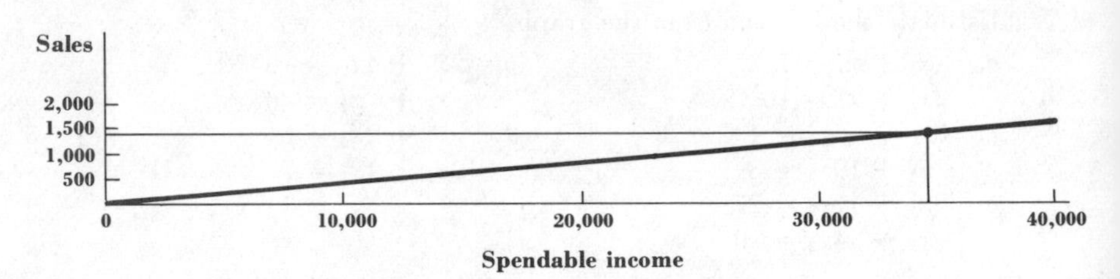

2. First
3. About $1,400
4. Slightly below expected sales

5.3 *Systems of two linear equations in two unknowns: matrices*

Problems

A. For the system of equations

$$\begin{aligned} x + 2y &= 20 \\ x + 5y &= 35 \end{aligned}$$

write the matrix of the system and perform the row operations necessary to make the matrix of coefficients an identity matrix. Read the solution from the column matrix of constants.

B. A furniture manufacturer packs his chairs and tables separately when he ships dinette sets. A chair which is disassembled and packed takes up 1 cu ft of space, and a table occupies 4 cu ft. The company truck has a capacity of 1,200 cu ft. How many chairs and tables should he pack if he wants four chairs with each table?

Two methods have been indicated to solve 2×2 systems of linear equations: algebraic and geometric. Both were presented by the elaborate use of examples, and in neither case was an *algorithm* proposed. An algorithm is a procedure—usually repetitive in nature—that will always work. For the geometric solution, the method is clear, but as was remarked, it is somewhat inaccurate since it calls for a visual reading of a graph. The two operations used in the algebraic solution—multiplying an equation by a constant and adding a constant multiple of one equation to the other—had their justification postponed until a later section. In this section, the algebraic procedure will be reexamined, some new notations and nomenclature will be introduced, and hopefully the algorithm—the repetitive procedure—will become more apparent. The important new word in this discussion is *matrix*, and the two operations discussed earlier will be interpreted as *row operations on matrices.*

Example 1. Consider once again the 2×2 system

$$\begin{aligned} (1)\quad 0.12x_1 + 0.60x_2 &= 240{,}000 \\ (2)\quad 0.08x_1 + 1.20x_2 &= 240{,}000 \end{aligned}$$

The *matrix of the coefficients* of the system is

$$\begin{pmatrix} 0.12 & 0.60 \\ 0.08 & 1.20 \end{pmatrix}$$

This matrix has two *rows:*

$$(0.12 \quad 0.60)$$

and

$$(0.08 \quad 1.20)$$

It also has two *columns:*

$$\begin{pmatrix} 0.12 \\ 0.08 \end{pmatrix}$$

and

$$\begin{pmatrix} 0.60 \\ 1.20 \end{pmatrix}$$

The first row is the set of coefficients of Eq. (1); the second row is the set of coefficients of Eq. (2). The first column is the set of coefficients of x_1; the second column is the set of coefficients of x_2. The matrix of coefficients is also 2×2 (rows $\times$ columns).

The *matrix of the system* is

$$\left(\begin{array}{cc|c} 0.12 & 0.60 & 240{,}000 \\ 0.08 & 1.20 & 240{,}000 \end{array}\right)$$

It is 2×3 (rows $\times$ columns), and the vertical bar is included to set off the constants of the equations. Recall from Example 3 of Sec. 5.1 that the first step in the algebraic solution is to *multiply* the first equation by $1/0.12 = {}^{100}\!/_{12}$ and *multiply* the second row by 100. If these operations are applied to the 3×2 matrix, we have a second matrix,

$$\left(\begin{array}{cc|c} 1 & 5 & 2{,}000{,}000 \\ 8 & 120 & 24{,}000{,}000 \end{array}\right)$$

Compare this matrix with the second step in the algebraic solution on page 315. The next step is to *add* the product of the first row and -8 to the second row. This operation yields a third 2×3 matrix,

$$\left(\begin{array}{cc|c} 1 & 5 & 2{,}000{,}000 \\ 0 & 80 & 8{,}000{,}000 \end{array}\right)$$

Again a comparison with the algebraic procedure on page 315 will be helpful. The third step was to multiply the second row by $1/80$. Applied to the third matrix, this gives a fourth 2×3 matrix:

$$\left(\begin{array}{cc|c} 1 & 5 & 2{,}000{,}000 \\ 0 & 1 & 100{,}000 \end{array}\right)$$

Finally, adding the product of the second row and -5 to the first row, we have

$$\left(\begin{array}{cc|c} 1 & 0 & 1{,}500{,}000 \\ 0 & 1 & 100{,}000 \end{array}\right)$$

The left-hand portion of this matrix is

$$\begin{pmatrix} 1 & 0 \\ 0 & 1 \end{pmatrix}$$

which is an *identity matrix*, and the column matrix of constants

$$\begin{pmatrix} 1{,}500{,}000 \\ 100{,}000 \end{pmatrix}$$

is the solution of the system. The identity matrix is so called because it has an element 1 in each position of the *main diagonal* and zeros elsewhere. For obvious reasons, this procedure is often called *diagonalizing a matrix*.

Example 2. Now we imitate this procedure on a new system, this time keeping the numbers smaller for convenience. Consider the system

$$\begin{aligned} 2x_1 + 3x_2 &= 8 \\ x_1 + 2x_2 &= 5 \end{aligned}$$

which has as its *coefficient matrix*

$$\begin{pmatrix} 2 & 3 \\ 1 & 2 \end{pmatrix}$$

and as its *system matrix*

$$\left(\begin{array}{cc|c} 2 & 3 & 8 \\ 1 & 2 & 5 \end{array}\right)$$

Multiplying the first row by ½ yields

$$\left(\begin{array}{cc|c} 1 & 3/2 & 4 \\ 1 & 2 & 5 \end{array}\right)$$

Adding the product of the first row and −1 to the second row yields

$$\left(\begin{array}{cc|c} 1 & 3/2 & 4 \\ 0 & 1/2 & 1 \end{array}\right)$$

Multiplying the second row by 2 gives

$$\left(\begin{array}{cc|c} 1 & 3/2 & 4 \\ 0 & 1 & 2 \end{array}\right)$$

Adding the product of $-3/2$ and the second row to the first row gives

$$\left(\begin{array}{cc|c} 1 & 0 & 1 \\ 0 & 1 & 2 \end{array}\right)$$

The left-hand, 2×2 portion of the matrix is an *identity matrix;* the solution of the system is $x_1 = 1$, $x_2 = 2$. This is verified by the computations

$$\begin{aligned} 2 \cdot 1 + 3 \cdot 2 &= 2 + 6 = 8 \\ 1 \cdot 1 + 2 \cdot 2 &= 1 + 4 = 5 \end{aligned}$$

Example 3. Another example should assist in making clear the procedure. In this example, the description of the row operations is omitted. See if you can supply them. The system to be solved is

$$\begin{aligned} 2x_1 + 3x_2 &= 1 \\ x_1 + 4x_2 &= 3 \end{aligned}$$

The matrix of the system is

$$\left(\begin{array}{cc|c} 2 & 3 & 1 \\ 1 & 4 & 3 \end{array}\right)$$

The matrices in the steps that lead to the solution are

$$\left(\begin{array}{cc|c} 1 & 3/2 & 1/2 \\ 1 & 4 & 3 \end{array}\right)$$

$$\left(\begin{array}{cc|c} 1 & 3/2 & 1/2 \\ 0 & 5/2 & 5/2 \end{array}\right)$$

$$\left(\begin{array}{cc|c} 1 & 3/2 & 1/2 \\ 0 & 1 & 1 \end{array}\right)$$

$$\left(\begin{array}{cc|c} 1 & 0 & -1 \\ 0 & 1 & 1 \end{array}\right)$$

That the solution is $x_1 = -1$, $x_2 = 1$ is verified by

$$2(-1) + 3 \cdot 1 = -2 + 3 = 1$$

and

$$1(-1) + 4 \cdot 1 = -1 + 4 = 3$$

One obvious advantage of omitting the variables and equals signs in the computations is that the computation is shorter and the symbols are more easily read. It is this advantage (along with other considerations) that will cause us to adopt this means of solution in future work. The operations on the matrices are called *row* operations because that is exactly what they are. The study of matrices is in itself a large branch of mathematics, but there will be little occasion here to develop rules for the addition or multiplication of matrices per se. Instead, only that manipulation essential to the solution of systems of equations will be discussed.

Exercises

1. For each of the following systems, write the matrix of the system and solve by use of row operations on the matrix. Verify each result by replacement of the variables in the equations of the system.

   a. $x_1 + 3x_2 = 13$, $x_1 + 2x_2 = 7$

   b. $x_1 + 8x_2 = 13$, $2x_1 + 6x_2 = 11$

   c. $2x_1 + x_2 = 5$, $x_1 + 2x_2 = 10$

   d. $2x_1 + x_2 = 4$, $4x_1 + 2x_2 = 6$

   e. $x_1 + 5x_2 = 12$, $2x_1 + 10x_2 = 14$

   f. $2x_1 + 6x_2 = 12$, $x_1 + x_2 = 3$

2. A creamery takes raw milk and separates the cream from the skim milk. If the process yields 4 percent cream and 96 percent skim milk and if the cream yields a profit of 20 cents per gal and skim milk yields a profit of 2 cents per gal, how many gallons of raw milk must be processed to yield a profit of $1,000?

Answers to problems

A.

$$\left(\begin{array}{cc|c} 1 & 2 & 20 \\ 1 & 5 & 35 \end{array}\right)$$

$$\left(\begin{array}{cc|c} 1 & 2 & 20 \\ 0 & 3 & 15 \end{array}\right)$$

$$\left(\begin{array}{cc|c} 1 & 2 & 20 \\ 0 & 1 & 5 \end{array}\right)$$

$$\left(\begin{array}{cc|c} 1 & 0 & 10 \\ 0 & 1 & 5 \end{array}\right)$$

$x_1 = 10$, $x_2 = 5$

B. Let x_1 be the number of chairs and x_2 the number of tables. Then the system of equations is

$$x_1 - 4x_2 = 0$$
$$x_1 + 4x_2 = 1{,}200$$

The solution is 600 chairs and 150 tables.

5.4 *Three linear equations in three unknowns*

Problems

A. Three truckers enter a joint venture of buying fresh produce from farmers and trucking it to the city market. They will each purchase the same amount of three products: grapefruit, lettuce, and asparagus. The first partner reports a profit of \$0.10 per lb on grapefruit, \$0.05 per lb on lettuce, and \$0.01 per lb on asparagus, for a profit of \$100. The second reports a profit of \$0.10 per lb on grapefruit, a loss of \$0.05 per lb on lettuce, and a profit of \$0.01 per lb on asparagus, for a profit of \$50. The corresponding figures for the third partner are a loss of \$0.10 per lb on grapefruit, a profit of \$0.05 per lb on lettuce, and a profit of \$0.02 per lb on asparagus, for a profit of \$20. Did they keep their agreement to purchase the same number of pounds of each commodity?

B. The Acme Company manufactures a feed product which can be made from three ingredients, x_1, x_2, and x_3, as shown in the following table:

Ingredient	*Weight per bushel*	*Vitamin content*	*Mineral content*
x_1	50 lb	5 units	2 units
x_2	40 lb	4 units	5 units
x_2	100 lb	2 units	6 units

1. The total product must weigh 1 ton and must contain 120 units of the vitamin and 168 units of the mineral. What combination of these three ingredients would provide a product which meets these three requirements exactly? Notice that

$$50x_1 + 40x_2 + 100x_3 = 2{,}000$$
$$5x_1 + 4x_2 + 2x_3 = 120$$
$$2x_1 + 5x_2 + 6x_3 = 168$$

2. What would be the effect on your answer if the vitamin and mineral requirements were *minimums* rather than exact requirements?

In the previous sections of this chapter, systems of equations with two equations in two unknowns were considered. It was shown that there are three different possibilities, namely, a unique solution, no solution, and an infinite number of solutions. In this section, the concepts of the previous sections will be extended in an obvious way to the consideration of systems of linear equations with three equations in three unknowns. Just as a coordinate system in a plane was used in Sec. 5.2 to illustrate the solutions of the 2×2 systems, a coordinate system in space can be used to illustrate the solutions of the systems of equations to be considered in this section. However, as we shall see, there are some serious handicaps in such an approach. The algebraic procedure using matrix notation will prove to be the most satisfactory method of solution.

Again, an example will be used to illustrate how systems of linear equations with three equations in three unknowns might arise out of problems connected with business.

Example 1. In Sec. 5.1, an example was given in which the board of directors of CCC considered how to determine the number of each of the products, sheets and blankets, it was necessary to sell in order to make a predetermined gross profit. They did not consider the price or profit on their third product, comforters, because they felt that the comforters should be considered separately. However, as a result of their considerations about the profit on sheets and blankets, they decide to consider three products rather than two. A study of production records and costs is made as an initial effort toward estimating the profit that can be made on each of the three products. The results of their deliberation are profit estimates of \$0.12 per sheet, \$0.60 per blanket, and \$1.80 per comforter. But now the profit to be made is based on the sales of three products rather than two, and they want the total gross profit to be \$360,000. The condition which expresses this situation (and which uses x_3 to represent the number of comforters that must be sold) is given by

$$(1) \quad 0.12x_1 + 0.60x_2 + 1.80x_3 = 360{,}000$$

An equation such as Eq. (1) has as a solution (x_1,x_2,x_3), which is an *ordered triple* of numbers. For example, see that (3,000,000, 0, 0) and (0, 600,000, 0) and (0, 0, 200,000) are each solutions of Eq. (1). (Verification of this is required in Exercise 2.) But each of these three solutions is immediately rejected because each of them implies that only one of the three products will be manufactured next year. Because the company has no intention of stopping production of any of the three products, these solutions to Eq. (1) are of no practical interest. Remember it is necessary to evaluate the mathematical solution with respect to the physical circumstance that generates the equation.

However, the solutions are helpful in illustrating geometrically the set of all possible solutions. Since Eq. (1) is an equation that involves three variables, the *solution set* is a plane in space, and a portion of that plane (with the three particular results given above) is indicated by the graph in Diagram 9. In this illustration, the coordinate system has three axes that relate to the three unknowns x_1, x_2, and x_3.

Note that one difficulty in using geometric diagrams for 3×3 systems has already been encountered. On a plane (the sheet of paper), which is two-dimensional, it is not possible to have three mutually perpendicular axes. The compromise necessary to indicate three axes can be seen in the diagram. We shall observe other shortcomings of the geometric method of solution as we proceed. The points which indicate the intersection of the solution set (a plane) with each of the coordinate axes are indicated in the diagram, and these are the three particular solutions rejected earlier. However, there are other points on this plane which are also solutions to Eq. (1). One such point is (1,000,000, 310,000, 30,000). The meaning of this ordered triple as applied to the original problem is that if 1,000,000 sheets, 310,000 blankets, and 30,000 comforters are sold with the estimated profits of $0.12, $0.60, and $1.80, respectively, then the desired total gross profit of $360,000 will be made. Verify by substitution in Eq. (1) that this ordered triple is a solution of the condition.

Diagram 9 shows only a part of the plane of all solutions. The solution set is a plane in space, and indicated in the diagram is only that portion of the plane which occurs when each of the three variables is positive. Since the company will sell either zero or positive quantities of each of their three products, this portion of the plane is the only part

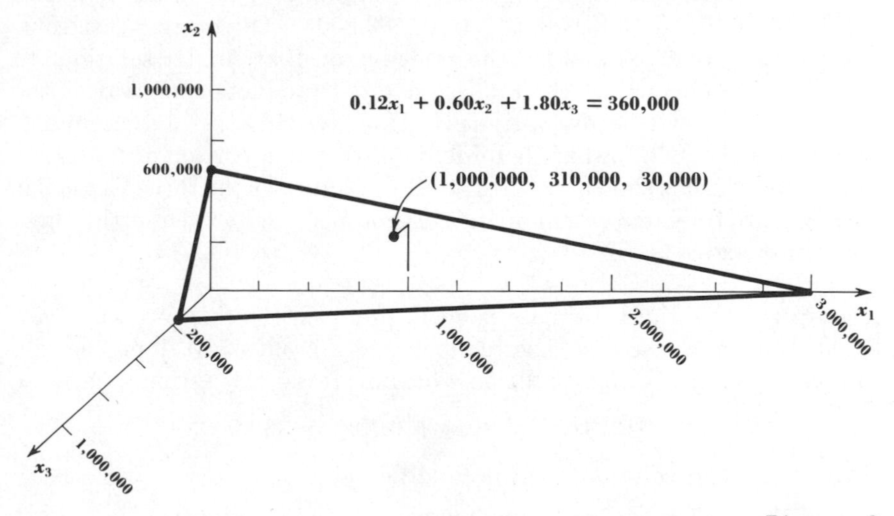

Diagram 9

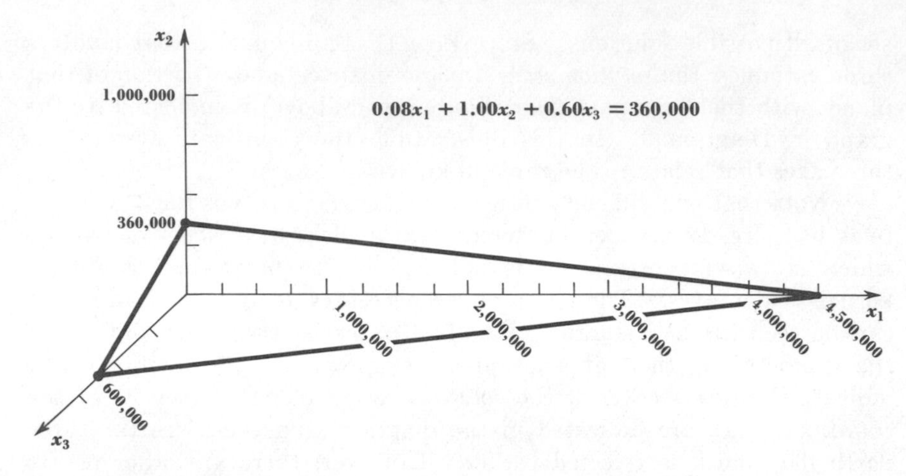

Diagram 10

that gives solutions which are feasible. Hence, the ability to exhibit only a portion of the plane is no serious handicap.

Example 2. The board of directors also decides to ask others in the company to estimate the profits that can be made if all three products are considered. Another estimate is that a profit of \$0.08 on sheets, \$1.00 on blankets, and \$0.60 on comforters can be made. If \$360,000 is the total gross profit to be made, then the condition which expresses the second estimate is

$$(2) \quad 0.08x_1 + 1.00x_2 + 0.60x_3 = 360{,}000$$

Three particular solutions for this condition are the *ordered triples* (4,500,000, 0, 0), (0, 360,000, 0), (0, 0, 600,000). Of course, these solutions are also rejected and for the same reason that similar solutions to Eq. (1) were rejected: each implies that the production of two of the three products will be discontinued. However, just as in Example 1, these particular solutions are helpful in illustrating a portion of the graph of Eq. (2), which is shown in Diagram 10. Note that the three particular solutions are the intersection of the solution plane with each of the three coordinate axes.

Example 3. A third estimate is \$0.10 profit on the sale of each sheet, \$0.90 profit on the sale of each blanket, and a profit of \$0.70 on the sale of each comforter. The condition which expresses this estimate is

$$(3) \quad 0.10x_1 + 0.90x_2 + 0.70x_3 = 360{,}000$$

Three particular solutions of this condition are

(3,600,000, 0, 0), (0, 400,000, 0), (0, 0, 514,286)

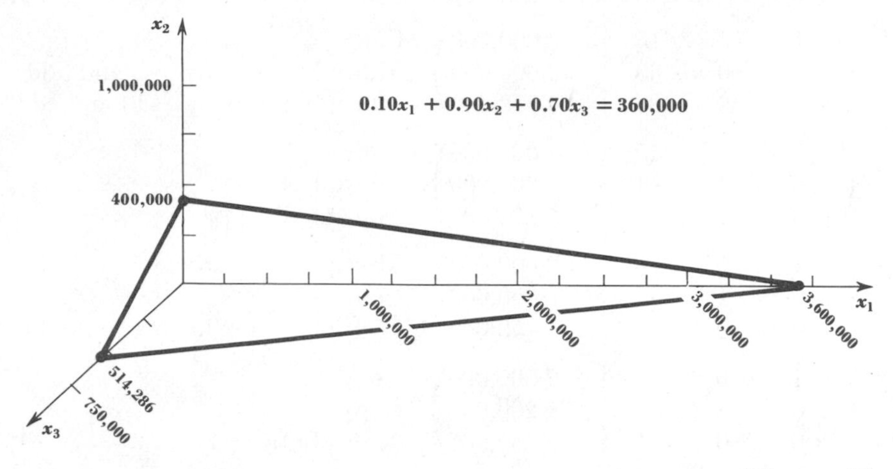

Diagram 11

The third solution is approximated. A partial graph of the solution plane for Eq. (3) is given in Diagram 11.

Example 4. After these preliminary estimates have been submitted, the board of directors meets again. The board now has for its consideration the estimate that they made and two others. The board decides that they will determine if there is any number of sales of the three products which satisfy Eqs. (1) to (3) simultaneously. The three conditions, written so that they can be considered simultaneously, are

$$\begin{aligned} 0.12x_1 + 0.60x_2 + 1.80x_3 &= 360{,}000 \\ 0.08x_1 + 1.00x_2 + 0.60x_3 &= 360{,}000 \\ 0.10x_1 + 0.90x_2 + 0.70x_3 &= 360{,}000 \end{aligned}$$

The method used to solve this system will be algebraic, and the matrix notation used for the solution of a 2×2 system will be generalized in an obvious fashion for a 3×3 system and used. This means that the symbols x_1, x_2, x_3 and the equals signs will be omitted from the intermediate steps of the computation. Listed below are the seven steps which lead to the solution of the three equations. Instructions for each of the steps in the solution have been written to the right of the computations. The operation is written beside the row to which it applies. Study the instructions to see if they have been followed correctly.

$$\left(\begin{array}{ccc|c} 0.12 & 0.60 & 1.80 & 360{,}000 \\ 0.08 & 1.00 & 0.60 & 360{,}000 \\ 0.10 & 0.90 & 0.70 & 360{,}000 \end{array}\right) \quad \text{Given}$$

$$\left(\begin{array}{ccc|c} 1 & 5 & 15 & 3{,}000{,}000 \\ 8 & 100 & 60 & 36{,}000{,}000 \\ 10 & 90 & 70 & 36{,}000{,}000 \end{array}\right) \quad \begin{array}{l} \text{Multiply by } 1/0.12 = 100/12 \\ \text{Multiply by } 100 \\ \text{Multiply by } 100 \end{array}$$

$$\left(\begin{array}{ccc|c} 1 & 5 & 15 & 3{,}000{,}000 \\ 0 & 60 & -60 & 12{,}000{,}000 \\ 0 & 40 & -80 & 6{,}000{,}000 \end{array}\right) \begin{array}{l} \text{Copy} \\ \text{Multiply row 1 by } -8 \text{ and add} \\ \text{Multiply row 1 by } -10 \text{ and add} \end{array}$$

$$\left(\begin{array}{ccc|c} 1 & 5 & 15 & 3{,}000{,}000 \\ 0 & 1 & -1 & 200{,}000 \\ 0 & 1 & -2 & 150{,}000 \end{array}\right) \begin{array}{l} \text{Copy} \\ \text{Multiply by } 1/60 \\ \text{Multiply by } 1/40 \end{array}$$

$$\left(\begin{array}{ccc|c} 1 & 0 & 20 & 2{,}000{,}000 \\ 0 & 1 & -1 & 200{,}000 \\ 0 & 0 & -1 & -50{,}000 \end{array}\right) \begin{array}{l} \text{Multiply row 2 by } -5 \text{ and add} \\ \text{Copy} \\ \text{Multiply row 2 by } -1 \text{ and add} \end{array}$$

$$\left(\begin{array}{ccc|c} 1 & 0 & 20 & 2{,}000{,}000 \\ 0 & 1 & -1 & 200{,}000 \\ 0 & 0 & 1 & 50{,}000 \end{array}\right) \begin{array}{l} \text{Copy} \\ \text{Copy} \\ \text{Multiply by } -1 \end{array}$$

$$\left(\begin{array}{ccc|c} 1 & 0 & 0 & 1{,}000{,}000 \\ 0 & 1 & 0 & 250{,}000 \\ 0 & 0 & 1 & 50{,}000 \end{array}\right) \begin{array}{l} \text{Multiply row 3 by } -20 \text{ and add} \\ \text{Multiply row 3 by 1 and add} \\ \text{Copy} \end{array}$$

$$\begin{aligned} x_1 \qquad\qquad &= 1{,}000{,}000 \\ x_2 \qquad &= 250{,}000 \\ x_3 &= 50{,}000 \end{aligned}$$

Note that the coefficient matrix of the final step is in *diagonal* form.

This unique solution, (1,000,000, 250,000, 50,000), interpreted with reference to the original question, means that if 1,000,000 sheets, 250,000 blankets, and 50,000 comforters are sold, the profit of $360,000 will be made. Furthermore, this gross profit can be achieved with the profits of any of the three estimates. If any one of the three (or any two or all three) estimates of profits are correct and if these sales are achieved, then the profit will be made.

Before another example is studied, some general comments about the method of solution are in order. First, the method was a generalization of the algebraic technique used in Sec. 5.1. But the matrix notation of Sec. 5.3 made it easier to keep clearly in mind what each succeeding step should be. Note that the final matrix is an *identity* matrix; there are 1s on the main diagonal and 0s elsewhere. Second, the geometric approach would have been extremely difficult to use. True, each solution plane has been drawn in Diagrams 9 to 11, but the difficulty of drawing all three planes on one set of axes and trying to decide where the planes intersect can probably only be appreciated if attempted. The reader should attempt the construction of the graph to understand the difficulty. The geometric interpretation is clear, however; the three planes meet in a point just as two walls and the floor of a room meet. *The geometric approach for systems more involved than* 2×2 *is impractical.*

Example 5. For another example, suppose a fourth estimate is that profits of \$0.24, \$1.20, and \$3.60 can be made on sheets, blankets, and comforters, respectively, with a resulting profit of \$720,000. The condition that expresses this estimate is

$$(4) \quad 0.24x_1 + 1.20x_2 + 3.60x_3 = 720{,}000$$

The board now decides to determine if there are sales consistent with the profit estimates of Eqs. (1), (2), and (4).

The three conditions in three unknowns designated by Eqs. (1), (2), and (4) are repeated below for simultaneous consideration:

$$\begin{aligned} 0.12x_1 + 0.60x_2 + 1.80x_3 &= 360{,}000 \\ 0.08x_1 + 1.00x_2 + 0.60x_3 &= 360{,}000 \\ 0.24x_1 + 1.20x_2 + 3.60x_3 &= 720{,}000 \end{aligned}$$

The matrix method of solution using operations on rows is applied again, with the instructions written to the right. The operation to be performed on each row is written beside the row.

$$\left(\begin{array}{ccc|r} 0.12 & 0.60 & 1.80 & 360{,}000 \\ 0.08 & 1.00 & 0.60 & 360{,}000 \\ 0.24 & 1.20 & 3.60 & 720{,}000 \end{array}\right) \quad \text{Given}$$

$$\left(\begin{array}{ccc|r} 1 & 5 & 15 & 3{,}000{,}000 \\ 8 & 100 & 60 & 36{,}000{,}000 \\ 24 & 120 & 360 & 72{,}000{,}000 \end{array}\right) \quad \begin{array}{l} \text{Multiply by } 1/0.12 = 100/12 \\ \text{Multiply by } 100 \\ \text{Multiply by } 100 \end{array}$$

$$\left(\begin{array}{ccc|r} 1 & 5 & 15 & 3{,}000{,}000 \\ 0 & 60 & -60 & 12{,}000{,}000 \\ 0 & 0 & 0 & 0 \end{array}\right) \quad \begin{array}{l} \text{Copy} \\ \text{Multiply row 1 by } -8 \text{ and add} \\ \text{Multiply row 1 by } -24 \text{ and add} \end{array}$$

The computations above show that for *any* numbers x_1, x_2, x_3, the third condition will be satisfied. This means that the solutions of the system are those solutions common to the first two equations, as represented in the last matrix. But the first row can be written as

$$x_1 + 5x_2 = 3{,}000{,}000 - 15x_3$$

and the second row can be written as

$$x_2 = 200{,}000 + x_3$$

Now the number x_3 can be assigned any value, say a, and the *general solution* of Eqs. (1), (2), and (4) is

$$(2{,}000{,}000 - 20a, \; 200{,}000 + a, \; a)$$

To determine *particular solutions*, the symbol a is replaced by a real number. For example, if the CCC will no longer make comforters (and consequently, $x_3 = 0$), a particular solution is

$$(2{,}000{,}000,\ 200{,}000,\ 0)$$

For a second particular solution, if comforters are manufactured but only 1,000 are sold, then the number of sheets and blankets it is necessary to sell to satisfy all three estimates and achieve the gross profit is

$$(1{,}980{,}000,\ 201{,}000,\ 1{,}000)$$

For a third particular solution, if the number of comforters sold is 5,000, then substitution of $a = 5{,}000$ into the general solution yields

$$(1{,}900{,}000,\ 205{,}000,\ 5{,}000)$$

Substitution of a positive integer for a in the general solution gives solutions for values 1, 2, . . . , 100,000. Why must replacements for a not exceed 100,000? Each of these solutions satisfies all three of the conditions of Eqs. (1), (2), and (4).

Example 6. Now a new and final situation about these production estimates will be presented. Suppose that a fifth estimate is $0.24 profit on each sheet, $1.20 profit for each blanket, and $3.60 profit for each comforter, with a total profit of $360,000. The system that uses Eqs. (1), (2), and (5) is

$$\begin{aligned}
0.12x_1 + 0.60x_2 + 1.80x_3 &= 360{,}000\\
0.08x_1 + 1.00x_2 + 0.60x_3 &= 360{,}000\\
0.24x_1 + 1.20x_2 + 3.60x_3 &= 360{,}000
\end{aligned}$$

If the first condition in this system is multiplied by -2 and added to the third condition, then the system of equations which results is expressed as

$$\begin{aligned}
0.12x_1 + 0.60x_2 + 1.80x_3 &= 360{,}000\\
0.80x_1 + 1.00x_2 + 0.60x_3 &= 360{,}000\\
0x_1 + 0x_2 + 0x_3 &= -360{,}000
\end{aligned}$$

Regardless of the value chosen for x_1, x_2, and x_3, the third equation can never be satisfied since the left-hand side is 0 and the right-hand side is $-360{,}000$. This means that the three conditions have no common solution.

The examples given in this section were of necessity very simple since the solutions were to be determined by ordinary arithmetical techniques. More complicated systems of linear equations lend themselves to solutions by the use of computers. But even so, the important points about a 3×3 system have been made.

The coefficients of the equations in the systems studied in this section have been chosen in such a way that the solutions are integers. Such a choice of coefficients need not have been made since nonintegral solutions could have led to reasonable approximations. It should be clear from the nature of the problem that no real harm can come from a close approximation.

Although the two previous paragraphs apologize for the kind of examples that have been given, the examples do illustrate the various possible cases. The three solution planes may intersect in a unique point (and the fourth example illustrates this); the three solution planes may intersect in a line (the fifth example illustrates this); or the three planes may not intersect at all (the sixth example illustrates this).

Exercises

1. For each of the systems below, determine (if possible) the solution set. Use row operations on matrices as the method of solution.

a.
$$\begin{aligned} 2x_1 - x_2 &= 2 \\ x_2 + x_3 &= 4 \\ x_1 - x_2 + x_3 &= 2 \end{aligned}$$

b.
$$\begin{aligned} 2x_1 + x_2 - x_3 &= 2 \\ -4x_1 - 2x_2 + 2x_3 &= -4 \\ -6x_1 - 3x_2 + 3x_3 &= -6 \end{aligned}$$

c.
$$\begin{aligned} 2x_1 - x_2 + x_3 &= 3 \\ 2x_1 + 2x_3 &= 6 \\ x_2 + x_3 &= 3 \end{aligned}$$

d.
$$\begin{aligned} x_1 + x_2 - x_3 &= 2 \\ x_1 + 2x_2 + x_3 &= 6 \\ x_1 - x_3 &= 0 \end{aligned}$$

e.
$$\begin{aligned} x_1 - 2x_2 + 4x_3 &= -2 \\ -3x_1 + 6x_2 - 12x_3 &= 6 \\ 4x_1 - 8x_2 + 16x_3 &= 5 \end{aligned}$$

f.
$$\begin{aligned} 5x_1 + x_2 - x_3 &= -4 \\ x_1 - x_2 - x_3 &= -6 \\ 8x_1 + x_2 - 2x_3 &= -9 \end{aligned}$$

g.
$$\begin{aligned} 2x_1 + 2x_2 + x_3 &= 5 \\ 3x_1 + 2x_2 - x_3 &= 4 \\ -2x_3 &= -1 \end{aligned}$$

h.
$$\begin{aligned} 2x_1 - x_2 + 3x_3 &= -3 \\ \tfrac{2}{3}x_1 - \tfrac{1}{3}x_2 + x_3 &= -1 \\ -4x_1 + 2x_2 - 6x_3 &= 6 \end{aligned}$$

i.
$$\begin{aligned} 3x_1 + x_2 + x_3 &= 12 \\ 6x_1 - x_2 + x_3 &= 9 \\ 3x_1 - 5x_2 - x_3 &= 14 \end{aligned}$$

2. Given the equation

$$0.12x_1 + 0.60x_2 + 1.80x_3 = 360{,}000$$

Verify that (3,000,000, 0, 0) and (0, 600,000, 0) and (0, 0, 200,000) are solutions to the given equations.

3. A lumber company wishes to make $500,000 during the next year. It can sell plywood for a profit of $1 per 1,000 board ft, lumber for a profit of $0.50 per 1,000 board ft, and pulp for a profit of $2 per ton. Sketch the solution set.

Answers to problems

A.
$$\begin{aligned} 10x_1 + 5x_2 + x_3 &= 10{,}000 \\ 10x_1 - 5x_2 + x_3 &= 5{,}000 \\ -10x_1 + 5x_2 + 2x_3 &= 2{,}000 \end{aligned}$$

$$\left(\begin{array}{rrr|r} 10 & 5 & 1 & 10{,}000 \\ 10 & -5 & 1 & 5{,}000 \\ -10 & 5 & 2 & 2{,}000 \end{array}\right)$$

$$\left(\begin{array}{rrr|r} 1 & \frac{1}{2} & \frac{1}{10} & 1{,}000 \\ 0 & -10 & 0 & -5{,}000 \\ 0 & 10 & 3 & 12{,}000 \end{array}\right)$$

$$\left(\begin{array}{rrr|r} 1 & \frac{1}{2} & \frac{1}{10} & 1{,}000 \\ 0 & 1 & 0 & 500 \\ 0 & 0 & 3 & 7{,}000 \end{array}\right)$$

$$\left(\begin{array}{rrr|r} 1 & \frac{1}{2} & \frac{1}{10} & 1{,}000 \\ 0 & 1 & 0 & 500 \\ 0 & 0 & 1 & 7{,}000/3 \end{array}\right)$$

$$\left(\begin{array}{rrr|r} 1 & 0 & \frac{1}{10} & 750 \\ 0 & 1 & 0 & 500 \\ 0 & 0 & 1 & 7{,}000/3 \end{array}\right)$$

$$\left(\begin{array}{rrr|r} 1 & 0 & 0 & 1{,}550/3 \\ 0 & 1 & 0 & 500 \\ 0 & 0 & 1 & 7{,}000/3 \end{array}\right)$$

Possible only if the pounds were 1,550/3, 500, and 7,000/3.

B. 1. $x_1 = 4$ bushels; $x_2 = 20$ bushels; $x_3 = 10$ bushels
2. There would be a large number of possible answers.

5.5 *Other systems of linear equations and row operations on matrices*

Problem

A company sells mixed nuts wholesale to such retail outlets as drugstores, dime stores, etc. They make their own mixture from peanuts, cashews, and chestnuts. They sell the mixture for \$0.50 per lb. Their costs are \$0.20 per lb for peanuts, \$0.80 per lb for cashews, and \$1.40 per lb for chestnuts. They wish to make \$20 per 100 lb. Is this possible? If so, how many pounds of each should be used in each 100 lb?

This section will present two ideas: the solution of new types of systems of linear equations and the mathematical justification for row operations on matrices in the solution of systems of equations. With regard to the first of these, the types of systems of linear equations studied thus far have been 2×2 and 3×3. It must be clear that the process of studying *square* systems could be extended indefinitely to the study of other systems—4×4, 5×5, etc. This will not be done. It should also be clear that had we chosen to study a 4×4 system, the geometric method of solution would be impossible. The technique of row operation on matrices, developed for the solution of 2×2 and 3×3

systems, does work for 4×4, 5×5, etc., systems. But instead of larger square systems, we shall study rectangular systems. The two examples of this section deal with a 2×3 system (the number of variables exceeds the number of equations) and a 3×2 system (the number of equations exceeds the number of variables). It will be shown that the method of solution of Secs. 5.3 and 5.4 works equally well for these new types of systems.

From examples, we proceed to the abstract. The latter portion of the section presents arguments to show that the row operations on matrices used in previous solutions are mathematically sound.

Example 1. Referring to the previous section, if only the first and third estimates on sales are considered, is there a common solution? Recall that the equations which express estimates 1 and 3 are

$$\begin{aligned} &(1) \quad 0.12x_1 + 0.60x_2 + 1.80x_3 = 360{,}000 \\ &(3) \quad 0.10x_1 + 0.90x_2 + 0.70x_3 = 360{,}000 \end{aligned}$$

Recall also that each of these conditions has a graph which is a plane in space and that the solution set of these two equations is the set of points in the intersection of the two planes in space. This set is either (1) empty, (2) a line, or (3) a plane. The intersection of the two planes of this system is determined in what follows. Note that once again the matrix of the system is used in the solution.

$$\left(\begin{array}{ccc|c} 0.12 & 0.60 & 1.80 & 360{,}000 \\ 0.10 & 0.90 & 0.70 & 360{,}000 \end{array}\right)$$

$$\left(\begin{array}{ccc|c} 1 & 5 & 15 & 3{,}000{,}000 \\ 10 & 90 & 70 & 36{,}000{,}000 \end{array}\right)$$

$$\left(\begin{array}{ccc|c} 1 & 5 & 15 & 3{,}000{,}000 \\ 0 & 40 & -80 & 6{,}000{,}000 \end{array}\right)$$

$$\left(\begin{array}{ccc|c} 1 & 5 & 15 & 3{,}000{,}000 \\ 0 & 1 & -2 & 150{,}000 \end{array}\right)$$

$$\left(\begin{array}{ccc|c} 1 & 0 & 25 & 2{,}250{,}000 \\ 0 & 1 & -2 & 150{,}000 \end{array}\right)$$

The last matrix, written as a system of equations, becomes

$$\begin{aligned} x_1 + 25x_3 &= 2{,}250{,}000 \\ x_2 - 2x_3 &= 150{,}000 \end{aligned}$$

or equivalently,

$$\begin{aligned} x_1 &= 2{,}250{,}000 - 25x_3 \\ x_2 &= 150{,}000 + 2x_3 \end{aligned}$$

The instructions for each of the computations have not been given, and this omission is deliberate since similar computations have already been made and should be familiar by now. As an exercise, the reader should write the instructions for each line of the computation above.

If the number of comforters sold, x_3, is the number k, then the *general solution* of the system is

$$(2{,}250{,}000 - 25k,\ 150{,}000 + 2k,\ k)$$

Any counting number (which is meaningful) can be substituted for k to give a *particular solution.* For example, if the company decides no longer to produce comforters, then the solution (2,225,000, 150,000, 0) makes possible the desired total gross profit and is consistent with both of the estimated profits. If it is decided that they can sell 5,000 comforters, then a particular solution satisfying both of the conditions imposed is

$$(2{,}125{,}000,\ 160{,}000,\ 5{,}000)$$

How is the number 2,125,000 derived? The number 160,000? Any counting number chosen as a replacement of x_3 will yield a particular solution of the system, but remember that $2{,}250{,}000 - 25x_3$ must be nonnegative. So what is the largest *meaningful* replacement for x_3? These conditions have as their graphs two planes which, in turn, have an intersection that is a line.

We promised one example in which the number of variables would exceed the number of equations, and this example has just been concluded. We shall now consider an example in which the number of equations exceeds the number of variables.

Example 2. The board of directors of CCC decides that they can estimate the gross profit in terms of the sale of sheets and blankets without considering the sale of comforters. They further decide that \$0.12 can be made on each sheet and \$0.60 on each blanket and that the desired profit is \$240,000. The equation that expresses this is

$$0.12x_1 + 0.60x_2 = 240{,}000$$

A second estimate can be stated in a similar fashion, and it is

$$0.08x_1 + 1.20x_2 = 240{,}000$$

Still a third estimate is expressed by the equation

$$0.18x_1 + 0.90x_2 = 300{,}000$$

The board wants to determine (if possible) the common solution to these three conditions. It has already been observed that the graph of these conditions is a line in a plane, and so it can be seen that the solution is the common point of the three lines (if it exists). Using the matrix techniques employed before, the solution is determined in what follows:

$$\begin{pmatrix} 0.12 & 0.60 & \vert & 240{,}000 \\ 0.08 & 1.20 & \vert & 240{,}000 \\ 0.18 & 0.90 & \vert & 300{,}000 \end{pmatrix}$$

$$\begin{pmatrix} 1 & 5 & \vert & 2{,}000{,}000 \\ 8 & 120 & \vert & 24{,}000{,}000 \\ 18 & 90 & \vert & 30{,}000{,}000 \end{pmatrix}$$

$$\begin{pmatrix} 1 & 5 & \vert & 2{,}000{,}000 \\ 0 & 80 & \vert & 8{,}000{,}000 \\ 0 & 0 & \vert & -6{,}000{,}000 \end{pmatrix}$$

The reader should supply reasons for the computations in the three steps. The third matrix written as a system of equations is

$$\begin{aligned} x_1 + 5x_2 &= 2{,}000{,}000 \\ 80x_2 &= 8{,}000{,}000 \\ 0x_1 + 0x_2 &= -6{,}000{,}000 \end{aligned}$$

From condition 3 of the final result, it is seen that regardless of the numbers chosen for x_1 and x_2, the left-hand side of the condition is 0 and the right-hand side of the condition is $-6{,}000{,}000$. So regardless of the number of sheets or blankets sold, the three conditions cannot be satisfied *simultaneously*.

That these three conditions have no simultaneous solution is also easily observed if the graph of each of the equations is considered. In Diagram 12, the graphs of the three conditions are represented, and it is

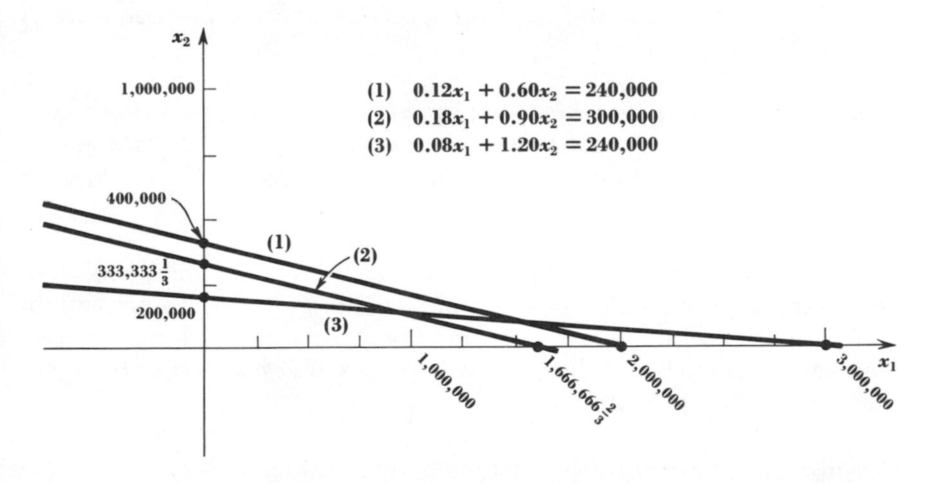

Diagram 12

clear from the graph that two of the lines are *parallel* (have no point in common). Hence, the three lines have no common point.

Only two examples have been given in this section. It should be clear to the reader, however, that the number of possible examples is strictly a function of the imagination and patience of the person who is making up the examples. There is an unlimited number of systems of such equations to be considered, but it should be comforting to know that such systems and their solutions have been analyzed and that a great deal is known about the circumstances under which solutions will exist. Although a detailed study of these systems will not be undertaken here, a list of properties that describe (at least partially) the situation is included in what follows.

To begin this general discussion, some terminology will be helpful. A general linear system is displayed below.

$$S: \quad \begin{array}{c} a_{11}x_1 + a_{12}x_2 + \cdots + a_{1n}x_n = b_1 \\ a_{21}x_1 + a_{22}x_2 + \cdots + a_{2n}x_n = b_2 \\ \cdots\cdots\cdots\cdots\cdots\cdots\cdots \\ a_{i1}x_1 + a_{i2}x_2 + \cdots + a_{in}x_n = b_i \\ \cdots\cdots\cdots\cdots\cdots\cdots\cdots \\ a_{m1}x_1 + a_{m2}x_2 + \cdots + a_{mn}x_n = b_n \end{array}$$

This is a *system of m equations in n variables*. The symbol a_{ij} is used to represent any one of the numbers which appear before the variables, and we have called these numbers *coefficients*. The symbol b_i is used to represent any of the numbers on the right-hand side of the equations, and we have called them the *constants* of the system.

There are some other expressions that should be given precise meaning, and the following definition does this for the expression *solution of a system of equations*.

Definition: **(*Solution of a system of equations*)** *The system S of m equations in n variables has a solution* $(a_1,a_2, \ldots ,a_n)$ *if and only if the ordered n-tuple* $(a_1,a_2, \ldots ,a_n)$ *satisfies each of the conditions in the system S.*

Now to determine a solution(s) of a system of equations, it is often necessary to change the form of the equations by performing certain arithmetical operations. These computations yield a different system (at least in appearance), but the two systems are *equivalent* in the sense that the following definition makes clear.

Definition: **(*Equivalent systems of equations*)** *If S and T are systems of equations, each of which involves the same number of equations* *and the same number of variables, and if S and T are such that every solution of S is also a solution of T and every solution of T is also a solution of S, then S and T are equivalent systems of equations.*

The next statement, offered without support now (but a supporting argument is given a little later), makes legitimate the operations that have been performed previously on systems of equations.

If S is a system of m equations in n variables and if each of the terms in the ith condition is multiplied by any nonzero real number a and substituted for the ith equation, then the resulting system of equations, T, is equivalent to S. Furthermore, if the ith equation in the system of equations S is added to any other equation in the system, then the resulting system, T, is equivalent to the system S.

These two operations are referred to as *row operations* on the system of equations.

Now that a vocabulary has been established, the next statement asserts that there is a method, already used extensively in this section, for determining the solution(s) of the system. An argument to support the statement follows the statement.

Any finite number of row operations applied to a system of conditions S yields an equivalent system of conditions.

An argument which shows that the two row operations when applied to a given system yield an equivalent system is easily given and follows.

Suppose that S is a system of linear conditions and that the ith equation is

$$c_{i1}x_1 + c_{i2}x_2 + \cdot \cdot \cdot + c_{in}x_n = b_i$$

Suppose, furthermore, that a is a nonzero number. If the ith equation is multiplied by a, the result is

$$ac_{i1}x_1 + ac_{i2}x_2 + \cdot \cdot \cdot + ac_{in}x_n = ab_i$$

Now let $X = (d_1, d_2, \ldots, d_n)$ be a solution of system S. It will be shown that X is also a solution of the system T (which is derived from S by multiplying the ith equation by the nonzero number a). It is clear that every equation in T is the same as the corresponding equation in S except the ith equation. Thus, X satisfies each of the equations of T with the possible exception of the ith equation. So it is sufficient to verify that the solution X satisfies the ith equation of the system T. However, upon replacing the variables x_i on the left-hand side of the equation with the corresponding numbers of the solution X of S assumed above, we have

$$ac_{i1}d_1 + ac_{i2}d + \cdot \cdot \cdot + ac_{in}d_n$$

By the distributive property of numbers, this can be written as

$$a(c_{i1}d_1 + c_{i2}d_2 + \cdots + c_{in}d_n)$$

But $X = (d_1,d_2, \ldots ,d_n)$ is a solution of S. So X satisfies the ith equation of S. That is, the equation

$$c_{i1}d_1 + c_{i2}d_2 + \cdots + c_{in}d_n = b_i$$

is satisfied. Therefore,

$$a(c_{i1}d_1 + c_{i2}d_2 + \cdots + c_{in}d_n) = ab_i$$

This verifies that $X = (d_1,d_2, \ldots ,d_n)$ is a solution of the system T.

It has been verified that any solution of S is also a solution of T. Now let $X = (e_1,e_2, \ldots ,e_n)$ be any solution of T for the purpose of establishing that it is also a solution of S. Since X satisfies all the equations in T, it will satisfy all the equations in S with the possible exception of the ith equation, since the two systems have precisely the same equations except for the ith equation. However, if $(e_1,e_2, \ldots ,e_n)$ is a solution of T, it follows that

$$ac_{i1}e_1 + ac_{i2}e_2 + \cdots + ac_{in}e_n = ab_i$$

But from the distributive property of numbers, it is also true that

$$a(c_{i1}e_1 + c_{i2}e_2 + \cdots + c_{in}e_n) = ab_i$$

Since $a \neq 0$, by the cancellation property of multiplication of real numbers, $(e_1,e_2, \ldots ,e_n)$ is a solution of the ith equation of S. (It is essential that a be nonzero, and that is why it was so stated on page 347.) This shows that the new system T does not have any solutions that are not solutions of S.

In summary, it has been verified that the operation of multiplying an equation of S by a nonzero number yields a system of equations T that is equivalent to S.

It will now be verified that adding one equation to a second equation gives an equivalent system of equations. Let

$$c_{i1}x_1 + c_{i2}x_2 + \cdots + c_{in}x_n = b_i$$

be the ith equation of S, and let

$$c_{j1}x_1 + c_{j2}x_2 + \cdots + c_{jn}x_n = b_j$$

be the jth equation of S. If the ith equation is added to the jth equation, the new equation is

$$(c_{i1} + c_{j1})x_1 + (c_{i2} + c_{j2})x_2 + \cdots + (c_{in} + c_{jn})x_n = b_i + b_j$$

The system T that is formed in this way will have every equation the same as in the original system except the jth equation. Now let

$$X = (d_1, d_2, \ . \ . \ . \ , d_n)$$

be a solution of S. Using the numbers of this solution as replacements for the variables in the left-hand side of the jth equation, we have

$$(c_{i1} + c_{j1})d_1 + (c_{i2} + c_{j2})d_2 + \cdots + (c_{in} + c_{jn})d_n$$

By the distributive, associative, and commutative properties of numbers, this can be written as

$$(c_{i1}d_1 + c_{i2}d_2 + \cdots + c_{in}d_n) + (c_{j1}d_1 + c_{j2}d_2 + \cdots + c_{jn}d_n)$$

But X is a solution of S, so in particular, X satisfies the ith and jth equations of T, that is,

$$(c_{i1}d_1 + c_{i2}d_2 + \cdots + c_{in}d_n) + (c_{j1}d_1 + c_{j2}d_2 + \cdots + c_{jn}d_n) = b_i + b_j$$

This verifies that any solution of the system S is also a solution of the system T.

Now let

$$X = (e_1, e_2, \ . \ . \ . \ , e_n)$$

be a solution of T. Since it satisfies every equation of T, it satisfies every equation of S with the possible exception of the jth equation. Since X is a solution of every equation of T,

$$(c_{i1} + c_{j1})e_1 + (c_{i2} + c_{j2})e_2 + \cdots + (c_{in} + c_{jn})e_n = b_i + b_j$$

This can be expressed as

$$(c_{i1}e_1 + c_{i2}e_2 + \cdots + c_{in}e_n) + (c_{j1}e_1 + c_{j2}e_2 + \cdots + c_{jn}e_n) = b_i + b_j$$

But X also satisfies the ith equation, so that

$$c_{i1}e_1 + c_{i2}e_2 + \cdots + c_{in}e_n = b_i$$

Upon subtracting the two equations above, the result is

$$c_{j1}e_1 + c_{j2}e_2 + \cdots + c_{jn}e_n = b_j$$

Hence $(e_1, e_2, \ . \ . \ . \ , e_n)$ is a solution of the jth condition of S. This verifies that any solution of T is also a solution of S.

It has now been verified that the operation of adding one equation to another equation in the system gives an equivalent system.

This concludes the discussion of linear conditions in this book, and it certainly is not a complete discussion. In actual practice, the number of equations is usually quite large, and the techniques that have been developed here are practical only when they are used to write a program for an electronic computer. That is why the reasons for the steps in the computations were referred to as *instructions*. Computers do almost instantly calculations that require large amounts of time if performed manually. The computations needed to solve systems of equations naturally lend themselves to computer programs because these are operations that such machines do well. A discussion of computer operations is not within the scope of this book, and the reader should refer to one of the numerous books which discuss this subject.

Exercises

1. For each of the systems below, determine (if possible) the solution set.

a.
$$\begin{aligned} 2x_1 - x_2 &= 2 \\ x_1 - x_2 + x_3 &= 2 \end{aligned}$$

b.
$$\begin{aligned} 2x_1 + x_2 - x_3 &= 2 \\ -4x_1 + 2x_2 + 2x_3 &= -4 \\ -6x_1 - 3x_2 + 3x_3 &= -6 \end{aligned}$$

c.
$$\begin{aligned} 2x_1 - x_2 + x_3 &= 3 \\ 2x_1 + 2x_3 &= 6 \\ x_2 + x_3 &= 3 \end{aligned}$$

d.
$$\begin{aligned} x_1 + x_2 - x_3 &= 2 \\ x_1 + 2x_2 - x_3 &= 6 \\ x_1 - x_3 &= 0 \end{aligned}$$

e.
$$\begin{aligned} x_1 - 2x_2 + 4x_3 &= -2 \\ -3x_1 + 6x_2 - 12x_3 &= 6 \\ 4x_1 - 8x_2 + 16x_3 &= 5 \end{aligned}$$

f.
$$\begin{aligned} 5x_1 + x_2 - x_3 &= -4 \\ x_1 - x_2 &= -6 \\ 8x_1 + x_2 - 2x_3 &= -9 \end{aligned}$$

g.
$$\begin{aligned} 2x_1 + 2x_2 + x_3 &= 5 \\ 3x_1 + 2x_2 + x_3 &= 4 \\ -2x_3 &= -1 \end{aligned}$$

h.
$$\begin{aligned} 2x_1 - x_2 + 3x_3 &= -3 \\ \tfrac{2}{3}x_1 - \tfrac{1}{3}x_2 + x_3 &= -1 \\ -4x_1 + 2x_2 - 6x_3 &= 6 \end{aligned}$$

i.
$$\begin{aligned} 3x_1 + x_2 + x_3 &= 12 \\ 6x_1 - x_2 + x_3 &= 9 \\ 3x_1 - 5x_2 - x_3 &= 14 \end{aligned}$$

2. The owners of a chain of three grocery stores are looking at sales reports which seem to them to be contradictory. Report 1 says that the total sales from all three stores in a typical hour of business are $1,000. Report 2 says that during a typical business hour, the total sales of stores *A* and *B* exceed the sales of store *C* by $600. Report 3 says that during a typical business hour, twice the sales of store *A* added to twice the sales of store *B* exceed the sales of store *C* by $1,800.
   a. Are these three reports consistent? That is, is there a common solution to all three reports?
   b. Is this solution unique?

3. Suppose that the owners of the chain of grocery stores (see Exercise 2) discover that report 3 was typed incorrectly and that it should have read that during a typical business hour, twice the sales of store *A* added to the sales of store *B* exceed the sales of store *C* by $1,600.
   a. Now are the three reports consistent?
   b. Is their solution unique?
   c. What is the solution?

4. Suppose that report 3 has been found to be incorrect once again (refer to Exercise 2) and that it should read that during a typical business hour, twice

the sales of store A added to twice the sales of store B total \$1,400. It has been determined that reports 1 and 2 are correct.

a. Are the three reports consistent?
b. What is a common solution?

Answer to problem

The equation for the cost is

$$20x_1 + 80x_2 + 140x_3 = 3{,}000$$

and the equation for the pounds to be used is

$$x_1 + x_2 + x_3 = 100$$

The general solution is

$$(^{250}/_3 + x_3,\ ^{50}/_3 - 2x_3,\ x_3)$$

with the restrictions that $0 \leq x_3 \leq {}^{25}/_3$. The answer is *yes*, in many ways.

5.6 *Linear programming: the maximum problem*

Problems

A. A company wants to make a combination of two products, A and B. The profit contribution is \$8 for each unit of A and \$10 for each unit of B. Assume that processing time and assembly time both restrict the number of units which can be produced in the following way:

$$4A + 3B \leq 130 \text{ (processing)}$$
$$5A + 2B \leq 110 \text{ (assembly)}$$

1. How many of A could be processed if no B were made?
2. How many of B could be processed if no A were made?
3. How many of A can be assembled if no B are made?
4. How many of B can be assembled if no A are made?
5. Plot your answers and find the combination of A and B which would completely use all processing and assembly time.
6. What is the maximum profit which can be made by a combination of A and B?

B. A manufacturing company has 360 hr of processing time and 420 units of raw material.

1. What combination of products A and B would maximize gross revenue, assuming:
   a. Each product sells for \$200?
   b. There is enough processing time to make 36 A's if no B's are made and 24 B's if no A's are made?
   c. There are enough raw materials to make 21 A's if no B's are made and 42 B's if no A's are made?

Summary

	Processing required	*Raw material required*	*Selling price*
Product A	10 hr	20 units	$200
Product B	15 hr	10 units	$200
Total available resources	360 hr	420 units	

2. Assuming that processing costs are $3 per hr and that raw materials cost $1 per unit, what combination of the two products would result in the greatest contribution to overhead and income?
3. Which objective, maximizing gross revenue or maximizing contribution to income, seems most logical to you?

The final two sections of this book touch lightly upon another large branch of mathematics—*linear programming*. The division of the treatment of this subject into two parts is based on the fact that there are two problems, *maximization problems* and *minimization problems*. This section deals with maximizing a linear objective function. A great deal of vocabulary has grown up in the field of linear programming around some rather simple mathematical ideas, but the vocabulary will be introduced as seems appropriate, in keeping with the theme of precision of language that has been adopted in this book. The vocabulary will also enable the reader to pursue the subject to greater depth if it is so desired.

Many compromises have been made in the selection of topics for these two sections. First and foremost, some compromise had to be made with regard to the number of variables involved in the problems to be solved. Because only the *graphical method* of solution is to be introduced here, the examples and exercises will contain only two variables. Second, the coefficients and solutions of the *inequalities* studied in the mathematical systems will involve, where possible, only counting numbers. This requires some judicious choices in the statement of the exercises and examples. In short, care has been taken so that the reader will not be distracted from the most important aspects of the objectives of the sections: the types of problems that lead to mathematical systems of inequalities and the techniques involved in the solution of these systems.

One other preliminary comment is that, while previous sections of Chap. 5 have dealt with systems of equations, in the final two sections we shall be concerned with *inequalities*. As usual, we proceed from examples to the general case, so read the next example to see what is meant by some of the technical terms introduced in these introductory paragraphs.

Example 1. The Ready Book Company manufactures both hardback and paperback books. They have two assembly lines for the books. On one line, it takes 3 min for each hardback book to be produced and 1 min for each paperback book. On the second line, a hardback book and a paperback book require 1 min each. The first production line works 40 hr (2,400 min) each week, but the second line works only $26\frac{2}{3}$ hr (1,600 min) each week. The profit expected on a hardback book is 5 cents and on a paperback book 1 cent. Assuming that the company can sell all of each type of book that it produces, how many books of each kind should be printed to make the maximum profit?

First, the real-life problem should be converted to mathematical notation. Two numbers must be determined: the number of hardback books, which we denote by x_1, and the number of paperback books, which we denote by x_2. Observe that a nonnegative number of each type is to be produced, so the first *constraints* imposed are

$$(1) \quad x_1 \geq 0$$

and

$$(2) \quad x_2 \geq 0$$

The symbol $x_1 \geq 0$ is read "x_1 is greater than 0" or "x_1 is equal to 0." In other words, the symbol $\geq$ is mathematical shorthand for the disjunction of two statements. Similar remarks apply for the second constraint.

Second, we need some expressions that reflect the production times in terms of the time available. For the first assembly line, the total time for hardback books is $3x_1$ min; and for paperback books, it is x_2 min. The sum of these two amounts of times cannot exceed 2,400 min per week. So the third condition is

$$(3) \quad 3x_1 + x_2 \leq 2{,}400$$

This is read "The sum of the time for hardback books and paperback books is less than or equal to 2,400 min." For the second assembly line, the condition is

$$(4) \quad x_1 + x_2 \leq 1{,}600$$

In the language of linear programming, inequalities (3) and (4) are called *structural constraints*. Be sure you understand how to read these expressions and how they are derived.

Third, the profit expected is the sum of the profit on hardback books,

$5x_1$, and the profit on paperback books, x_2. This sum is

$$(5) \quad 5x_1 + x_2$$

It is the profit expressed by (5) that is to be maximized. This expression is sometimes called an *objective function.*

In summary, the mathematical statement of the problem is given by the expressions

$$\begin{aligned}
&(1) \quad x_1 \geq 0 \\
&(2) \quad x_2 \geq 0 \\
&(3) \quad 3x_1 + x_2 \leq 2{,}400 \\
&(4) \quad x_1 + x_2 \leq 1{,}600 \\
&(5) \quad 5x_1 + x_2 \text{ is to be a maximum}
\end{aligned}$$

The method of solution will be graphic; that is, we shall use a graph to illustrate the procedure to find the desired numbers.

From what was explained in Sec. 5.2, it should be clear that the inequality (1) imposes the condition that the solution must be such that x_1 is nonnegative. See Diagram 13 to verify that this means that all solutions must be in the first or fourth quadrants (to the right of the x_2 axis). Diagram 13 shows a graphical display of this permissible set of solutions.

From inequality (2) we conclude that all solutions for x_2 must be nonnegative and hence in the first or second quadrant. See Diagram 14, which is shaded in the area that represents this permissible set of solutions.

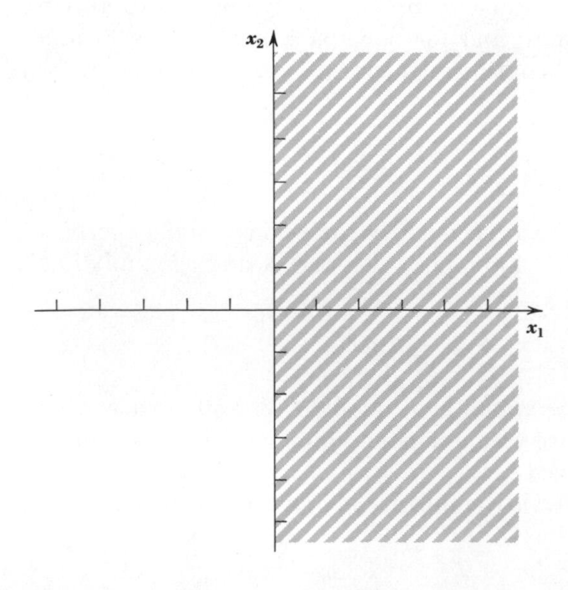

Diagram 13

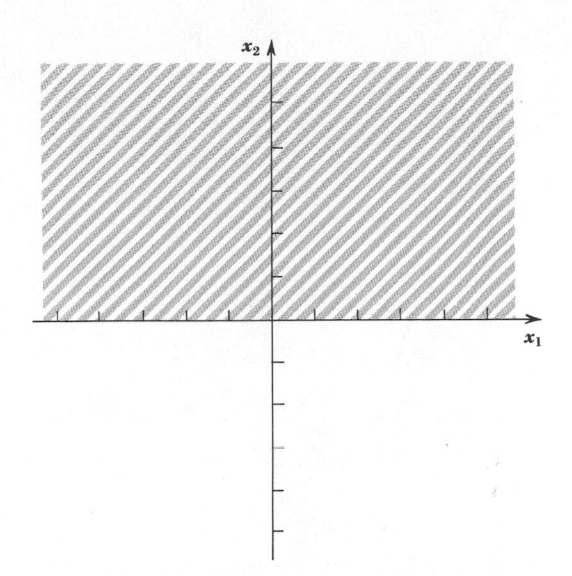

Diagram 14

Since conditions (1) *and* (2) must both be satisfied, the solution must be in the intersection of these two sets, which is the first quadrant. Recall that the word *and* corresponds to conjunction and the intersection of sets (Sec. 1.3). See Diagram 15 for a graphic display of the intersection of the sets. The common set of points is crosshatched in the diagram.

This situation is familiar to you. We have already seen in earlier

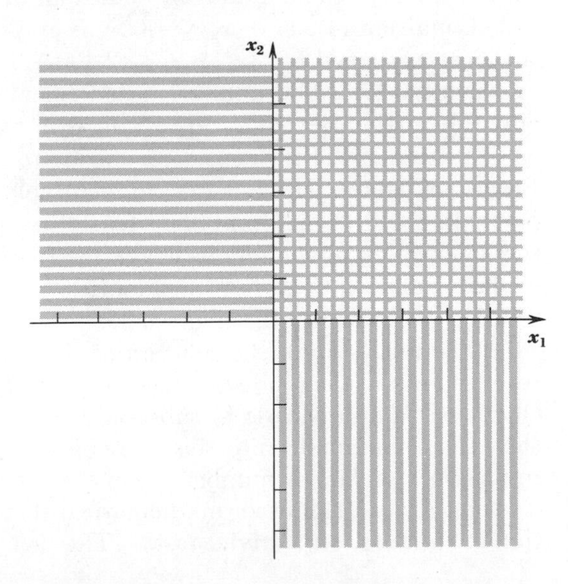

Diagram 15

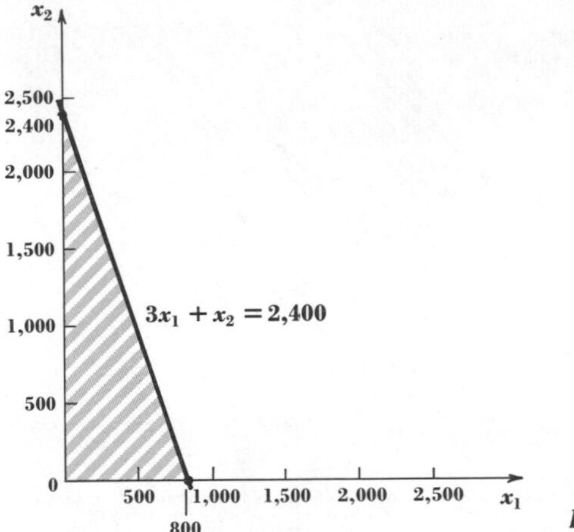

Diagram 16

work with systems of equations that often the feasible (meaningful) solutions to real-life problems are only those in the first quadrant.

The third condition is a disjunction of $3x_1 + x_2 = 2{,}400$ or $3x_1 + x_2 < 2{,}400$. Consider $3x_1 + x_2 = 2{,}400$. The intercepts are (800, 0) and (0, 2,400). These two points determine the line (Sec. 5.2), but the interest here is restricted, for the reason given above, to that portion of the line in the first quadrant. The other part of the disjunction, $3x_1 + x_2 < 2{,}400$, has as its graph those points "under" the line $3x_1 + x_2 = 2{,}400$. The union of the sets of points on the line and under the line is shown in Diagram 16. The unit of measurement chosen is 500.

Condition (4), $x_1 + x_2 \leq 1{,}600$, is graphed in a fashion exactly like that just explained for $3x_1 + x_2 \leq 2{,}400$. The graph of (4) is displayed in Diagram 17. The unit of measurement is again 500. Also again, only the portion in the first quadrant is in the feasibility area.

A summary of the considerations up to this point will now be made. From inequalities (1) and (2) we were able to deduce that only those points in the first quadrant were feasible solutions. As a result of this, the intersection of the sets in Diagrams 16 and 17 is the set which is the common solution to all four inequalities. The summary graph, with the solution set, is shown in Diagram 18.

The final step in the solution of the problem is to choose from the set of points in the shaded area of Diagram 18 those which maximize (5). One possible approach is to substitute each ordered pair whose graph is shown in Diagram 18 into the expression $5x_1 + x_2$ and determine which results in the larger number. This is, of course, an impossible task! Fortunately, we can use a mathematical theorem which makes the selection of the solution a trivial task. The fact which will be used, but not

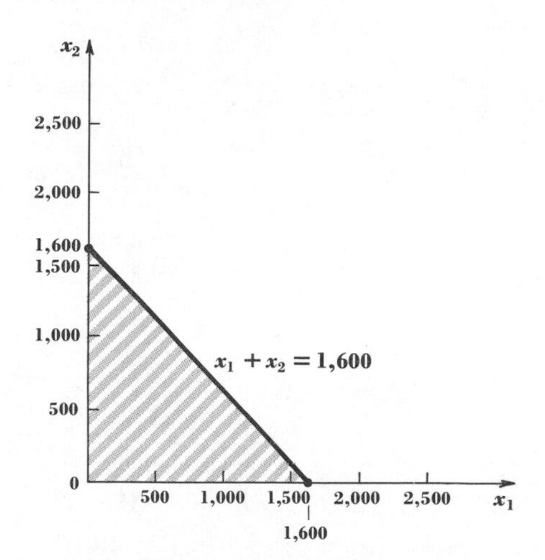

Diagram 17

proved here, is stated next. Some liberties are taken in its statement since the concern here is with techniques rather than with theory.

If there is a unique ordered pair, (x_1,x_2), which maximizes (minimizes) a linear expression of the form $Ax_1 + Bx_2$, then (x_1,x_2) is a vertex (corner) point of the polygon of permissible solutions.

The corner points of the permissible solutions are (0,0), (800, 0), (400, 1,200), (0, 1,600). These are all read from the graph except (400, 1,200). We have already shown in Sec. 5.2 how to determine the common solution of two linear equations. Nevertheless, the computation follows to remind you how to determine the coordinates of this point

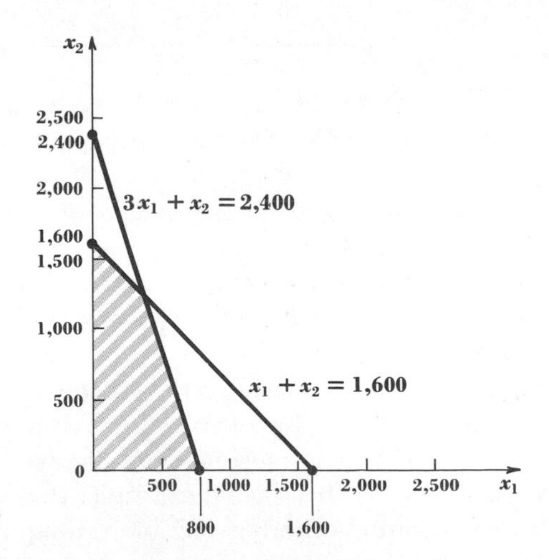

Diagram 18

of intersection. The reader should supply the reasons for the row operations on the matrices.

$$3x_1 + x_2 = 2{,}400$$
$$x_1 + x_2 = 1{,}600$$
$$\left(\begin{array}{cc|c} 3 & 1 & 2{,}400 \\ 1 & 1 & 1{,}600 \end{array}\right)$$
$$\left(\begin{array}{cc|c} 1 & \frac{1}{3} & 800 \\ 1 & 1 & 1{,}600 \end{array}\right)$$
$$\left(\begin{array}{cc|c} 1 & \frac{1}{3} & 800 \\ 0 & \frac{2}{3} & 800 \end{array}\right)$$
$$\left(\begin{array}{cc|c} 1 & \frac{1}{3} & 800 \\ 0 & 1 & 1{,}200 \end{array}\right)$$
$$\left(\begin{array}{cc|c} 1 & 0 & 400 \\ 0 & 1 & 1{,}200 \end{array}\right)$$
$$x_1 = 400 \qquad x_2 = 1{,}200$$

Now the final computations can be made. In Table 1, the first column shows the corner points and the second column is the number obtained when the corner points are used as replacements for the variables in (5).

Table 1

Corner points	*Objective function*
(0,0)	$5 \cdot 0 + 1 \cdot 0 = 0$
(800, 0)	$5 \cdot 800 + 1 \cdot 0 = 4{,}000$
(400, 1,200)	$5 \cdot 400 + 1 \cdot 1{,}200 = 3{,}200$
(0, 1,600)	$5 \cdot 0 + 1 \cdot 1{,}600 = 1{,}600$

The table shows that the maximum profit is $4,000 and that it is achieved by producing 800 hardback books and no paperback books.

Before the second example is offered, it is in order to remark that all the major ideas about linear programming have already been presented. Physical situations lead to systems of inequalities which, like systems of equalities, can under "good" circumstances be solved geometrically. Some help was needed from previous sections; intercepts, graphs, and intersections of sets were a few of these. It is true that some vocabulary was introduced but hopefully in such a way that the meaning was clear.

We now give a second example, which differs but little from the previous one. In this second example, there are still only two variables, but for this example there is one more inequality.

Example 2. On the basis of past records, the CCC knows that it can sell no more than 36,000 sheets, 10,000 blankets, and 20,000 comforters per week. It has two production lines, and it knows the production capacity of each of these two lines. The first assembly line can produce 5,000 sheets, 1,000 blankets, and 3,000 comforters per workday, and the second production line can produce 3,000 sheets, 1,000 blankets, and 1,000 comforters per workday. The directors decide that each week they will manufacture an amount of each product which is less than or equal to the amount that they know they can sell per week. In other words, they decide they will not overproduce and have to store merchandise in a warehouse. How many days per week should each of the two production lines work on manufacturing the three products?

Let x_1 and x_2 represent the number of days per week that assembly lines 1 and 2, respectively, are to work. The inequalities that express the productions of the products are

$$
\begin{aligned}
&(1) \quad x_1 \geq 0 \\
&(2) \quad x_2 \geq 0 \\
&(3) \quad 5{,}000x_1 + 3{,}000x_2 \leq 36{,}000 \\
&(4) \quad 1{,}000x_1 + 1{,}000x_2 \leq 10{,}000 \\
&(5) \quad 3{,}000x_1 + 1{,}000x_2 \leq 20{,}000
\end{aligned}
$$

Inequalities (1) and (2) must be satisfied because the number of days to be worked is nonnegative. Condition (3) states that the sum of the number of sheets produced on line 1, $5{,}000x_1$, and the number of sheets produced on line 2, $3{,}000x_2$, is to be less than or equal to the maximum possible sales per week, 36,000. Inequality (4) states a similar fact about the production of blankets. The sum of blankets produced on line 1, $1{,}000x_1$, and on line 2, $1{,}000x_2$, must be less than or equal to 10,000. Condition (5) is arrived at in a similar fashion. In the language introduced earlier, inequalities (3) to (5) are structural constraints.

There is one other consideration of prime importance: The CCC wishes to maximize its profit. This brings us to the *objective function.* Suppose that the CCC can make $0.09 on each sheet sold, $0.80 on each blanket sold, and $1.10 on each comforter sold. The total profit is obtained by adding the profit on each of these products. The number of sheets produced is

$$5{,}000x_1 + 3{,}000x_2$$

so the profit on sheets is

$$(5{,}000x_1 + 3{,}000x_2) \cdot 0.09$$

Similarly,

$$(1{,}000x_1 + 1{,}000x_2) \cdot 0.80$$

and

$$(3{,}000x_1 + 1{,}000x_2) \cdot 1.10$$

are the profits on blankets and comforters. The total profit is given by each of the next four expressions, which appear in simpler and simpler form, but it is the last expression that will be used as the profit function to be maximized.

$$\begin{aligned}(5{,}000x_1 + 3{,}000x_2) \cdot 0.09 &+ (1{,}000x_1 + 1{,}000x_2) \cdot 0.80 \\ &+ (3{,}000x_1 + 1{,}000x_2) \cdot 1.10 \\ &= (450x_1 + 270x_2) + (800x_1 + 800x_2) + (3{,}300x_1 + 1{,}100x_2) \\ &= (450 + 800 + 3{,}300)x_1 + (270 + 800 + 1{,}100)x_2 \\ &= 4{,}550x_1 + 2{,}170x_2\end{aligned}$$

The profit objective function to be maximized is

$$(6) \quad 4{,}550x_1 + 2{,}170x_2$$

See if it is clear how each of the simplifications above is achieved by use of some of the properties of numbers. The area which is shaded in Diagram 19 illustrates the solutions which will satisfy the first five inequalities that describe this problem.

Since a fact about maximizing a profit function was accepted (without proof) in the first example, the corner points of the solution set illustrated in Diagram 19 are crucial. But the methods of solution for determining these points was established in Sec. 5.2. One of the solutions will be computed in detail here; the others are left for the reader. Two of the lines are

$$(3) \quad 5{,}000x_1 + 3{,}000x_2 = 36{,}000$$

and

$$(4) \quad 1{,}000x_1 + 1{,}000x_2 = 10{,}000$$

Their intersection is arrived at by the following matrix computation:

$$\left(\begin{array}{cc|c} 5{,}000 & 3{,}000 & 36{,}000 \\ 1{,}000 & 1{,}000 & 10{,}000 \end{array}\right)$$

$$\left(\begin{array}{cc|c} 1 & 3/5 & 36/5 \\ 1 & 1 & 10 \end{array}\right)$$

$$\left(\begin{array}{cc|c} 1 & 3/5 & 36/5 \\ 0 & 2/5 & 14/5 \end{array}\right)$$

$$\left(\begin{array}{cc|c} 1 & 3/5 & 36/5 \\ 0 & 1 & 7 \end{array}\right)$$

$$\left(\begin{array}{cc|c} 1 & 0 & 3 \\ 0 & 1 & 7 \end{array}\right)$$

$$x_1 = 3 \qquad x_2 = 7$$

The corner points are (0,0), (0,10), (3,7), (6,2), and (20/3,0). We repeat that the profit function to be maximized is

$$4{,}550x_1 + 2{,}170x_2$$

In Table 2, the corner points are in the first column and the numbers

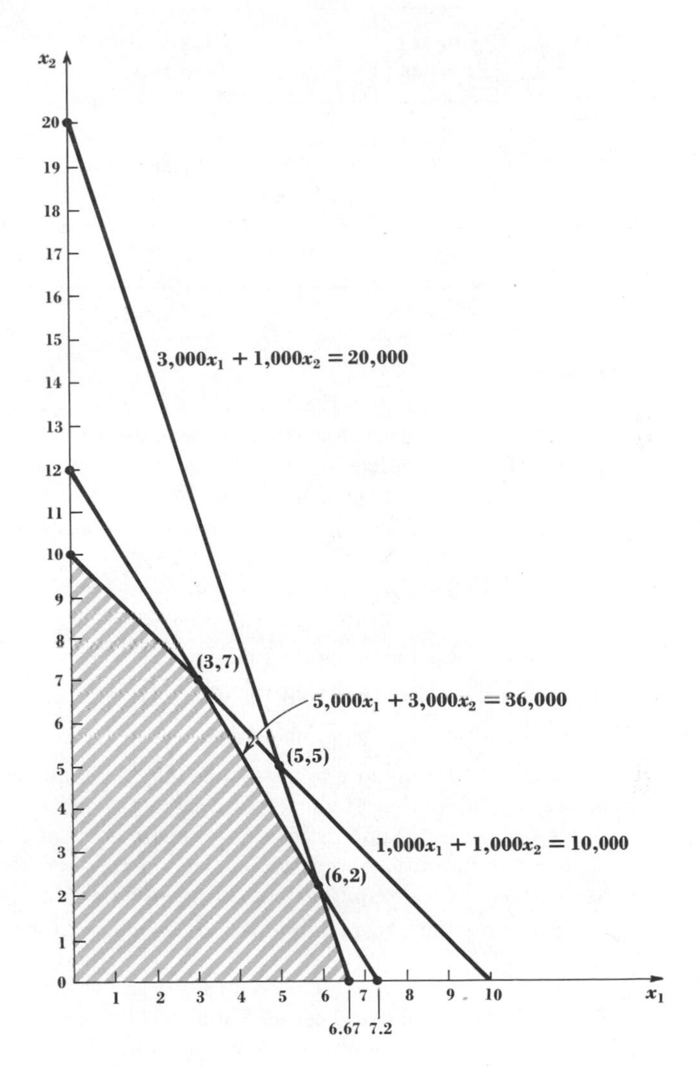

Diagram 19

obtained by the replacement of the coordinates of the corner points are in the second column.

The maximum amount of profit which can be obtained is $31,640 per week. It is desirable for CCC to choose six and two (6,2) as the number of days to be worked by assembly lines 1 and 2, respectively.

Two examples of linear programming which require a maximization have been given. The graphic method was used in the solution. The exercises that follow are patterned after these examples.

Table 2

Corner points	*Objective function*
(0,0)	$4{,}550 \cdot 0 + 2{,}170 \cdot 0 = 0$
(0,10)	$4{,}550 \cdot 0 + 2{,}170 \cdot 10 = 21{,}700$
(3,7)	$4{,}550 \cdot 3 + 2{,}170 \cdot 7 = 28{,}840$
(6,2)	$4{,}550 \cdot 6 + 2{,}170 \cdot 2 = 31{,}640$
$(6\frac{2}{3},0)$	$4{,}550 \cdot 6\frac{2}{3} + 2{,}170 \cdot 0 = 30{,}333$

Exercises

1. In each of the following, a system of inequalities and a profit function are given. Find the ordered pair such that the profit is maximized. For all these, assume $x_1 \geq 0$ and $x_2 \geq 0$ and that the profit function is $x_1 + 2x_2$.

   a. $3x_1 + 2x_2 = 5$
   $x_1 + 3x_2 = 4$

   b. $2x_1 + 4x_2 = 16$
   $x_1 + x_2 = 6$

   c. $3x_1 + 15x_2 = 18$
   $x_1 + 5x_2 = 6$

   d. $2x_1 + 6x_2 = 8$
   $x_1 + 3x_2 = 5$

2. Rework Exercise 1 for the profit function $2x_1 + x_2$.
3. Rework Exercise 1 for the profit function $10x_1 + x_2$.
4. Rework Exercise 1 for the profit function $x_1 + 10x_2$.
5. Assume that your company wants to schedule for manufacture product K or L or some combination of the two to make the most profitable use of the firm's resources. You are asked to prepare a set of equations which will express the limitations imposed on the company's decision, given the following restrictions, requirements, and production data:

   a. Sales restriction: Maximum units which can be sold are $K = 80$, $L = 70$.

   b. Selling price: $K = \$140$, $L = \$165$.

   c. Raw materials: Costs per unit of finished product are $K = \$6$, $L = \$4$. Adequate supplies of raw materials are readily available.

   d. Shipping facilities: Can handle up to 500 finished units.

   e. Capital: Adequate for all possible sales.

   f. Sales commitments: 10 units of K already contracted for.

   g. Processing time available: Dept. A, 300 hr; Dept. B, 450 hr; Dept. C, 90 hr.

   h. Processing time required: For product K, 3 hr in A, 5 hr in B, and 1.2 hr in C; for product L, 5 hr in A, 9 hr in B, and 1.5 hr in C.

   i. Processing costs: Dept. A, \$3 per hr; Dept. B, \$5 per hr; Dept. C, \$10 per hr.

   j. The real-life restriction: There cannot be a negative number of either product produced.

k. Profit per unit produced and sold:

	Product *K*		Product *L*	
Selling price per unit		\$140.00		\$165.00
Cost per unit:				
Raw materials	\$ 6.00		\$ 4.00	
Processing—*A*	9.00		15.00	
B	25.00		45.00	
C	12.00		15.00	
Total variable cost per unit		\$ 52.00		\$ 79.00
Contribution toward overhead and income		\$ 88.00		\$ 86.00

What is the most profitable combination of the two products?

6. A farmer in the Rio Grande Valley raises cotton. He also raises geese because they will eat the weeds between the cotton plants but not the cotton itself. He has his own farm, which will grow 1¼ bales of cotton to the acre. At the same time, he can feed 75 geese per acre. He also works a nearby farm, which will raise only ¾ bale of cotton per acre, but owing to the sandy soil, there are more weeds, so that 110 geese can be raised per acre. The government regulates cotton production, and his total cotton yield cannot exceed 60 bales. He is also limited by the number of goose eggs available to hatch, and he estimates that he can hatch at the most 4,950 goslings. He anticipates a net profit of \$75 per acre on the first farm and \$60 per acre on the second. How many acres should he plant on each farm to maximize his profit?

Answers to problems

A. 1. 32.5 *A*'s
2. 43⅓ *B*'s
3. 22 *A*'s
4. 55 *B*'s
5. 10 *A*'s and 30 *B*'s
6. 380

B. 1. 13 *A*'s and 15 *B*'s
2. All *B*, $24[200 - (3 \cdot 15) - (10 \cdot 1)] = 3{,}480$
All *A*, $21[200 - (3 \cdot 10) - (20 \cdot 1)] = 3{,}150$
13 *A*'s and 15 *B*'s $(13 \cdot 150) + (15 \cdot 145) = 4{,}125$. This is the best.
3. Maximizing contribution to income

5.7 Linear programming: the minimum problem

Problems

A. A small factory receives a contract for 80 receivers and 90 amplifiers. This factory has two men who can work in the section to assemble these items, Mr. Adams and Mr. Smith. Mr. Adams can assemble 10 receivers and 5 amplifiers in an hour, while Mr. Smith can assemble 5 receivers and 15 amplifiers in an hour. Mr. Adams works for $2 per hr, and Mr. Smith works for $3 per hr. Quite naturally, the factory wants to get the most work for the least expenditure. It is not required that Mr. Adams work the same number of hours on this project as Mr. Smith. What is the minimum labor expenditure that the factory will have to make to get these items assembled? How many hours will each man work?

B. Assume that your company makes a product which must meet certain rigid specifications with regard to weight, vitamin content, and mineral content. The product can be made by combining two different raw materials, as summarized in the table.

Raw material	*Units of vitamin*	*Units of mineral*	*Weight per bushel*	*Cost per bushel*
A	5	2	40 lb	$1.00
B	4	5	20 lb	$1.50

1. Assume that the finished product must contain at least 16 units of the vitamin and at least 12 units of the mineral. It must not contain more than 25 units of the vitamin, and it must weigh at least 100 lb. What combination of the two ingredients would produce the required amount of the product at the least cost?
2. What would have been the answer to part 1 if the cost of ingredient *B* had been $3 per bushel instead of $1.50?

The examples of this section closely parallel those of Sec. 5.6. The only important change is one of emphasis—an objective function is to be minimized rather than maximized. Just as business firms try to maximize profits, they also attempt to minimize costs. Therefore, the examples are chosen to be representative of that class of problem.

Example 1. Mr. Gate is the head of the shipping department of Acme Clothing Company. He has been told that for the next few months, his department must prepare for shipping, and ship, at least 1,600 cartons of socks and 4,900 cartons of shirts per month. Mr. Gate has two work crews that he can use for preparing the cartons for shipping, and he knows that it costs $60 per day to use the first work crew and $80 per day to use the second work crew. He also knows (on the basis of past performances) that the first crew can prepare and ship 200 cartons of socks and 500 cartons of shirts per day, while the second crew can package and ship

100 cartons of socks and 700 cartons of shirts per day. Naturally, Mr. Gate wishes to *minimize costs* connected with the shipping department because, for one reason, his department is partially judged on this factor. How many days of work should he assign to each crew in order to meet his minimum requirements of shipping and at the same time keep the cost at the lowest possible amount?

This question can be answered by the determination of two variables. The first variable is the number of days per month that the first crew is to be assigned to this task, and x_1 will be used throughout to represent the first unknown. The other variable is the number of days per month that the second crew should be assigned to this task, and x_2 will be used to represent the second variable. Now that names have been assigned to the two variables which are to be determined, the problem can be stated by a sequence of five conditions:

(1) $200x_1 + 100x_2 \geq 1{,}600$
(2) $500x_1 + 700x_2 \geq 4{,}900$
(3) $x_1 \geq 0$
(4) $x_2 \geq 0$
(5) $60x_1 + 80x_2$ is to be a minimum

An explanation will be given to show how each of these expressions was determined.

The first crew can package and ship 200 cartons of socks per workday, so if they work x_1 days per month, they will prepare $200x_1$ cartons of socks. Similarly, if the second crew works x_2 days, they will ship $100x_2$ cartons of socks. The sum of these two quantities must be equal to or greater than the minimum amount of socks that Mr. Gate's department is to ship each month, which is 1,600 cartons. Hence the inequality (1).

The second inequality is arrived at by similar reasoning. The expressions $500x_1$ and $700x_2$ represent the number of cartons of shirts prepared and shipped each month by the first and second crews, respectively. The sum of the amounts that each can ship must meet or exceed the number of cartons of shirts in the quota given to Mr. Gate. Hence, the inequality (2).

Other conditions on the variables x_1 and x_2 are implied in the statements that describe the problem. It must be assumed (but it was not explicitly stated) that the *least number* of days that either crew will labor *is zero*. They cannot, of course, work a negative number of days. Hence inequalities (3) and (4).

Mr. Gate's problem is to minimize the amount of money required to prepare and ship the quotas that his department has been assigned. Since it costs \$60 per workday for crew 1 and \$80 per workday for crew 2, the total amount that it will cost for the two crews is $60x_1 + 80x_2$. It is this quantity which Mr. Gate wishes to keep to a minimum.

On the basis of his experience, Mr. Gate feels that he can keep this cost below \$480 and meet his quota. This estimate is questioned by his supervisor, who feels that it will cost at least twice that much. Which (if either) of the two men is correct?

The reader will note the strong similarity between this example and the two examples considered in Sec. 5.6. The number of variables in this example is again two, so that the graphic method of solution can be used.

In the language of set notation, inequalities (1) to (4) can be combined into two expressions, which are

(6) $S_1 = \{(x_1,x_2)|200x_1 + 100x_2 \geq 1{,}600 \text{ and } x_1 \geq 0 \text{ and } x_2 \geq 0\}$
(7) $S_2 = \{(x_1,x_2)|500x_1 + 700x_2 \geq 4{,}900 \text{ and } x_1 \geq 0 \text{ and } x_2 \geq 0\}$

The intersection of these two sets, $S_1 \cap S_2$, must be determined, and the element (elements) chosen such that $60x_1 + 80x_2$ is a minimum. The graph of $200x_1 + 100x_2 \geq 1{,}600$ is the union of the sets of points such that $200x_1 + 100x_2 = 1{,}600$ or $200x_1 + 100x_2 > 1{,}600$. Indeed the symbol $\geq$ means "either $>$ or $=$." The first of these sets is a line in the plane, and it has intercepts (8,0) and (0,16). The second set is those points above (greater than) the line. Note that *less than* means "below" and *greater than* means "above" the line. The union (*or*) of the two sets is graphed in Diagram 20.

In a similar fashion, the points that correspond to the ordered pairs (x_1,x_2) that satisfy $500x_1 + 700x_2 = 4{,}900$ are on the line $500x_1 + 700x_2 = 4{,}900$ or are above the line. The intercepts are

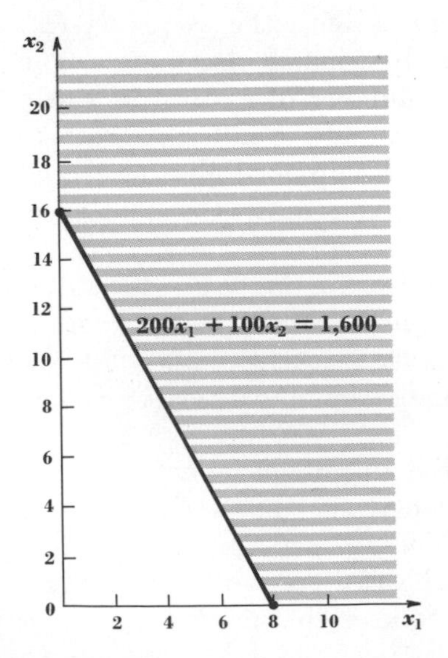

Diagram 20

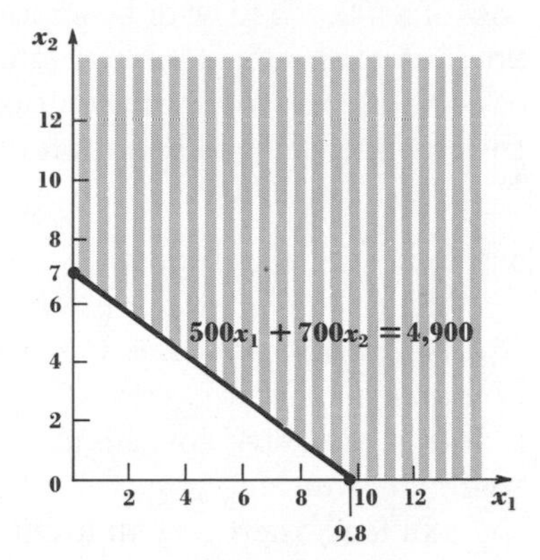

Diagram 21

($^{49}\!/_5$,0) and (0,7). The set of points on the line unioned with those above it are shown in Diagram 21.

Each of the sets S_1 and S_2 has been illustrated, but on separate coordinate systems. Since $S_1 \cap S_2$ is the solution set of (6) and (7), one coordinate system is used in Diagram 22, and $S_1 \cap S_2$ is crosshatched. Observe that there are points in S_1 not in S_2 (these are indicated with horizontal lines) and points in S_2 not in S_1 (these are indicated with vertical lines). The solution set of (6) and (7) has been illustrated, but the problem has not been solved. From this solution set is to be chosen those ordered pairs (x_1,x_2) such that the numbers represented by $60x_1 + 80x_2$ is a minimum. This question will be reconsidered shortly.

Mr. Gate has estimated that this work can be done each month at a

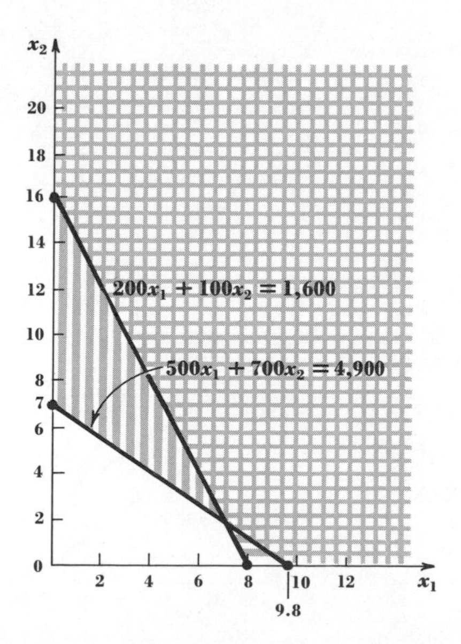

Diagram 22

cost of \$480. This can be expressed by the condition $60x_1 + 80x_2 = 480$. So it is the intersection of this linear condition with the intersection of $S_1 \cap S_2$ that will yield solutions which satisfy both (6) and (7) and keep the cost at \$480. The graph of $S_1 \cap S_2$ and the condition

$$60x_1 + 80x_2 = 480$$

are shown in Diagram 23.

Since the line $60x_1 + 80x_2 = 480$ does not intersect the set of points $S_1 \cap S_2$, it follows that it is not possible to do the work with any combination of the two crews for \$480 per month. However, if \$960 per month is allocated for the job, then the graph in Diagram 24 shows that there are solutions.

In fact, there are an infinite number of solutions to the two inequalities (6) and (7) and the condition $60x_1 + 80x_2 = 960$. So it is possible to do the job for \$960 per month, and furthermore, there are an infinite number of different ways that the two crews can be chosen to do the job. It is likely that in practice most of these solutions would be rejected

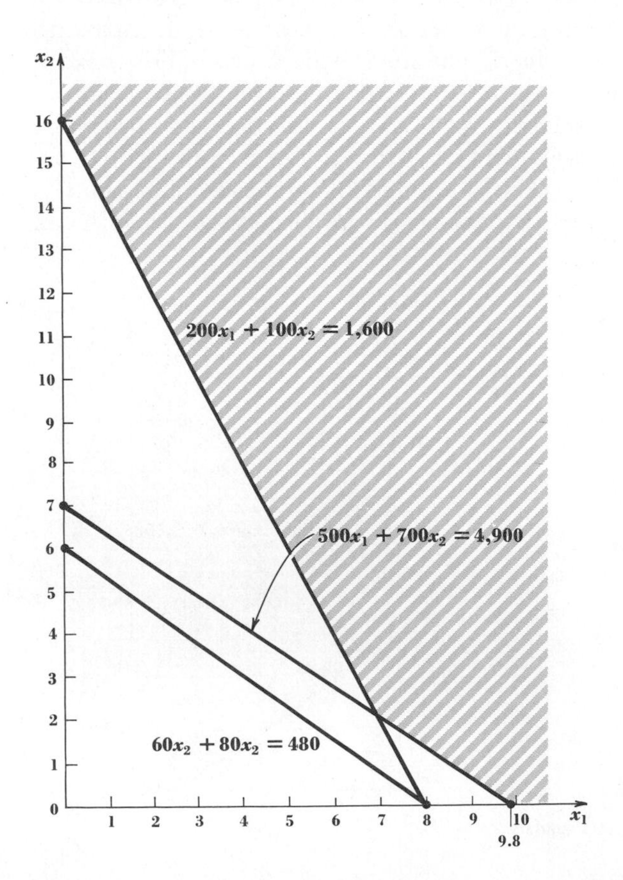

Diagram 23

since the crews would be used only full days and hence only *counting-number solutions are acceptable.* Particular solutions that satisfy (1) to (4) and $60x_1 + 80x_2 = 960$ are (8,6), (12,3), (6,7½). Each of these solutions will meet (in fact, exceed) the minimum shipping quotas, but each will cost \$960 per month. To fully understand these solutions, they should be used as replacements in (1) and (2) to see how many cartons are shipped.

It may be possible to meet minimum shipping quotas at a lower cost, but this has not yet been determined. The point of intersection of the two structural constraints on the graph is the point corresponding to (7,2). When these values for (x_1,x_2) are substituted into $60x_1 + 80x_2$, it becomes

$$(60 \cdot 7) + (80 \cdot 2) = 420 + 160 = 580$$

It can be shown that the replacement of (x_1,x_2) by any point in the shaded area representing the intersection of S_1 and S_2 into the expression $60x_1 + 80x_2$ will yield a number greater than (or equal to) the amount that has just been determined. This is not an accident; the minimum amount for $60x_1 + 80x_2$ is achieved by substitution of the corner points of the *convex set* of the solution points. Although it was not proved (in this book), it has been accepted as a fact that the minimum and maximum values of the objective function are obtained at the corner points of the solution set.

The answer to the question first posed is: Work the first crew seven days and the second crew two days at a cost of \$580.

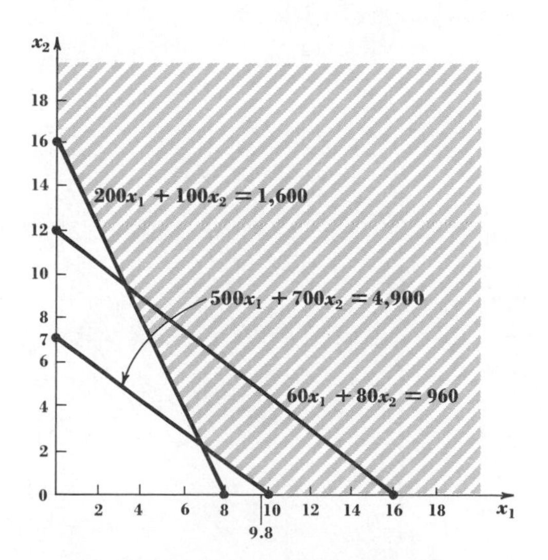

Diagram 24

The second, and final, example that will be given in this section also has to do with problems about minimums. Again we refer to a business situation to give the data which lead to a problem of minimization.

Example 2. A company has two warehouses, with trucks at each warehouse to deliver refrigerators and washing machines to a nearby city. The trucks at the first warehouse are built to carry 10 refrigerators and 4 washers, while the trucks at the second warehouse are so constructed that they carry 5 refrigerators and 10 washers. On any day, 120 refrigerators and 80 washers must be delivered. Owing to the union contract, at least five trucks must be sent from the first warehouse and at least two trucks must leave from the second warehouse each day.

If it costs \$50 for each truck from the first warehouse to make a delivery and \$30 for each truck from the second warehouse, how many trucks should be sent from each warehouse to minimize the cost?

Let x_1 be the number of trucks to make deliveries from the first warehouse and x_2 the number of trucks to be used from the second warehouse. Because of the union contracts, it follows that

$$(1) \quad x_1 \geq 5$$
$$(2) \quad x_2 \geq 2$$

Note that these constraints differ from those of previous examples, which required only that the variables be nonnegative. Because of these differences, we pause to draw the graphs before establishing the structural constraints. The graphs are in Diagrams 25 and 26.

Since both constraints must be met simultaneously, it can already be stated that the solutions must be in that portion of the first quadrant that is shaded in Diagram 27.

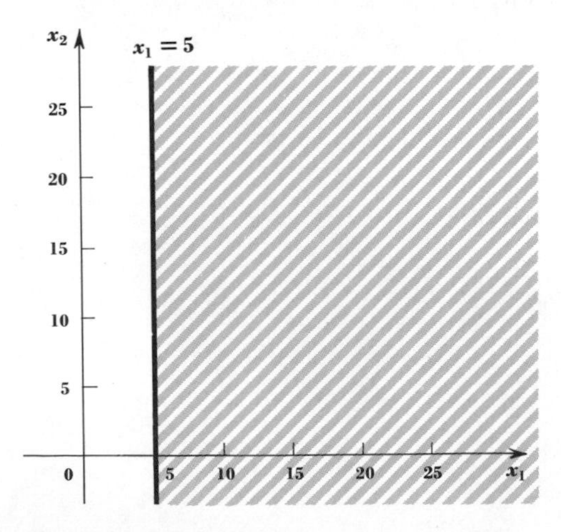

Diagram 25

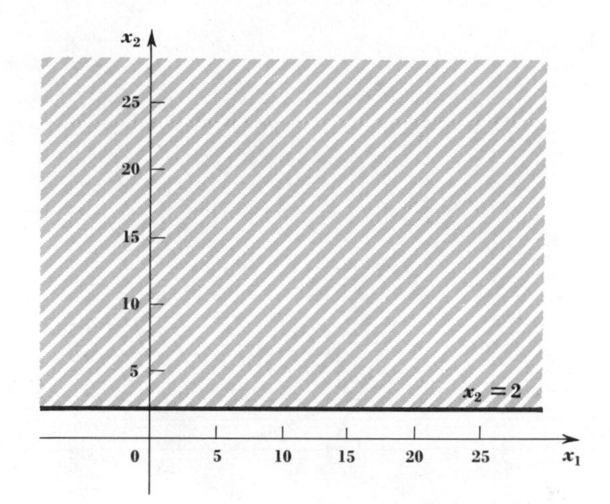

Diagram 26

Now we return to the statement of the problem. The number of refrigerators to be transported from warehouse 1 is $10x_1$ and from warehouse 2 is $5x_1$. Therefore, the third inequality is

$$(3) \quad 10x_1 + 5x_2 \geq 120$$

By now it should be clear that the fourth constraint is expressed by

$$(4) \quad 4x_1 + 10x_2 \geq 80$$

The solution set for the four inequalities is shaded in Diagram 28.

The corner points are located by solving simultaneously the equations of the lines. One corner point is the solution of

$$(1) \quad x_1 = 5$$

and

$$(3) \quad 10x_1 + 5x_2 = 120$$

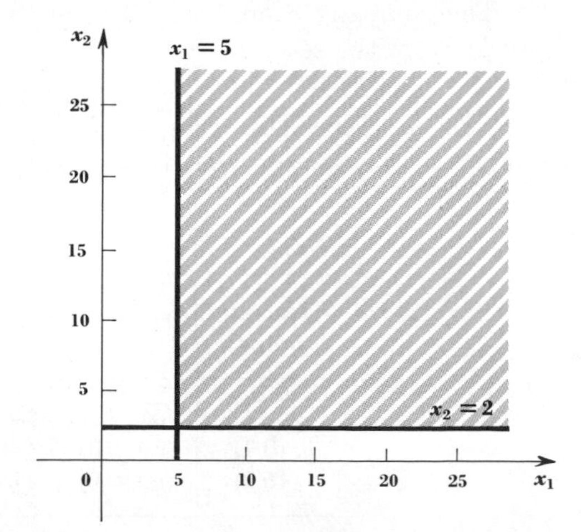

Diagram 27

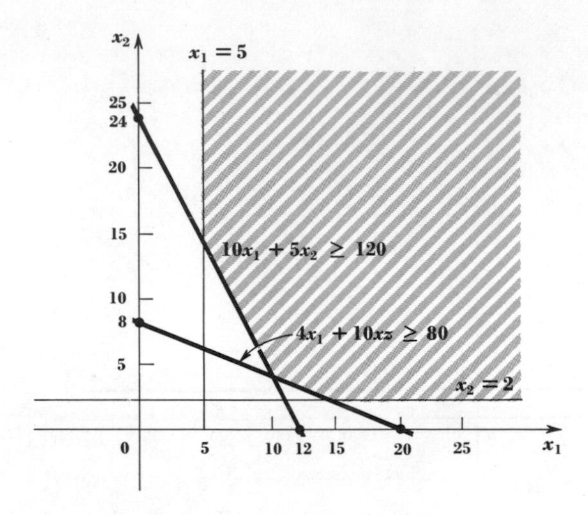

Diagram 28

which is (5,14). Similarly, the solution of

$$(3) \quad 10x_1 + 5x_2 = 120$$

and

$$(4) \quad 4x_1 + 10x_2 = 80$$

is (10,4). Finally, the solution of

$$(2) \quad x_2 = 2$$

and

$$(4) \quad 4x_1 + 10x_2 = 80$$

is (15,2). These three points are tested in the objective function

$$50x_1 + 30x_2$$

and the results are given in Table 3.

The company should send 10 trucks from the first warehouse and 4 trucks from the second.

This concludes the examples, but a word of explanation about the examples of this section and the previous one should be given. As was

Table 3

Corner points	*Objective function*
(5,14)	$(50 \cdot 5) + (30 \cdot 14) = 670$
(10,4)	$(50 \cdot 10) + (30 \cdot 4) = 620$
(15,2)	$(50 \cdot 15) + (30 \cdot 2) = 810$

explained, the examples are very simple. Rarely in practice is it possible to keep the number of variables to two. We have already seen in the sections on equalities that for three variables the graphs become difficult to draw and for four or more variables graphs are not possible. The graphic method becomes impractical for most real-life problems. So why has it been dwelt upon for so long? First, it makes clear the nature of linear-programming problems. Other problems may be more complex, but they follow the pattern that you have seen. Second, for more complex problems it is still essential to write in mathematics the problem furnished in words. You have been given some experience in that phase of the solution. In practice, the inequalities are programmed and turned over to a computer. As a final comment, it is the invention and use of electronic computers that have made practical the solution of many business problems.

Exercises

1. Find values of x_1 and x_2 satisfying

$$\begin{aligned} 2x_1 + 5x_2 &\geq 10 \\ 4x_1 + 3x_2 &\geq 12 \\ x_1 &\geq 0 \\ x_2 &\geq 0 \end{aligned}$$

   and minimizing $3x_1 + 4x_2$.

2. A company knows that they have to produce at least 25,000 cans and 14,500 metal trays. They have two machines which they can utilize to fill the order. The first costs $1,000 per hr to operate and can produce 1,000 cans and 300 trays per hour. The second costs $1,200 per hr to operate and can produce 600 cans and 600 trays per hr.
   - *a.* How many hours should each machine operate to minimize the cost of production?
   - *b.* What will this minimum cost be?

3. A tire company has gotten a contract to supply at least 1,000,000 black-wall tires and at least 1,000,000 white-wall tires to an auto manufacturing company. The company has two assembly lines to use to fill this particular contract. Assembly line 1 can produce 1,000 black-wall and 500 white-wall tires in an hour, and assembly line 2 can produce 750 black-wall and 1,000 white-wall tires in an hour. It costs $12,000 per hr to operate line 1 and $15,000 per hr to operate line 2.
   - *a.* How many hours should each line operate to minimize production costs?
   - *b.* What will the minimum production cost be?

4. The CCC has two assembly lines. The amounts that each line can produce per day are: 30,000 sheets, 20,000 blankets, and 300 comforters on the first line and 20,000 sheets, 20,000 blankets, and 400 comforters on the second line.

   a. How many days must each line work so that the total numbers produced are at least 240,000 sheets, 210,000 blankets, and 36,000 comforters?

   b. If it costs $100,000 per day to run line 1 and $85,000 per day to run line 2, how many days should each line operate to minimize the cost?

5. Work Example 1 of this section with the added restriction that $x_1 \leq 5$.

Answers to problems

A. Let x_1 be the number of hours worked by Mr. Adams and x_2 be the number of hours worked by Mr. Smith.

$$10x_1 + 5x_2 \geq 80$$
$$5x_1 + 15x_2 \geq 90$$
$$x_1 \geq 0$$
$$x_2 \geq 0$$
$$2x_1 + 3x_2 \text{ is to be minimum}$$

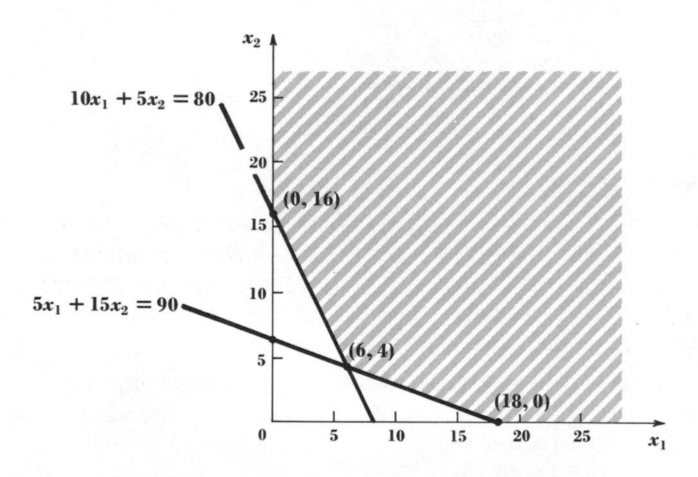

$$(2 \cdot 0) + (3 \cdot 16) = 48$$
$$(2 \cdot 6) + (3 \cdot 4) = 24$$
$$(2 \cdot 18) + (3 \cdot 0) = 36$$

Work Mr. Adams 6 hr and Mr. Smith 4 hr.

B. Let x_1 and x_2 be the number of bushels of A and B.

Weight:	$40x_1 + 20x_2 \geq 100$	
Vitamin minimum:	$5x_1 + 4x_2 \geq 16$	$x_1 \geq 0$
Vitamin maximum:	$5x_1 + 4x_2 \leq 25$	$x_2 \geq 0$
Mineral minimum:	$2x_1 + 5x_2 \geq 12$	

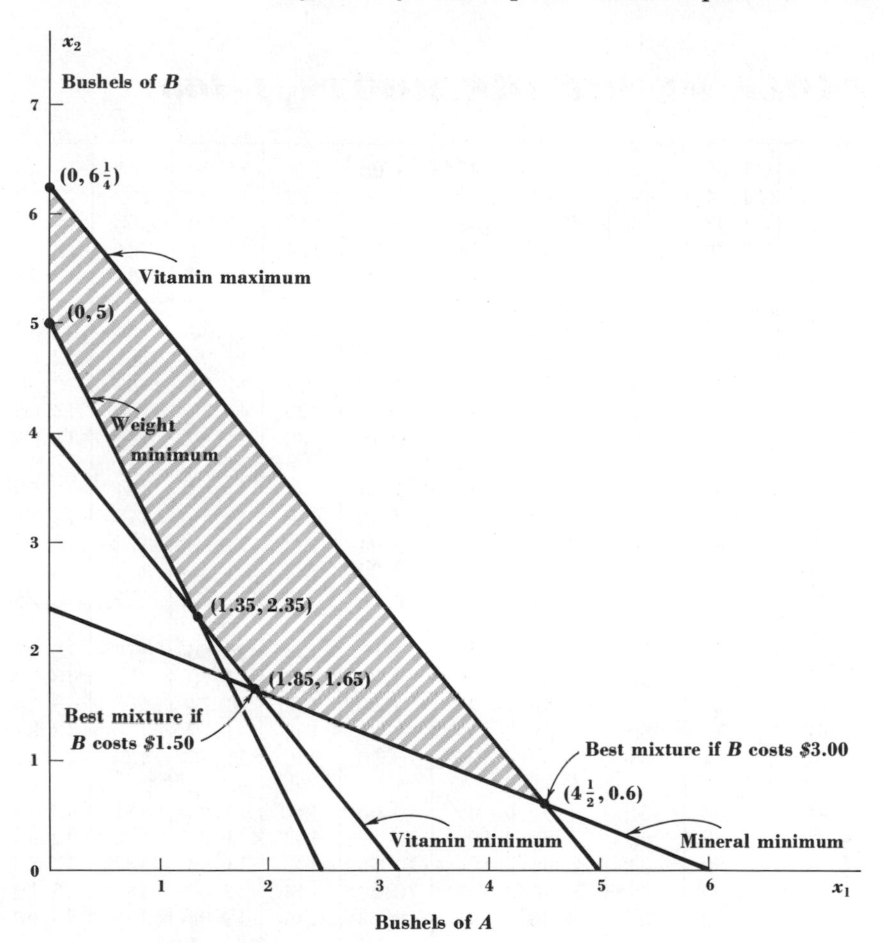

Costs at	*A*	*B at $1.50*	*Total*
(0,6.25)	0	$9.35	$9.35
(0,5)	0	$7.50	$7.50
(1.35,2.35)	$1.35	$3.53	$4.88
(1.85,1.65)	$1.85	$2.48	$4.33*
(4.5,0.6)	$4.50	$0.90	$5.40

* Best cost if *B* costs $1.50.

Costs at	*A*	*B at $3*	*Total*
(0,6.25)	0	$18.75	$18.75
(0,5)	0	$15.00	$15.00
(1.35,2.35)	$1.35	$ 7.05	$ 8.40
(1.85,1.65)	$1.85	$ 4.95	$ 6.80
(4.5,0.6)	$4.50	$ 1.80	$ 6.30*

* Best cost if *B* costs $3.

TABLE OF SQUARE ROOTS (1-400)

1	1.00	41	6.40	81	9.00	121	11.00	161	12.69
2	1.41	42	6.48	82	9.06	122	11.05	162	12.73
3	1.73	43	6.56	83	9.11	123	11.09	163	12.77
4	2.00	44	6.63	84	9.17	124	11.14	164	12.81
5	2.24	45	6.71	85	9.22	125	11.18	165	12.85
6	2.45	46	6.78	86	9.27	126	11.23	166	12.88
7	2.65	47	6.86	87	9.33	127	11.27	167	12.92
8	2.83	48	6.93	88	9.38	128	11.31	168	12.96
9	3.00	49	7.00	89	9.43	129	11.36	169	13.00
10	3.16	50	7.07	90	9.49	130	11.40	170	13.04
11	3.32	51	7.14	91	9.54	131	11.45	171	13.08
12	3.46	52	7.21	92	9.59	132	11.49	172	13.11
13	3.61	53	7.28	93	9.64	133	11.53	173	13.15
14	3.74	54	7.35	94	9.70	134	11.58	174	13.19
15	3.87	55	7.42	95	9.75	135	11.62	175	13.23
16	4.00	56	7.48	96	9.80	136	11.66	176	13.27
17	4.12	57	7.55	97	9.85	137	11.70	177	13.30
18	4.24	58	7.62	98	9.90	138	11.74	178	13.34
19	4.36	59	7.68	99	9.95	139	11.79	179	13.38
20	4.47	60	7.75	100	10.00	140	11.83	180	13.42
21	4.58	61	7.81	101	10.05	141	11.87	181	13.45
22	4.69	62	7.87	102	10.10	142	11.92	182	13.49
23	4.80	63	7.94	103	10.15	143	11.96	183	13.53
24	4.90	64	8.00	104	10.20	144	12.00	184	13.56
25	5.00	65	8.06	105	10.25	145	12.04	185	13.60
26	5.10	66	8.12	106	10.30	146	12.08	186	13.64
27	5.20	67	8.19	107	10.34	147	12.12	187	13.67
28	5.29	68	8.25	108	10.39	148	12.17	188	13.71
29	5.39	69	8.31	109	10.44	149	12.21	189	13.75
30	5.48	70	8.37	110	10.49	150	12.25	190	13.78
31	5.57	71	8.43	111	10.54	151	12.29	191	13.82
32	5.66	72	8.49	112	10.58	152	12.33	192	13.86
33	5.74	73	8.54	113	10.63	153	12.37	193	13.89
34	5.83	74	8.60	114	10.68	154	12.41	194	13.93
35	5.92	75	8.66	115	10.72	155	12.45	195	13.96
36	6.00	76	8.72	116	10.77	156	12.49	196	14.00
37	6.08	77	8.77	117	10.82	157	12.53	197	14.04
38	6.16	78	8.83	118	10.86	158	12.57	198	14.07
39	6.25	79	8.89	119	10.91	159	12.61	199	14.11
40	6.32	80	8.94	120	10.95	160	12.65	200	14.14

201	14.18	241	15.52	281	16.76	321	17.92	361	19.00
202	14.21	242	15.56	282	16.79	322	17.94	362	19.03
203	14.25	243	15.59	283	16.82	323	17.97	363	19.05
204	14.28	244	15.62	284	16.85	324	18.00	364	19.08
205	14.32	245	15.65	285	16.88	325	18.03	365	19.11
206	14.35	246	15.68	286	16.91	326	18.06	366	19.13
207	14.39	247	15.72	287	16.94	327	18.08	367	19.16
208	14.42	248	15.75	288	16.97	328	18.11	368	19.18
209	14.46	249	15.78	289	17.00	329	18.14	369	19.21
210	14.49	250	15.81	290	17.03	330	18.17	370	19.24
211	14.53	251	15.84	291	17.06	331	18.19	371	19.26
212	14.56	252	15.87	292	17.09	332	18.22	372	19.29
213	14.59	253	15.91	293	17.12	333	18.25	373	19.31
214	14.63	254	15.94	294	17.15	334	18.28	374	19.34
215	14.66	255	15.97	295	17.18	335	18.30	375	19.36
216	14.70	256	16.00	296	17.20	336	18.33	376	19.39
217	14.73	257	16.03	297	17.23	337	18.36	377	19.42
218	14.76	258	16.06	298	17.26	338	18.38	378	19.44
219	14.80	259	16.09	299	17.29	339	18.41	379	19.47
220	14.83	260	16.12	300	17.32	340	18.44	380	19.49
221	14.87	261	16.16	301	17.35	341	18.47	381	19.52
222	14.90	262	16.19	302	17.38	342	18.49	382	19.54
223	14.93	263	16.22	303	17.41	343	18.52	383	19.57
224	14.97	264	16.25	304	17.44	344	18.55	384	19.60
225	15.00	265	16.28	305	17.46	345	18.57	385	19.62
226	15.03	266	16.31	306	17.49	346	18.60	386	19.65
227	15.07	267	16.34	307	17.52	347	18.63	387	19.67
228	15.10	268	16.37	308	17.55	348	18.65	388	19.70
229	15.13	269	16.40	309	17.58	349	18.68	389	18.72
230	15.17	270	16.43	310	17.61	350	18.71	390	19.75
231	15.20	271	16.46	311	17.64	351	18.74	391	19.77
232	15.23	272	16.49	312	17.66	352	18.76	392	19.80
233	15.26	273	16.52	313	17.69	353	18.79	393	19.82
234	15.30	274	16.55	314	17.72	354	18.81	394	19.85
235	15.33	275	16.58	315	17.75	355	18.84	395	19.87
236	15.36	276	16.61	316	17.78	356	18.87	396	19.90
237	15.39	277	16.64	317	17.80	357	18.89	397	19.92
238	15.43	278	16.67	318	17.83	358	18.92	398	19.95
239	15.46	279	16.70	319	17.86	359	18.95	399	19.98
240	15.49	280	16.73	320	17.89	360	18.97	400	20.00

TABLE OF NORMAL DISTRIBUTION

	.00	.01	.02	.03	.04	.05	.06	.07	.08	.09
0.0	.50000	.50399	.50798	.51197	.51595	.51994	.52392	.52790	.53188	.53586
0.1	.53983	.54380	.54776	.55172	.55567	.55962	.56356	.56749	.57142	.57535
0.2	.57926	.58317	.58706	.59095	.59483	.59871	.60257	.60642	.61026	.61409
0.3	.61791	.62172	.62552	.62930	.63307	.63683	.64058	.64431	.64803	.65173
0.4	.65542	.65910	.66276	.66640	.67003	.67364	.67724	.68082	.68439	.68793
0.5	.69146	.69497	.69847	.70194	.70540	.70884	.71226	.71566	.71904	.72240
0.6	.72575	.72907	.73237	.73536	.73891	.74215	.74537	.74857	.75175	.75490
0.7	.75804	.76115	.76424	.76730	.77035	.77337	.77637	.77935	.78230	.78524
0.8	.78814	.79103	.79389	.79673	.79955	.80234	.80511	.80785	.81057	.81327
0.9	.81594	.81859	.82121	.82381	.82639	.82894	.83147	.83398	.83646	.83891
1.0	.84134	.84375	.84614	.84849	.85083	.85314	.85543	.85769	.85993	.86214
1.1	.86433	.86650	.86864	.87076	.87286	.87493	.87698	.87900	.88100	.88298
1.2	.88493	.88686	.88877	.89065	.89251	.89435	.89617	.89796	.89973	.90147
1.3	.90320	.90490	.90658	.90824	.90988	.91149	.91309	.91466	.91621	.91774
1.4	.91924	.92073	.92220	.92364	.92507	.92647	.92785	.92922	.93056	.93189
1.5	.93319	.93448	.93574	.93699	.93822	.93943	.94062	.94179	.94295	.94408
1.6	.94520	.94630	.94738	.94845	.94950	.95053	.95154	.95254	.95352	.95449
1.7	.95543	.95637	.95728	.95818	.95907	.95994	.96080	.96164	.96246	.96327
1.8	.96407	.96485	.96562	.96638	.96712	.96784	.96856	.96926	.96995	.97062
1.9	.97128	.97193	.97257	.97320	.97381	.97441	.97500	.97558	.97615	.97670

2.0	.97725	.97784	.97831	.97882	.97932	.97982	.98030	.98077	.98124	.98169
2.1	.98214	.98257	.98300	.98341	.98382	.98422	.98461	.98500	.98537	.98574
2.2	.98610	.98645	.98679	.98713	.98745	.98778	.98809	.98840	.98870	.98899
2.3	.98928	.98956	.98983	.99010	.99036	.99061	.99086	.99111	.99134	.99158
2.4	.99180	.99202	.99224	.99245	.99266	.99286	.99305	.99324	.99343	.99361
2.5	.99379	.99396	.99413	.99430	.99446	.99461	.99477	.99492	.99506	.99520
2.6	.99534	.99547	.99560	.99573	.99585	.99598	.99609	.99621	.99632	.99643
2.7	.99653	.99664	.99674	.99683	.99693	.99702	.99711	.99720	.99728	.99736
2.8	.99744	.99752	.99760	.99767	.99774	.99781	.99788	.99795	.99801	.99807
2.9	.99813	.99819	.99825	.99831	.99836	.99841	.99846	.99851	.99856	.99861
3.0	.99865	.99869	.99874	.99878	.99882	.99886	.99899	.99893	.99896	.99900
3.1	.99903	.99906	.99910	.99913	.99916	.99918	.99921	.99924	.99926	.99929
3.2	.99931	.99934	.99936	.99938	.99940	.99942	.99944	.99946	.99948	.99950
3.3	.99952	.99953	.99955	.99957	.99958	.99960	.99961	.99962	.99964	.99965
3.4	.99966	.99968	.99969	.99970	.99971	.99972	.99973	.99974	.99975	.99976
3.5	.99977	.99978	.99978	.99979	.99980	.99981	.99981	.99982	.99983	.99983
3.6	.99984	.99985	.99985	.99986	.99986	.99987	.99987	.99988	.99988	.99989
3.7	.99989	.99990	.99990	.99990	.99991	.99991	.99992	.99992	.99992	.99992
3.8	.99993	.99993	.99993	.99994	.99994	.99994	.99994	.99995	.99995	.99995
3.9	.99995	.99995	.99996	.99996	.99996	99996	.99996	.99996	.99997	.99997

Directions: To find the area under the curve between the left-hand end and any point, determine how many standard deviations that point is to the right of the average, then read the area directly from the body of the table. *Example:* The area under the curve from the left-hand end and a point 1.81 standard deviations to the right of the average is 0.96485 of the total area under the curve.

BIBLIOGRAPHY

Blalock, Hubert M.: "Social Statistics," McGraw-Hill Book Company, New York, 1960.

Chung, An-Min: "Linear Programming," Charles E. Merrill Books, Inc., Columbus, Ohio, 1963.

Croxton, Frederick E., and Dudley J. Cowden: "Applied General Statistics," 2d ed., Prentice-Hall, Inc., Englewood Cliffs, N.J., 1956.

Dornbusch, Sanford M., and Calvin F. Schmid: "A Primer of Social Statistics," McGraw-Hill Book Company, New York, 1955.

Ekeblad, Frederick A.: "The Statistical Method in Business Applications of Probability and Inference to Business and Other Problems," John Wiley & Sons, Inc., New York, 1962.

Ficken, F. A.: "The Simplex Method of Linear Programming," Holt, Rinehart and Winston, Inc., New York, 1961.

Fowler, Parker, and E. W. Sandberg: "Basic Mathematics for Administration," John Wiley & Sons, Inc., New York, 1962.

Fraser, D. A. S.: "Statistics: An Introduction," John Wiley & Sons, Inc., New York, 1958.

Freund, J. E.: "Modern Elementary Statistics," 3d ed., Prentice-Hall, Inc., Englewood Cliffs, N.J., 1967.

Gale, David: "The Theory of Linear Economic Models," McGraw-Hill Book Company, New York, 1960.

Hoel, P. G.: "Elementary Statistics," John Wiley & Sons, Inc., 20th ed., New York, 1966.

Howell, J. E., and D. Teichroew: "Mathematical Analysis for Business Decisions," Richard D. Irwin, Inc., Homewood, Ill., 1963.

Huff, Darrell, and I. Geis: "How to Lie with Statistics," W. W. Norton & Company, Inc., New York, 1954.

Irwin, W. C.: "Digital Computer Principles," D. Van Nostrand Company, Inc., Princeton, N.J., 1960.

Johnston, John B., Baley Price, and Frederick Von Vleck: "Linear Equations and Matrices," Addison-Wesley Publishing Company, Inc., Reading, Mass., 1966.

Kattsoff, Louis O., and Albert J. Simone: "Finite Mathematics with Applications in the Social and Management Sciences," McGraw-Hill Book Company, New York, 1965.

Kemeny, John G., et al.: "Finite Mathematics with Business Applications," Prentice-Hall, Inc., Englewood Cliffs, N.J., 1957.

Lipschultz, Seymour: "Outline of Theory and Problems of Finite Mathematics," Schaum Publishing Company, New York, 1966.

McCracken, D. D.: "Digital Computer Programming," John Wiley & Sons, Inc., New York, 1957.

Mack, S. F.: "Elementary Statistics," Holt, Rinehart and Winston, Inc., New York, 1960.

Mode, E. B.: "Elements of Statistics," 30th ed., Prentice-Hall, Inc., Englewood Cliffs, N.J., 1961.

Mood, A. M., and F. A. Graybill: "Introduction to the Theory of Statistics," 2d ed., McGraw-Hill Book Company, New York, 1963.

Mosteller, F., R. E. K. Rourke, and G. B. Thomas, Jr.: "Probability with Statistical Applications," Addison-Wesley Publishing Company, Inc., Reading, Mass., 1961.

National Bureau of Standards: Tables of the Binomial Probability Distribution, Government Printing Office, 1949.

Rutledge, W. E., and T. W. Cains: "Mathematics for Business Analysis," Holt, Rinehart and Winston, Inc., New York, 1963.

Stockton, John R.: "Business Statistics," South-Western Publishing Company, Cincinnati, 1958.

Stockton, R. S.: "Introduction to Linear Programming," 20th ed., Allyn and Bacon, Inc., Boston, 1960.

Suppes, Patrick: "Introduction to Logic," D. Van Nostrand Company, Inc., Princeton, N.J., 1957.

Theodore, Chris A.: "Applied Mathematics: An Introduction," Richard D. Irwin, Inc., Homewood, Ill., 1965.

Weinberg, George H., and John A. Schumaker: "Statistics: An Intuitive Approach," Wadsworth Publishing Company, Inc., Belmont, Calif., 1962.

ANSWERS TO SELECTED EXERCISES

Chapter 1

Section 1.1

1. *a.* Simple statement *b.* Compound statement
   c. Not a statement *d.* Compound statement

2. *a.* p *b.* q *c.* Not p *d.* Not q *e.* p and q

3. *a.* p: Profits are decreasing.
   q: Market price of the stock goes down.
   r: Stockholders are unhappy.

4. *a.* p and q *b.* Not p *c.* If p, then q.

Section 1.2

1. *a.* False *b.* True *c.* True *d.* True *e.* False

2. *a.* Not possible *b.* $\{\ \}, \{0\}, \{1\}, \{2\}, \{0,1\}$
   c. Not possible *d.* $\{\ \}, \{0\}, \{1\}, \{2\}, \{3\}, \{4\}$

3. *a.* False *b.* False *c.* False

4. $2^5 = 32$

5. *a.* $2^9 = 256$

6. Yes; $2^{12} = 4{,}096$

7. *a.* $\{P_{10}, P_{11}, P_{12}\}$

8. *a.* $\{P_3, P_6, P_9, P_{12}\}$ *b.* $2^4 = 16$

9. *a.* $\{\ \}$, $\{a\}$, $\{b\}$, $\{c\}$, $\{a,b\}$, $\{a,c\}$, $\{b,c\}$, $\{a,b,c\}$
   b. The first seven of 9*a*

Section 1.3

1. $S \cap T = \{a,e\}$

2. *a.*

p	q	$p \wedge q$
T	T	T
T	F	F
F	T	F
F	F	F

b.

p	q	$q \wedge p$
T	T	T
T	F	F
F	T	F
F	F	F

c.

p	q	r	$q \wedge r$	$p \wedge (q \wedge r)$
T	T	T	T	T
T	T	F	F	F
T	F	T	F	F
T	F	F	F	F
F	T	T	T	F
F	T	F	F	F
F	F	T	F	F
F	F	F	F	F

d.

p	q	r	$p \wedge q$	$(p \wedge q) \wedge r$
T	T	T	T	T
T	T	F	T	F
T	F	T	F	F
T	F	F	F	F
F	T	T	F	F
F	T	F	F	F
F	F	T	F	F
F	F	F	F	F

3.

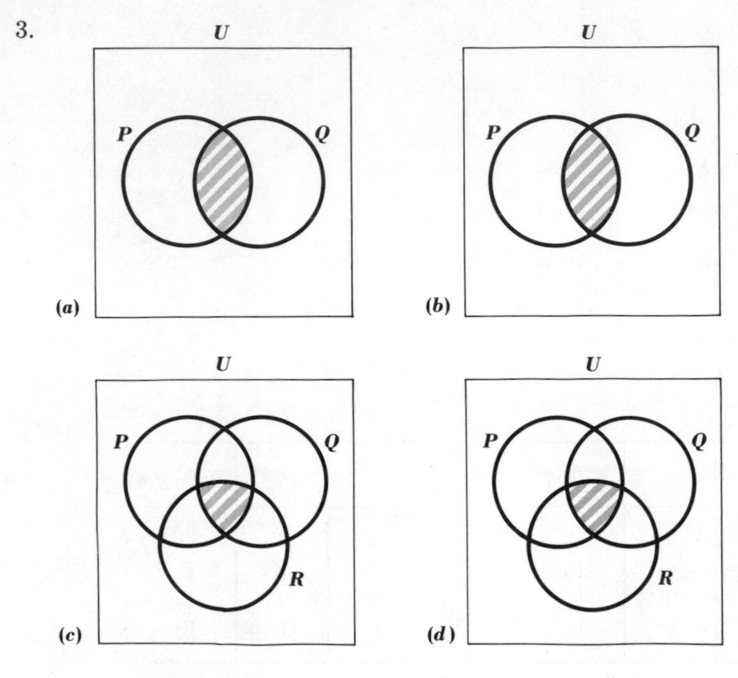

4. $S = \{0,1,2,3,4,5,6\}$
5. No
6. *a.* False
   b. It is true if the other simple statement is true, but it is false if the other statement is false.
8. The set $P \cap Q \cap R$ is the set of all elements in P and in Q and in R.

Section 1.4

1. *a.* Exclusive *b.* Inclusive *c.* Inclusive *d.* Exclusive
4. *a.* $\{0,1,2,5,6\}$ *b.* $\{0,1,5,6\}$ *c.* $\{0,1,2,3,5\}$
6.

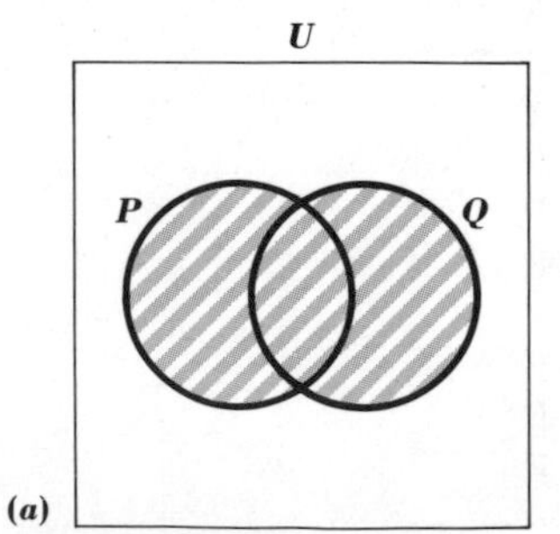

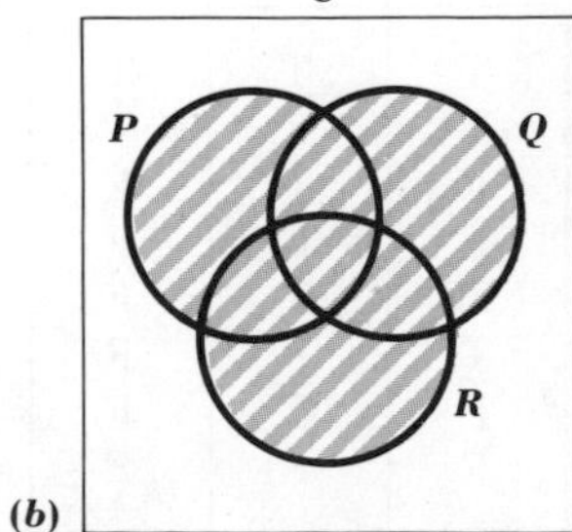

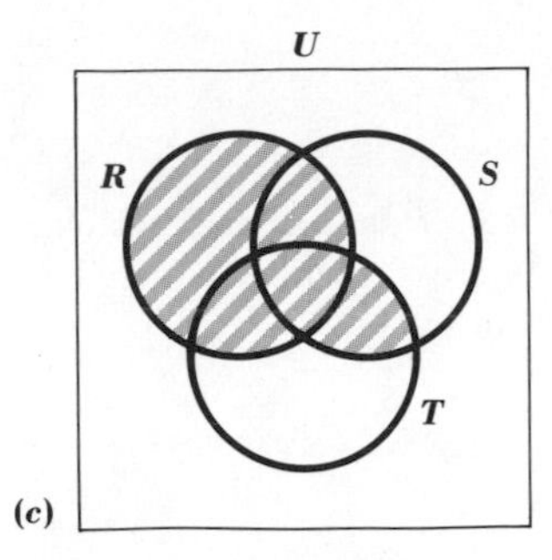

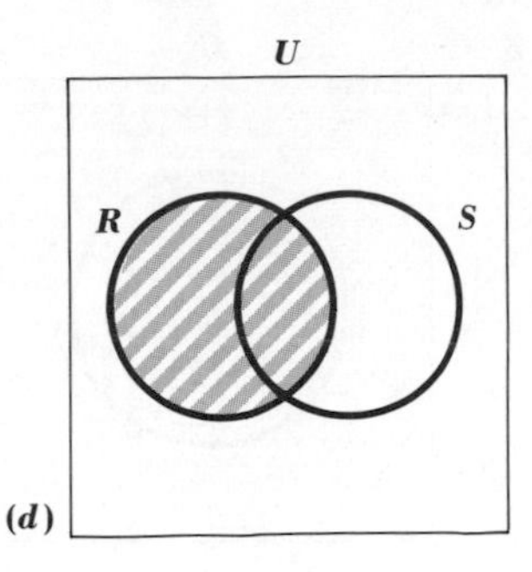

7. *a.* Let p: The union is demanding that Washington's birthday be a holiday.
   q: The union is demanding that the day after Thanksgiving be a holiday.
   The original sentence becomes $p \vee q$ (p or q is acceptable).
   b. Let p: Television is a more expensive advertising medium than is radio.
   q: Television is more effective than radio.
   r: Television reaches a wider audience than radio.
   The original sentence becomes $p \wedge (q \wedge r)$.
   c. Let p: Industry and the university must work together.
   q: The student is wasting his money coming to college.
   The original statement becomes $p \underline{\vee} q$ ($p \vee q$ is not acceptable).

Section 1.5

1. *a.*

p	$\sim p$
T	F
F	T

b.

p	q	$\sim p$	$\sim p \vee q$
T	T	F	T
T	F	F	F
F	T	T	T
F	F	T	T

c.

p	q	$\sim q$	$p \wedge \sim q$
T	T	F	F
T	F	T	T
F	T	F	F
F	F	T	F

d.

p	q	$p \wedge q$	$\sim(p \wedge q)$
T	T	T	F
T	F	F	T
F	T	F	T
F	F	F	T

2. $\tilde{S} = \{d,e\}$

3.

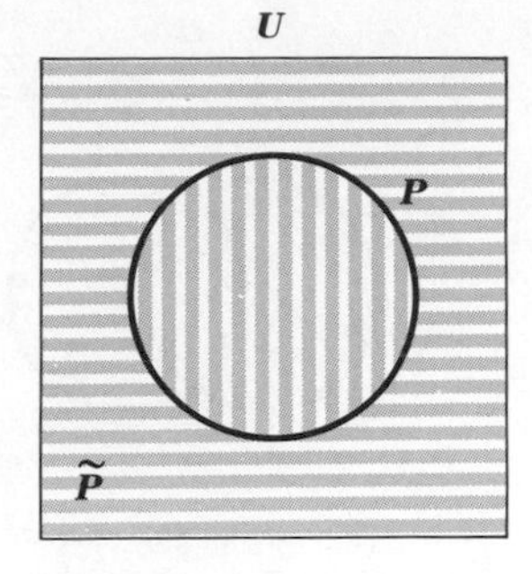

$P \cap \tilde{P}$ is that area which is crosshatched, which is the set { }.

4.

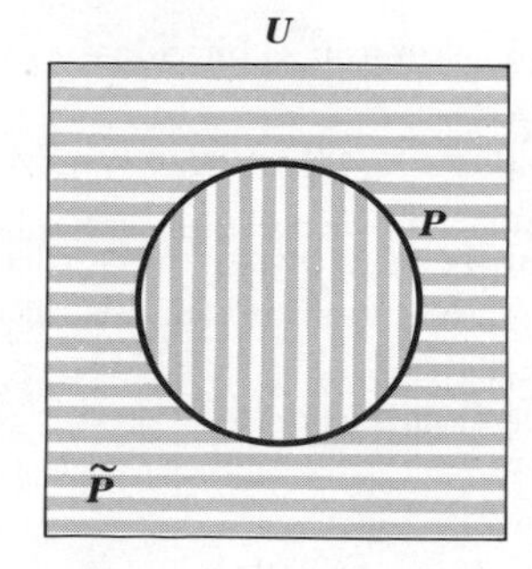

$P \cup \tilde{P}$ is that area which is shaded, which is U.

5. Profits are not rising.

6. All are negations. Let p = productivity goes up, q = wages go up, and r = the union will strike. The given statement becomes $p \wedge (q \vee r)$.

   a. Becomes $\sim[p \wedge (q \vee r)]$
   b. Becomes $\sim p \vee \sim(q \vee r)$
   c. Becomes $\sim p \vee (\sim q \wedge \sim r)$

				Given:	*a*			*b*
p	q	r	$q \vee r$	$p \wedge (q \vee r)$	$\sim[p \wedge (q \vee r)]$	$\sim p$	$\sim(q \vee r)$	$\sim p \vee \sim(q \vee r)$
T	T	T	T	T	F	F	F	F
T	T	F	T	T	F	F	F	F
T	F	T	T	T	F	F	F	F
T	F	F	F	F	T	F	T	T
F	T	T	T	F	T	T	F	T
F	T	F	T	F	T	T	F	T
F	F	T	T	F	T	T	F	T
F	F	F	F	F	T	T	T	T

c

p	q	r	$\sim p$	$\sim q$	$\sim r$	$\sim q \wedge \sim r$	$\sim p \vee (\sim q \wedge \sim r)$
T	T	T	F	F	F	F	F
T	T	F	F	F	T	F	F
T	F	T	F	T	F	F	F
T	F	F	F	T	T	T	T
F	T	T	T	F	F	F	T
F	T	F	T	F	T	F	T
F	F	T	T	T	F	F	T
F	F	F	T	T	T	T	T

The truth tables for parts *a*, *b*, and *c* show that their statements are true when the given statement is false and false when the given statement is true.

Section 1.6

1. *a.* Profits are rising and dividends are rising.
   b. It is not true that profits are rising and dividends are rising, and stockholders are not satisfied.
   c. Profits are not rising or dividends are not rising, and stockholders are not satisfied.
   d. Profits are rising and dividends are rising, whether stockholders are satisfied or not.

7. *a.* {1,2,5,8,9} *b.* {5} *c.* {2,4,6,8,10}

9. *a.* {1,5,9,17,18,19,20}

Section 1.7

1. *a.* Hypothesis: The sales forecast is accurate.
   Conclusion: We shall make a million dollars.
   b. Hypothesis: We are to succeed.
   Conclusion: We must have a plan of action.
   c. Hypothesis: Foreign cars become popular.
   Conclusion: The American automobile industry is hurt.

2. The statement is true.
   p: Productivity increases.
   q: Wages will increase.
   The statement is $p \rightarrow q$.

p	q	$p \rightarrow q$
T	T	T
T	F	F
F	T	T
F	F	T

The third case, p false, q true, is the case when wages increase and productivity does not increase. In the third case, the statement $p \rightarrow q$ is true.

3. If people with fixed incomes do not suffer, then prices are not rising.

5. *a.*

p	q	$p \rightarrow q$
T	T	T
T	F	F
F	T	T
F	F	T

b.

p	q	$\sim p$	$\sim q$	$\sim p \rightarrow \sim q$
T	T	F	F	T
T	F	F	T	T
F	T	T	F	F
F	F	T	T	T

c.

p	q	$\sim p$	$q \rightarrow \sim p$	$p \vee (q \rightarrow \sim p)$
T	T	F	F	T
T	F	F	T	T
F	T	T	T	T
F	F	T	T	T

d.

p	q	$\sim p$	$p \vee q$	$\sim p \rightarrow (p \vee q)$
T	T	F	T	T
T	F	F	T	T
F	T	T	T	T
F	F	T	F	F

e.

p	q	$q \rightarrow p$	$(q \rightarrow p) \wedge p$
T	T	T	T
T	F	T	T
F	T	F	F
F	F	T	F

6. Let

   p: Our high prices cause customers to buy from our competitors.
   q: The stockholders will be unhappy.
   r: The general manager will be fired.

   The statement becomes $p \rightarrow (q \wedge r)$.

9. Yes. As proof, let

   p: Sales will fall.
   q: Profits will fall.

   a. $\sim(p \vee q)$

b. $\sim p \wedge \sim q$
and these are logically equivalent.

p	q	$p \vee q$	$\sim(p \vee q)$	$\sim p$	$\sim q$	$\sim p \wedge \sim q$
T	T	T	F	F	F	F
T	F	T	F	F	T	F
F	T	T	F	T	F	F
F	F	F	T	T	T	T

The corresponding truth elements are identical.

Section 1.8

1. *a.* If the organization suffers, then a good replacement will not be found.
   b. If actual profits are less than potential profits, then the sales forecast is too low.
   c. If we must have accurate accounting records, then we are to borrow money from a bank.
   d. If costs should go down, then sales will go down.

2. *a.*

p	q	$p \leftrightarrow q$
T	T	T
T	F	F
F	T	F
T	T	T

b.

p	q	$p \wedge q$	$p \vee q$	$(p \wedge q) \leftrightarrow (p \vee q)$
T	T	T	T	T
T	F	F	T	F
F	T	F	T	F
F	F	F	F	T

c.

p	q	$q \leftrightarrow p$	$p \vee (q \leftrightarrow p)$
T	T	T	T
T	F	F	T
F	T	F	F
F	F	T	T

d.

p	q	$p \vee q$	$\sim(p \vee q)$	$p \wedge q$	$\sim(p \vee q) \leftrightarrow (p \wedge q)$
T	T	T	F	T	F
T	F	T	F	F	T
F	T	T	F	F	T
F	F	F	T	F	F

5. *a.* $p \wedge q$ *b.* $p \rightarrow q$ *c.* $p \veebar q$ *d.* $q \leftrightarrow p$

Section 1.9

1. *a.* No *b.* No
3. *a.* Yes *b.* Yes
7. *a.* $\sim p \vee \sim q,\ p \wedge q$
8. *a.* $p \rightarrow q,\ q \rightarrow p$
13. *a.* Valid *b.* Valid
14. *a.* Valid

Section 1.10

2. *a.* This sheet has fitted corners. *d.* This sheet is white.
3. *b.* $q \leftrightarrow \sim(p \wedge r)$
4. *c.* $r \leftrightarrow (\sim p) \wedge q$ *e.* $(p \wedge q) \rightarrow r$

Chapter 2

Section 2.1

1. *a.* 111 *b.* 99 *c.* 82 *d.* Not possible
2. *a.* 20 *c.* 18
5. *a.* $(\{a_1\}, \{a_3\}, \{a_2,a_4\}, \{a_5,a_6\})$
   c. The prospects who have purchased previously and those expected to purchase in the future
6. *a.* $(P,\tilde{P}), (Q,\tilde{Q}), (R,\tilde{R})$
   b. $(P \cap Q \cap R,\ P \cap Q \cap \tilde{R},\ P \cap \tilde{Q} \cap R,\ P \cap \tilde{Q} \cap \tilde{R},\ \tilde{P} \cap Q \cap R,\ \tilde{P} \cap Q \cap \tilde{R},\ \tilde{P} \cap \tilde{Q} \cap R,\ \tilde{P} \cap \tilde{Q} \cap \tilde{R})$
9. *a.* Let U be the set of customers. Let P_1 be the subset who use the credit card zero times per month, P_2 the subset who use the credit card one time per month, and P_3 the subset who use the credit card two or more times per month (P_1,P_2,P_3). Let Q_1 be the subset who purchase less than \$7 per month and Q_2 the subset who purchase \$7 or more per month (Q_1,Q_2). Let R_1 be the subset who pay the bill when due and R_2 the subset who do not pay the bill when due (R_1,R_2).
   b. $(P_1 \cap Q_1,\ P_1 \cap Q_2,\ P_2 \cap Q_1,\ P_2 \cap Q_2,\ P_3 \cap Q_1,\ P_3 \cap Q_2)$
   c. $(P_1 \cap Q_1 \cap R_1,\ P_1 \cap Q_2 \cap R_1,\ P_2 \cap Q_1 \cap R_1,\ P_2 \cap Q_2 \cap R_1,\ P_3 \cap Q_1 \cap R_1,\ P_3 \cap Q_2 \cap R_1,\ P_1 \cap Q_1 \cap R_2,\ P_1 \cap Q_2 \cap R_2,\ P_2 \cap Q_1 \cap R_2,\ P_2 \cap Q_2 \cap R_2,\ P_3 \cap Q_1 \cap R_2,\ P_3 \cap Q_2 \cap R_2)$
   e. 80,000 *f.* 60,000 *g.* 40,000

Section 2.2

1. *a.* $10^7 = 10{,}000{,}000$ *b.* 9,000,000 *c.* 8,100,000
2. $7! = 5{,}040$
3. $7 \cdot 6 \cdot 5 \cdot 4 = 840$

6. *a.* $5! = 120$ *b.* 120
7. $14! \cdot 45 = 3{,}923{,}023{,}104{,}000$ min
9. $2^{14} = 16{,}384$

Section 2.3

1. $_nC_n = \dfrac{_nP_n}{_nP_n} = 1$
3. $_6C_2 = \dfrac{6!}{2!4!} = \dfrac{6 \cdot 5}{1 \cdot 2} = \dfrac{30}{2} = 15$
4. $_{100}C_2 = \dfrac{100!}{2!98!} = \dfrac{100 \cdot 99}{1 \cdot 2} = 4{,}950 = {}_{100}C_{98}$
5. $_4C_2\ {}_2C_1 = \dfrac{4!}{2!2!}2 = 12$
8. $_7C_2\ {}_5C_1\ {}_3C_2 = 21 \cdot 5 \cdot 3 = 315$

Section 2.4

1. $_{18}C_{13} = \dfrac{18!}{13!5!} = \dfrac{18 \cdot 17 \cdot 16 \cdot 15 \cdot 14}{1 \cdot 2 \cdot 3 \cdot 4 \cdot 5} = 8{,}568$
3. $_{20}C_{14}(\frac{1}{2})^6(\frac{1}{2})^{14}$
7. $(2u - 3w)^4 = (2u)^4 + 4(2u)^3(-3w) + 6(2u)^2(-3w)^2 + 6(2u)(-3w)^3 + (-3w)^4$
   $= 16u^4 - 96u^3w + 216u^2w^2 - 216uw^3 + 81w^4$
9. *a.* $_7C_2(2x)^2(-3y)^5 = 21 \cdot 4x^2(-243y^5) = -20{,}412x^2y^5$
   The answer is $-20{,}412$.

Section 2.5

1. *a.* 1 *b.* 0
2. *a.* $_9C_{3,3,3} = \dfrac{9!}{3!3!3!} = 1{,}680$ *b.* $_9C_{4,4,1} = \dfrac{9!}{4!4!1!} = 630$
3. $_5C_{2,1,2} = \dfrac{5!}{2!1!2!} = 30$
4. *a.* $6! = 720$ *b.* $_6C_{2,2,2} = \dfrac{6!}{2!2!2!} = 90$ *c.* $_6C_{2,2,1,1} = \dfrac{6!}{2!2!1!1!} = 180$
5. *a.* $15!$ *b.* $_{15}C_{5,5,5} = \dfrac{15!}{5!5!5!} = 756{,}756$

Chapter 3

Section 3.1

1. *a.* $\frac{1}{2}$ *b.* $\frac{1}{4}$ *c.* $\frac{1}{13}$ *d.* $\frac{1}{13}$ *e.* $\frac{3}{13}$
2. *a.* $\frac{1}{36}$ *b.* $\frac{2}{36}$ *c.* $\frac{3}{36}$

3. $^{25}/_{100}$
7. *b.* $^1/_8$
8. *a.* $^1/_6$ *b.* $^4/_6$ *c.* $^2/_6$
9. $^1/_7$
10. $^1/_7$
11. $^1/_2$
12. *a.* $20/2{,}000 = {}^1/_{100}$ *b.* $^0/_{400} = 0$ *c.* 1,540/1,700
d. 1,540/1,600

Section 3.2

1. *a.* $^1/_4$ *b.* $^1/_{26}$ *c.* $^1/_{26}$ *d.* $^3/_{26}$ *e.* $^1/_{52}$
2. *a.* $^5/_{100}$ *b.* $^9/_{100}$ *c.* $^{26}/_{100}$ *d.* $^3/_{100}$
3. *a.* 1,700/2,000 *b.* 260/2,000 *c.* 40/2,000 *d.* 0
e. 160/2,000 *f.* 1,540/2,000
4. *a.* 1,540/1,600 *b.* 20/1,600 *c.* 40/1,600
5. *a.* 770/1,100 *b.* 330/1,100 *c.* 485/1,100 *d.* 320/1,100
e. 165/1,100
6. *a.* 160/2,000 *b.* $^{240}/_{400}$ *c.* $^{240}/_{260}$

Section 3.3

1. *a.* $^1/_2$ *b.* $^1/_{13}$ *c.* $^1/_{26}$ *d.* $^{28}/_{56}$
2. *a.* $^3/_{36}$ *b.* $^{15}/_{36}$ *c.* $^{30}/_{36}$ *d.* $^{17}/_{36}$
4. *a.* $^2/_8$ *b.* $^3/_8$ *c.* $^1/_8$
7. *a.* $^{15}/_{20}$ *b.* $^5/_{20}$ *c.* $^{10}/_{20}$
8. *a.* 50 percent *b.* 180 *c.* $^{300}/_{800}$ *d.* 480/2,000
e. 1,100/2,000

Section 3.4

1. *a.* $1 - {}^{26}/_{52} = {}^{26}/_{52}$ *b.* $1 - {}^{13}/_{52} = {}^{39}/_{52}$ *c.* $1 - {}^4/_{52} = {}^{48}/_{52}$
d. $1 - {}^{13}/_{52} = {}^{39}/_{52}$
2. *a.* $1 - {}^1/_{36} = {}^{35}/_{36}$ *b.* $1 - {}^2/_{36} = {}^{34}/_{36}$ *c.* $1 - {}^3/_{36} = {}^{33}/_{36}$
3. *a.* $1 - {}^{25}/_{100} = {}^{75}/_{100}$ *b.* $1 - {}^5/_{100} = {}^{95}/_{100}$ *c.* $1 - {}^9/_{100} = {}^{91}/_{100}$
d. $1 - {}^{26}/_{100} = {}^{74}/_{100}$ *e.* $1 - {}^3/_{100} = {}^{97}/_{100}$
4. *a.* $^5/_6$ *b.* $^1/_6$ *c.* $^5/_6$ *d.* $^1/_2$
5. *a.* $^3/_4$ *b.* $^1/_2$ *c.* $^1/_2$
6. *a.* $1 - 770/1{,}100 = 330/1{,}100$ *b.* $1 - 300/1{,}100 = 800/1{,}100$
c. $1 - 0 = 1$ *d.* 330/1,100

Section 3.5

2. *a.* { }, {*a*}, {*b*}, {*c*}, {*a,b*}, {*a,c*}, {*b,c*}, {*a,b,c*}
b. 0, $^1/_2$, $^3/_8$, $^1/_8$, $^7/_8$, $^5/_8$, $^1/_2$, 1

4. *a.* $m(R \cup S \cup T) = m(R) + m(S) + m(T)$
   b. $m(R \cup S \cup T) = m(R) + m(S) + m(T) - m(R \cap S) - m(R \cap T) - m(S \cap T) + m(R \cap S \cap T)$
5. *a.* $m(R) \leq m(T)$ *b.* No
6. *a.* $(P \cap Q) \cup (\tilde{P} \cap Q) = (U \cap Q) = Q$
   $Pr(p \wedge q) + Pr(\sim p \wedge q) = Pr(q)$
   c. $P \cap Q = Q \cap P$
   $Pr(p \wedge q) = Pr(q \wedge p)$
7. E_1: H
   E_2: T H
   E_3: T T H
   E_4: T T T H
   . . . .
8. *a.* $5 \cdot 4 = 20$ *b.* $3/5$
9. *a.* 12/10,000 *b.* 188/10,000 *c.* 9,662/10,000
   d. 338/10,000 *e.* 200/10,000 *f.* 150/10,000

Section 3.6

1. *a.* $\dfrac{{}_6C_2\,{}_2C_1}{{}_8C_3} = 30/56$ *b.* $\dfrac{{}_6C_1\,{}_2C_2}{{}_8C_3} = 6/56$ *c.* $\dfrac{{}_6C_3\,{}_2C_0}{{}_8C_3} = 20/56$
   d. Not possible
4. *a.* $\dfrac{{}_3C_1\,{}_1C_1}{{}_4C_2} = 3/6$ *b.* $5/6$ *c.* $1 - 3/6 = 3/6$ *d.* $1 - 5/6 = 1/6$
5. $1/20$
7. *a.* ${}_{20}C_5 = 15{,}504$ *b.* 2,002/15,504 *c.* 6,006/15,504 *e.* 1
8. *a.*
$$Pr(18) = \frac{20!}{18!2!}\,.5^{18} \cdot .5^2 = 0.0002$$
$$Pr(19) = \frac{20!}{19!1!}\,.5^{19} \cdot .5^1 = 0.0000$$
$$Pr(20) = \frac{20!}{20!0!}\,.5^{20} \cdot .5^0 = 0.0000$$
   Total $= Pr(18 \text{ or more}) = 0.0002$

   There are only about 2 chances in 10,000 that 18 or more of the panel would be right if there is no difference.

   b. There is a detectable taste difference.

Section 3.7

1. *a.* ${}_3C_2 \cdot 1/5(4/5)^2 = 48/125$
2. *a.* ${}_{12}C_1 0.97^{11} \cdot 0.03^1$ *b.* ${}_{12}C_2 0.97^{10} \cdot 0.03^2$
   c. ${}_{12}C_7 0.97^5 \cdot 0.03^7 + {}_{12}C_8 0.97^4 \cdot 0.03^8 + {}_{12}C_9 0.97^3 \cdot 0.03^9 + {}_{12}C_{10} 0.97^2 \cdot 0.03^{10} + {}_{12}C_{11} 0.97^1 \cdot 0.03^{11} + {}_{12}C_{12} 0.97^0 \cdot 0.03^{12}$
5. ${}_3C_2(3/4)^2(1/4)^1$
6. Yes; $1/100$
7. *a.* ${}_{20}C_{15} 0.5^{15} \cdot 0.5^5$ *b.* ${}_{20}C_{16} 0.5^{16} \cdot 0.5^4$

c. Add the probabilities of 15, 16, 17, 18, 19, and 20.
d. No. It is, however, unlikely.

8. *a.* ${}_{10}C_1 0.03^1 \cdot 0.97^9$ *b.* ${}_{10}C_2 0.03^2 \cdot 0.97^8 = 45 \cdot 0.03^2 \cdot 0.97^8$
c. $0.03^2 \cdot 0.97^8$. Only one of the 45 combinations meets the requirements of the statement.

9. *a.* ${}_8C_5 0.1^5 \cdot 0.9^3 + {}_8C_6 0.1^6 \cdot 0.9^2 + {}_8C_7 0.1^7 \cdot 0.9^1 + {}_8C_8 0.1^8$
b. ${}_8C_0 0.1^0 \cdot 0.9^8$

10. Zero

11. $1 - 0.01 \cdot 0.01 = 1 - 0.0001 = 0.9999$

Section 3.8

1. ${}_{10}C_{8,1,1} 0.92^8 \cdot 0.06^1 \cdot 0.02^1$

5. *a.* ${}_5C_{0,3,1,1}(0.9)^3(0.1)^2(0.005)^1(0.995)^4$
b. ${}_5C_{0,4,1,0} 0.9^4 \cdot 0.1^1 \cdot 0.005^1 \cdot 0.995^4$

6. *a.* ${}_4C_{1,2,0,1} 0.2^1 \cdot 0.6^2 \cdot 0.1^0 \cdot 0.1^1$ *b.* ${}_4C_{0,0,0,4} 0.2^0 \cdot 0.6^0 \cdot 0.1^0 \cdot 0.1^4$

7. ${}_{80}C_{25,30,10,15} 0.7^{55} \cdot 0.3^{25} \cdot 0.2^{35} \cdot 0.8^{45}$

Section 3.9

1. $\dfrac{18/38}{1 - 18/38} = \dfrac{18/38}{20/38} = 18/20 = 9/10$

2. $\dfrac{1/7}{1 - 1/7} = \dfrac{1/7}{6/7} = 1/6$

3. No; $\dfrac{14/36}{1 - 14/36} = \dfrac{14/36}{22/36} = 14/22 = 7/11$

6. $E = 1/4 \cdot \$76{,}000 - 3/4 \cdot \$24{,}000 = \$19{,}000 - \$18{,}000 = \$1{,}000$
$1/4 w = 3/4 \cdot \$24{,}000$, $w = \$72{,}000$
Total return $24,000 + $72,000 = $96,000

7. $1/4 \cdot 1/5 \cdot \$1{,}000 + 1/2 \cdot 1/10 \cdot \$1{,}000 + 1/4 \cdot 8/100 \cdot \$1{,}000$
$= \$50 + \$50 + \$20 = \$120$. Gain $20.

Section 3.10

1. *a.* 13/65 *b.* 13/48
2. *a.* 110/580 *b.* 110/200 *c.* 110/260 *d.* 70/580 *e.* 70/200 *f.* 70/110
3. *a.* 5/10 *b.* 10/50 *c.* 5/30 *d.* 3/50 *e.* 7/25 *f.* 25/95, 5/95, 40/95
4. *a.* 11/20 *b.* 3/50 *c.* 87/400 *d.* 120/220 *e.* 13/22 *f.* 13/153
6. *a.* 10/32 *b.* 1/4 *c.* 10/31

Chapter 4

Section 4.1

1. *a.* The range of A is $220 - 2 = 218$. The range of B is $1{,}800 - 2 = 1{,}798$.
3. *a.* 4 *b.* 20 *c.* 4

Section 4.2

4. *a.* No *b.* 10
5. *a.* $91/7$ *b.* $2 \cdot 91/7 + 2 = 196/7$
6. *a.* x_{r+1} *b.* $x_{r+1} + c$ *c.* Yes *d.* Yes
8. *a.* x_{r+1} *b.* Yes *c.* $x_{r+1}c$ *d.* Yes *e.* Yes, for $c > 0$
13. *a.* 2 *b.* 6 *c.* 8 *d.* $2 + 6 = 8$
15. Let P have n elements. Then Q has $2n$ elements. The overall mean is $(nx_1 + 2nx_2)/3n$.
16. The number of times at bat in each of the five years.

Section 4.3

1. *a.* 0 *b.* $10/6$ *c.* $20/6$ *d.* $80/6$ *f.* $2\sqrt{10}/\sqrt{3}$
4. *a.* 9 *b.* 7 *c.* $3/4$ *d.* 12 *e.* 16 *f.* 0
9. *a.* $(100\sqrt{10})/(7\sqrt{3})$

Section 4.4

1. *a.* $15\frac{2}{9}$ *b.* $15\frac{2}{9}$
2. *a.* $64/6$ *b.* $64/6$

Section 4.5

1. *a.* 24 *b.* $897\frac{1}{9}$ *c.* Approximately 29.9 *d.* 17 *e.* 18
   f. 18 *g.* $13\frac{1}{2}$ or 14, 16, 17
4. Approximately $21/30 = 7/10$
5. Approximately $1/30$
10. *a.* 2 *b.* 3 *c.* 4 *d.* 5 *e.* 10

Section 4.6

1. Using 30 as an approximation of the standard deviation, the scores are $\{-233/30, -233/30, -211/30, -200/30, -199/30, -188/30, -188/30, -166/30, -155/30, -144/30, -100/30, -2/30, -1/30, 1/30, 8/30, 566/30, 599/30, 766/30\}$.
4. $17/18$ or approximately 94 percent
7. *a.* 75 *b.* 89
8. *a.* 93 *b.* 97
9. John

Section 4.7

1. $65\frac{5}{9}$, $67\frac{1}{2}$
2. $66\frac{2}{3}$, $64\frac{7}{12}$
3. *a.* The fourth, 61 to 80 *b.* The seventh, 61 to 70

Section 4.8

1. $\$500(-7/495) \cong -\$7$
3. *a.* 4.5 *b.* 3.75
4. *a.* 5.4 *b.* 4.5
5. *a.* 3.6 *b.* 3
6. *a.* 0.45 *b.* $15/16$
7. *a.* 0.54 *b.* $18/16$
8. *a.* 0.36 *b.* 0.75
9. $32/16$

Section 4.10

1. *a.* 10
   b. {8,9}, {8,10}, {8,11}, {8,12}, {9,10}, {9,11}, {9,12}, {10,11}, {10,12}, {11,12}
   c. $\{8\frac{1}{2}, 9, 9\frac{1}{2}, 10, 10, 9\frac{1}{2}, 10, 10\frac{1}{2}, 10\frac{1}{2}, 11, 11\frac{1}{2}\}$
   f. 10 *g.* $\sqrt{3/2}$
3. *a.* Do not reject. *b.* Reject. *c.* Reject.
4. *a.* Do not reject. *b.* Reject. *c.* Reject.
5. *a.* Do not reject. *b.* Do not reject. *c.* Reject.

Section 4.11

1. $79\frac{1}{3} < \mu < 82\frac{2}{3}$
2. $72\frac{2}{3} < \mu < 77\frac{1}{3}$
3. $91 < \mu < 93$
4. $78\frac{1}{2} < \mu < 83\frac{1}{2}$
5. $71\frac{1}{2} < \mu < 78\frac{1}{2}$
6. $90\frac{1}{2} < \mu < 93\frac{1}{2}$

Section 4.12

1. *a.* 0.0987 *d.* 0.1554 *e.* 0.4332 *f.* $0.0987 + 0.3413 = 0.44$
   g. 0.95 *h.* 0.9544 *i.* 0.9974
3. *a.* 0.1587 *b.* 0.0228 *d.* 0.1587
5. *a.* 0.176 *b.* 0.03 *c.* 0.001

Section 4.13

1. *a.* $49\frac{5}{8} < \mu < 50\frac{3}{8}$ *b.* $119\frac{1}{8} < \mu < 120\frac{7}{8}$ *c.* $79\frac{3}{8} < \mu < 80\frac{5}{8}$
   d. $1{,}238\frac{1}{2} < \mu < 1{,}241\frac{1}{2}$ *e.* $45\frac{3}{4} < \mu < 46\frac{1}{4}$
2. *a.* $49\frac{1}{4} < \mu < 50\frac{3}{4}$ *b.* $118\frac{1}{4} < \mu < 121\frac{3}{4}$ *c.* $78\frac{3}{4} < \mu < 81\frac{1}{4}$
   d. $1{,}237 < \mu < 1{,}243$ *e.* $45\frac{1}{2} < \mu < 46\frac{1}{2}$
3. *a.* $48\frac{7}{8} < \mu < 51\frac{1}{8}$ *b.* $117\frac{3}{8} < \mu < 122\frac{5}{8}$ *c.* $78\frac{1}{8} < \mu < 81\frac{7}{8}$
   d. $1{,}235\frac{1}{2} < \mu < 1{,}244\frac{1}{2}$ *e.* $45\frac{1}{4} < \mu < 46\frac{3}{4}$

Chapter 5

Section 5.1

1. *a.* 1,500,000 *b.* 1,000,000
2. *a.* 150,000 *b.* 0
5. *a.* Any number except $8/3$ *b.* Not possible *c.* $8/3$
6. *a.* Any number except 3 *b.* $a = 3$ *c.* Not possible
8. The only solution is 0 lb of each.
9. He should buy 2,100 lb of chickens and 2,200 lb of ducks.
11. $a = 2.4$, $b = 20$
12. $a = 120$, $b = 420$; sales $= 120 + 420 \cdot 300 = 126{,}120$.
    Number of clerks is 126,120/12,000 or 11.

Section 5.2

2. *a.* (3,4) *b.* (4,2)
3. *a.* (13,0), $(0, 13/3)$ *b.* (7,0), $(0, 7/2)$
4. *a.* (−5,6) *b.* $(1, 3/2)$

Section 5.3

1. *a.* (−5,6) *b.* $(1, 3/2)$
2. Let x_1 be the number of gallons of cream and x_2 the number of gallons of skim milk. Then the system of equations is

$$\begin{aligned} 24x_1 - x_2 &= 0 \\ 0.20x_1 + 0.02x_2 &= 1{,}000 \end{aligned}$$

   The solution is (1,471, 35,294).

Section 5.4

1. *a.*

$$\left(\begin{array}{ccc|c} 2 & -1 & 0 & 2 \\ 0 & 1 & 1 & 4 \\ 1 & -1 & 1 & 2 \end{array}\right) \quad \left(\begin{array}{ccc|c} 1 & -\frac{1}{2} & 0 & 1 \\ 0 & 1 & 1 & 4 \\ 1 & -1 & 1 & 2 \end{array}\right)$$

$$\left(\begin{array}{ccc|c} 1 & -\frac{1}{2} & 0 & 1 \\ 0 & 1 & 1 & 4 \\ 0 & -\frac{1}{2} & 1 & 1 \end{array}\right) \quad \left(\begin{array}{ccc|c} 1 & -\frac{1}{2} & 0 & 1 \\ 0 & 1 & 1 & 4 \\ 0 & 0 & \frac{3}{2} & 3 \end{array}\right)$$

$$\left(\begin{array}{ccc|c} 1 & -\frac{1}{2} & 0 & 1 \\ 0 & 1 & 1 & 4 \\ 0 & 0 & 1 & 2 \end{array}\right) \quad \left(\begin{array}{ccc|c} 1 & -\frac{1}{2} & 0 & 1 \\ 0 & 1 & 0 & 2 \\ 0 & 0 & 1 & 2 \end{array}\right)$$

$$\left(\begin{array}{ccc|c} 1 & 0 & 0 & 2 \\ 0 & 1 & 0 & 2 \\ 0 & 0 & 1 & 2 \end{array}\right)$$

Section 5.5

1. *a.* $\left(\dfrac{x_2 + 2}{2}, x_2, \dfrac{x_2 + 2}{2}\right)$ *b.* $(x_1, 0, 2x_1 - 2)$ *c.* $(3 - x_3, 3 - x_3, x_3)$
2. No. Reports 1 and 2 imply that the sales of *C* are \$200. This makes reports 2 and 3 inconsistent.
3. *a.* Yes *b.* No *c.* $(800 - x_2, x_2, 200)$

Section 5.6

1. *a.* Corner points are (0,0), (0,$\frac{4}{3}$), (1,1), ($\frac{5}{3}$,0). The profit-function values are 0, $\frac{8}{3}$, 3, $\frac{5}{3}$, respectively. The maximum is 3 at (1,1).
2. *a.* The profit-function values are 0, $\frac{4}{3}$, 3, $\frac{10}{3}$, respectively. The maximum is $\frac{10}{3}$ at ($\frac{5}{3}$,0).
3. *a.* The profit-function values are 0, $\frac{4}{3}$, 11, $\frac{50}{3}$, respectively. The maximum is $\frac{50}{3}$ at ($\frac{5}{3}$,0).
4. *a.* The profit-function values are 0, $\frac{40}{3}$, 11, $\frac{5}{3}$, respectively. The maximum is $\frac{40}{3}$ at (0,$\frac{4}{3}$).

Section 5.7

1. The corner points are (0,4), ($\frac{15}{7}$,$\frac{8}{7}$), (5,0). The function values are 16, $\frac{73}{7}$, 15. The minimum value is $\frac{73}{7}$ at ($\frac{15}{7}$,$\frac{8}{7}$).

INDEX